Eugene Stefanovich
Elementary Particle Theory

De Gruyter Studies in Mathematical Physics

Volume 45

Eugene Stefanovich

Elementary Particle Theory

Volume 1: Quantum Mechanics

DE GRUYTER

Mathematics Subject Classification 2010
Primary: 81-02, 81P10, 22E43; Secondary: 47A40, 81V10

Author
Dr. Eugene Stefanovich
San Jose, California
USA
eugene_stefanovich@usa.net

ISBN 978-3-11-049088-6
e-ISBN (PDF) 978-3-11-049213-2
e-ISBN (EPUB) 978-3-11-049103-6
ISSN 2194-3532

Library of Congress Control Number: 2018016481

Bibliographic information published by the Deutsche Nationalbibliothek
The Deutsche Nationalbibliothek lists this publication in the Deutsche Nationalbibliografie; detailed bibliographic data are available on the Internet at http://dnb.dnb.de.

Typesetting: VTeX UAB, Lithuania
Printing and binding: CPI books GmbH, Leck

www.degruyter.com

To Regina

Contents

List of figures

https://doi.org/10.1515/9783110492132-201

List of tables

https://doi.org/10.1515/9783110492132-202

Postulates, statements, theorems, assertions

https://doi.org/10.1515/9783110492132-203

Conventional notation

Quantum logic

$\mathcal{L}$	set of all propositions (logic) (page 10)
$\mathcal{A}, \mathcal{B}, \mathcal{X}, \mathcal{Y}, \mathcal{Z}$	logical propositions (page 9)
$\mathcal{A} < \mathcal{B}$	$\mathcal{B}$ follows from $\mathcal{A}$ and $\mathcal{B} \neq \mathcal{A}$ (page 11)
$\mathcal{A} \leq \mathcal{B}$	$\mathcal{B}$ follows from $\mathcal{A}$ (page 11)
$\mathcal{I}$	maximal proposition (page 11)
$\emptyset$	minimal proposition (page 11)
$\mathcal{A}^{\perp}$	orthocomplement to $\mathcal{A}$ (NOT $\mathcal{A}$) (page 11)
$\mathcal{A} \vee \mathcal{B}$	join ($\mathcal{A}$ OR $\mathcal{B}$) (page 10)
$\mathcal{A} \wedge \mathcal{B}$	meet ($\mathcal{A}$ AND $\mathcal{B}$) (page 10)
$\mathcal{A} \leftrightarrow \mathcal{B}$	$\mathcal{A}$ compatible with $\mathcal{B}$ (page 21)
$\mathfrak{S}$	set of all states (page 10)
ϕ	state (page 9)
$(\phi\vert\mathcal{X})$	probability measure (page 16)

Hilbert space

$\mathscr{H}$	Hilbert space (page 209)
$\mathscr{H}^*$	dual space (page 210)
$\mathscr{H}_1 \otimes \mathscr{H}_2$	tensor product (page 212)
$\vert x\rangle$	ket vector in $\mathscr{H}$ (page 210)
$\langle x\vert$	bra vector in $\mathscr{H}^*$ (page 210)
$\mathscr{A} \cap \mathscr{B}$	intersection of subspaces $\mathscr{A}$ and $\mathscr{B}$ (page 23)
$\mathscr{A} \subseteq \mathscr{B}$	$\mathscr{A}$ is a subspace in $\mathscr{B}$ (page 189)
$\mathscr{A} \uplus \mathscr{B}$	linear span of $\mathscr{A}$ and $\mathscr{B}$ (page 23)
$\mathscr{A}'$	orthogonal complement of $\mathscr{A}$ (page 225)
$\mathscr{A} \perp \mathscr{B}$	orthogonal subspaces (page 225)
$\mathscr{A} \oplus \mathscr{B}$	direct sum of subspaces (page 225)
$[A, B]$	commutator of operators $AB - BA$ (page 217)
$\{A, B\}$	anticommutator $AB + BA$ (page 217)
$\mathrm{Tr}(F)$	trace of operator F (page 217)
$\hat{F}$	operator acting on wave functions (page 93)
$P_{\mathscr{A}}$	projection on the subspace $\mathscr{A}$ (page 225)
P_f	spectral projection for eigenvalue f (page 26)
ϕ	phase of a wave packet (page 151)

https://doi.org/10.1515/9783110492132-204

3D space

$\mathbb{R}$	real numbers (page 24)
$\mathbb{C}$	complex numbers (page 24)
$i, j, k, \ldots$	3-vector indices $(1, 2, 3 = x, y, z)$ (page 192)
$\mathbf{i}, \mathbf{j}, \mathbf{k}$	basis vectors in $\mathbb{R}^3$ (page 191)
$\boldsymbol{a} \cdot \boldsymbol{b}$	scalar product of two 3-vectors (page 191)
$[\boldsymbol{a} \times \boldsymbol{b}]$	vector product of two 3-vectors (page 197)
δ_{ij}	Kronecker delta (page 193)
ϵ_{ijk}	Levi-Civita tensor (page 195)

Minkowski space–time

$\mathcal{M}$	Minkowski space–time (page 237)
$\tilde{A}$	4-vector (page 237)
$\tilde{A} \cdot \tilde{B}$	pseudoscalar product (page 237)
$\mu, \nu, \sigma, \ldots$	4-vector indices (0, 1, 2, 3) (page 237)
$\eta_{\mu\nu}$	metric tensor (page 237)

Poincaré group

$\{\boldsymbol{\eta}\}$	element of a Lie group (page 203)
$\{\boldsymbol{v}(\boldsymbol{\theta}); \boldsymbol{\varphi}; \boldsymbol{r}; t\}$	inertial transformation (page 38)
$\mathcal{H}$	generator of time translations (page 41)
$\mathcal{J}$	generator of rotations (page 41)
$\mathcal{P}$	generator of space translations (page 41)
$\mathcal{K}$	generator of boosts (page 41)
$R_{\boldsymbol{\varphi}}$	matrix of rotation (page 192)
$\boldsymbol{\phi}, \boldsymbol{\varphi}$	rotation vectors (page 197)
$\boldsymbol{\theta}, \boldsymbol{\vartheta}$	boost rapidity vectors (page 50)
$\boldsymbol{\theta} \circ \boldsymbol{\varphi}$	composition of transformations from the Lorentz subgroup (page 69)
C^a_{bc}	structure constants (page 204)

Group representations

$\mathbb{K}$	isomorphism of subspaces in $\mathscr{H}$ (pages 56, 125)
$\mathbb{k}$	unitary or antiunitary generator of $\mathbb{K}$ (pages 56, 125)
$[U_g]$	ray of unitary transformations (page 57)
U_g	unitary representation of a group (page 231)
$\mathfrak{U}_g$	unitary projective representation of a group (page 61)

$\mathfrak{T}$	generator of a projective representation (page 62)
$\mathfrak{Q}$	central charge (page 63)
$[A,B]_L$	Lie bracket (page 205)
$\mathfrak{h}_n$	Heisenberg algebra (page 232)
$\boldsymbol{\kappa}$	standard momentum (page 102)
$\mathscr{D}^s$	irreducible representation of the rotation group (page 235)
$\varphi_W(\boldsymbol{p},\boldsymbol{\theta})$	Wigner angle (page 101)
$U(\Lambda;\tilde{a})$	unitary representation of the Poincaré group (page 69)
$\tilde{\boldsymbol{\theta}}$	4×4 pseudo-orthogonal matrices of boosts (page 239)

Observables

H	Hamiltonian (energy) (page 72)
$\boldsymbol{K}$	boost operator (page 72)
$\boldsymbol{J}$	total angular momentum (page 72)
$\boldsymbol{P}$	total linear momentum (page 72)
$H_0, \boldsymbol{P}_0, \boldsymbol{J}_0, \boldsymbol{K}_0$	generators of the noninteracting representation (page 128)
V	potential energy (page 132)
$\boldsymbol{Z}$	potential boost (page 132)
M	operator of mass (page 74)
N	interaction in the mass operator (page 133)
$\boldsymbol{R}$	center-of-energy position (page 77)
$\boldsymbol{S}$	spin (page 77)
$\boldsymbol{V}$	velocity (page 73)
$\tilde{W}$	$\equiv (W_0, \boldsymbol{W})$ Pauli–Lubanski operator (page 75)
$\boldsymbol{r}_i$	particle position (page 125)
$\lvert\boldsymbol{p}\rangle^\tau$	photon state with momentum $\boldsymbol{p}$ and helicity τ (page 118)
$\omega_{\boldsymbol{p}}$	$\equiv \sqrt{m^2c^4 + p^2c^2}$ one-particle energy (page 96)
$\mathrm{pow}(A)$	power of operator A (page 85)
S	S-operator (page 162)
Φ	scattering phase operator (page 166)
$\underline{Y(t)}$	$\equiv -\frac{i}{\hbar}\int_{-\infty}^{t} Y(t')dt'$ (page 165)
$\underbrace{Y(t)}$	$\equiv -\frac{i}{\hbar}\int_{-\infty}^{+\infty} Y(t')dt'$ (page 165)
$\langle F\rangle$	expectation value of observable F (page 29)
$<x>$	physical dimension of observable x (page 49)
$\mathcal{S}$	phase space (page 9)
$[A,B]_P$	Poisson bracket (page 155)
m_e	electron mass (page 92)
m_p	proton mass (page 92)

Preface

Theoretical physics, as a science, began with Newton. His ideas were based on particles – corpuscles,[1] and the first realistic model of interactions was the Newtonian theory of gravity, in which the planets and the Sun were attracted to each other by instantaneous forces at a distance.

Newton himself was very unhappy about this model. He wrote [61]

> That Gravity should be innate, inherent and essential to Matter, so that one body may act upon another at a distance thro' a Vacuum, without the Mediation of any thing else, by and through which their Action and Force may be conveyed from one to another, is to me so great an Absurdity that I believe no Man who has in philosophical Matters a competent Faculty of thinking can ever fall into it.

Indeed, over time, the idea of Newtonian corpuscles began to lose its appeal. The first blow was caused by the wave theory of light by Young and Fresnel. The second blow was Maxwell's theory of electromagnetic phenomena. The culmination of these misfortunes was Einstein's theory of relativity. By 1905, a harmonious system of views had developed, which denied the Newtonian action-at-a-distance. The theory of relativity forbade the superluminal transmission of any signals and interactions. The Maxwell–Liénard–Wiechert theory explained that the carrier of the retarded interaction between charges is the electromagnetic field propagating at the speed of light. Energy and momentum flowing between charges are temporarily stored in the field, so that conservation laws are not violated even in the case of such a retarded transmission of forces.

For a short period of time this field picture was shaken by the arrival of quantum mechanics. In particular, to explain the photoelectric effect, Einstein revived the Newtonian corpuscles of light – photons [4]. It turned out that these corpuscles (their wave functions) can also interfere, and to explain the structure of the atom it was sufficient to solve the Schrödinger equation for particles interacting via the instantaneous Coulomb potential.

However, early quantum theory was soon criticized for its alleged incompatibility with the principle of relativity and replaced with quantum field theory (QFT). The fantastic agreement of this theory with experiments, it would seem, has forever discouraged the return to the corpuscular past. It is enough to go over titles of some articles in respected journals,[2] to understand that in today's physics particles are in deep disgrace.

---

1 Even light was understood by Newton as a stream of a huge number of microscopic particles.

2 "No place for particles in relativistic quantum theories?" [35], "There are no particles, there are only fields" [37], "Why there cannot be a relativistic quantum mechanics of (localizable) particles" [53].

https://doi.org/10.1515/9783110492132-205

> In its mature form, the idea of quantum field theory is that quantum fields are the basic ingredients of the universe and particles are just bundles of energy and momentum of the fields – S. Weinberg [96].

However, the "particles vs. fields" argument is still far from a happy resolution. Modern field theories face two difficult problems.

The first problem is ultraviolet divergences. All realistic quantum field theories suffer from divergent loop integrals occurring in calculations of scattering amplitudes. These divergent theories are "renormalized" by adding infinite counterterms to their Hamiltonians. In fact, the renormalization sweeps the problem of ultraviolet divergences "under the carpet," because it results in a poorly defined formally infinite energy operator, which is not suitable for describing the time-dependent dynamics of states. On closer examination, it turns out that the problem of divergences is related to the self-interaction of particles in QFT. In this theory, the electron interacts with itself, which is often depicted by diagrams in which an electron absorbs its own emitted virtual photons.

In the third volume of our book, we shall see that the problems of self-interaction and renormalization can be solved by introducing the so-called dressed interaction theory. This will bring us back to Newton's corpuscles, interacting with each other through instantaneous potentials. But how can one reconcile this action-at-a-distance with the theory of relativity, which prohibits superluminal propagation of interactions?

To answer this question, we turn to the second important problem of theoretical physics. It is sometimes formulated as the problem of quantum gravity, although, in fact, quantum mechanics is poorly compatible even with Einstein's special relativity theory. In special relativity, positions and time are treated on an equal basis as coordinates in the four-dimensional Minkowski space–time. However, in quantum mechanics these two quantities play quite different roles. The spatial coordinate (like any other physical observable) is described by an Hermitian operator, whereas time is simply a numerical parameter that cannot be converted into an operator without contradictions.

Our main goal is to understand the essence of contradictions between quantum mechanics and the special theory of relativity. For this, we will have to return to the very foundations of theoretical physics. We begin with indisputable postulates of *quantum mechanics* and the *principle of relativity*. Strict adherence to these postulates will lead us to the idea of unitary representations of the Poincaré group in a Hilbert space of states as the basis of the entire mathematical apparatus of our theory. Although applications of this approach to interacting systems are well known since the fundamental work of Dirac [23], it was not recognized that Dirac's interacting generators of boosts[3] imply that Lorentz transformations cannot be exact and universal,

3 The generators of boosts are interaction-dependent in the instant form of Dirac dynamics. In Volume 3, we will argue that only this form should be used to describe nature.

as required by special relativity. Boost transformations of observables must depend on the particular physical system and forces acting therein. This important observation will enable us to lift the prohibition on superluminal propagation of interactions and formulate a theory of particles acting on each other by means of instantaneous potentials. At the same time, we will be able to avoid conflicts with the unshakable principles of relativity and causality.

In the third volume, we will analyze in detail the recent experiment [21] conducted by the team of professor Pizzella at the Frascati Research Center. With this experiment they discovered the superluminal propagation of Coulomb forces, which, in our opinion, is the most convincing validation of the theory presented in this book. In some sense, the ultimate goal of the entire book is to demonstrate that Pizzella's unusual results are naturally expected in a rigorous approach to quantum relativistic physics.

In this book, we will focus on systems of charged particles and photons as well as on electromagnetic forces acting in such systems. Traditionally, these phenomena are described by *quantum electrodynamics* (QED). Our approach will lead us to another theory, which we call *relativistic quantum dynamics*, or RQD. This theory is exactly equivalent to the renormalized QED as long as one is interested in properties related to the S-matrix (scattering cross sections, lifetimes, energies of bound states, etc.). However, unlike QED, our approach can also describe the time evolution and boost transformations in interacting systems.

This book is divided into three volumes.[4] This is Volume 1, where we will try to avoid contradictory issues and will, basically, adhere to the generally accepted views on relativistic quantum theory. We will define our basic assumptions, notation, and terminology and also try to trace a logical path starting from the postulates of relativity and probability and leading to relativistic quantum theory of interacting systems. In this volume, we confine ourselves to interactions that do not change the number of particles in the system, which is an acceptable approximation for low-energy processes within the framework of elementary *quantum mechanics*.

Volume 1 consists of seven chapters.

In Chapter 1, *Quantum logic*, we derive the basic laws of quantum theory from simple axioms of measurements and probability (= quantum logic). We turn to the old, but not yet very popular idea that in order to understand quantum laws it is necessary to replace some of the postulates of classical logic. Despite the apparent radicalism of this approach, it leads to the well-known quantum formalism with wave functions and Hermitian operators in the Hilbert space. For us it will be important to emphasize that, being rooted in logic, the foundations of quantum mechanics are solid and unshak-

4 This work is based partially on our earlier publications [82, 83], which were rewritten, updated and improved in significant ways.

able. Therefore, we do not expect any modification of the laws of quantum mechanics[5] in the foreseeable future.

In Chapter 2, *Poincaré group*, we introduce the Poincaré group as a set of transformations connecting different (but equivalent) inertial reference frames. This chapter is central to understanding the principle of relativity. In our approach, the group properties of inertial transformations are at the core of the relativistic description of nature.

Chapter 3, *Quantum mechanics and relativity*, will combine the two theories presented above and establish unitary representations of the Poincaré group as the most general and complete mathematical description of any isolated physical system. This is the most adequate language for a relativistic quantum description of nature. One can even say that the rest of this book is simply an exercise in constructing and analyzing various unitary representations of the Poincaré group.

In Chapter 4, *Observables*, we examine the correspondence between known physical quantities (such as mass, energy, momentum, spin, position, etc.) and specific Hermitian operators in the Hilbert space of states. The most important point is the connection between physical observables and generators of the Poincaré group representation. From this connection we derive the commutation relations of observables and how these operators change with respect to inertial transformations of the observers.

Chapter 5, *Elementary particles*, is devoted to the Wigner theory of unitary irreducible representations of the Poincaré group. This theory fully describes the basic properties and dynamics of isolated stable elementary particles. For us, the special importance of this chapter is that Wigner's elementary particles are the most fundamental ingredients in our model of the world. As we explain in Volume 3, quantum fields are just formal technical constructions, and real physical systems are composed of elementary particles that interact directly with each other.

In Chapter 6, *Interaction*, we discuss relativistically invariant interactions in many-particle systems. Here we emphasize the most important conclusion of Dirac [23], that relativistically invariant interactions require modification not only of the Hamiltonian (as in the familiar non-relativistic theory) but also of other generators of the Poincaré group. We will base our theories on the Dirac instant form of dynamics, where interaction is present in both the Hamiltonian and the boost generators. In Volume 3 this will lead us to the conclusion that Lorentz transformations of special relativity are, strictly speaking, inapplicable to interacting systems.

Chapter 7, *Scattering*, is devoted to the quantum-mechanical description of particle collisions. Scattering is important first because it is the most informative experimental method for studying subatomic phenomena and second because the scattering

5 Such modifications are sometimes contemplated in attempts to develop a quantum theory of gravity. See, for example, [47] and references therein.

matrix is the main target of QFT (see Volume 2). In this book, we will build our theory (RQD) by modifying QFT, so for us the scattering matrix is of central importance. We will pay special attention to the notion of scattering equivalence, when two different Hamiltonians lead to the same S-operator. This property will play an important role in the derivation of the "dressed" Hamiltonian in Volume 3.

Some useful mathematical facts and technical calculations are collected in *Appendix*.

In Volume 2 [84], we will formulate the foundations of the most successful quantum field theory – QED, explain the causes of ultraviolet divergences and demonstrate the renormalization of the S-matrix by introducing counterterms into the Hamiltonian. There is no new physics introduced in Volumes 1 and 2. They present textbook quantum mechanics and QFT, perhaps sometimes viewed from unusual angles, but still rather orthodox. The main goal of the first two volumes is to prepare the ground for the formulation of our unconventional approach, based on the notion of physical particles and "dynamical" relativity, in Volume 3 [85] of this book.

We use the Heaviside–Lorentz system of units,[6] in which the potential energy of the electron–proton interaction has the form $V = -e^2/(4\pi r)$, and the proton charge is $e = 2\sqrt{\pi}\times 4.803\times 10^{-10}$ statcoulomb. The speed of light is $c = 2.998\times 10^{10}$ cm/s, and the Planck constant is $\hbar = 1.055\times 10^{-27}$ erg·s $= 6.582\cdot 10^{-16}$ eV·s, so that the *fine structure constant* is equal to $\alpha \equiv e^2/(4\pi\hbar c) \approx 1/137$.

I would like to express my sincere gratitude to Peter Enders, Rainer Grobe, Theo Ruijgrok, Alexander Shebeko and Boris Zapol, who read the draft of this book and gave me many priceless comments and much advice, which I tried to take into account in the final manuscript. I also thank Harvey R. Brown, William Klink, Vladimir Korda, Chris Oakley, Federico Piazza, Guido Pizzella, Wayne Polyzou, Mikhail Shirokov and Charles Su for enlightening discussions at various stages of this work. I enjoyed online communications with Juan Bernard Chaverondier, Wolfgang Engelhardt, Juan R. González-Álvarez, Bill Hobba, Igor Khavkine, Mike Mowbray, Arnold Neumaier and Dan Solomon. All these contacts and exchanges of ideas have formed my understanding of relativistic quantum physics and, ultimately, led to the writing of this book. However, this does not at all mean that the mentioned researchers share or approve my views. For all the misconceptions and errors contained in this book, the author bears full responsibility.

---

6 See Appendix in [39].

Introduction

As a result, it was almost three o'clock in the morning before the final result of my computations lay before me. The energy principle held for all the terms, and I could no longer doubt the mathematical consistency and coherence of the kind of quantum mechanics to which my calculations pointed. At first, I was deeply alarmed. I had the feeling that, through the surface of atomic phenomena, I was looking at a strangely beautiful interior, and felt almost giddy at the thought that I now had to probe this wealth of mathematical structures nature had so generously spread out before me.
Werner Heisenberg

In this Introduction, we will try to formulate more precisely what is the goal of theoretical physics, what are the fundamental concepts of this science and the relationship between them. Some of our statements may look self-evident or even trivial. However, it seems important to us to spell out these definitions and clarify our positions here and now, in order to avoid misunderstandings in further parts of the book.

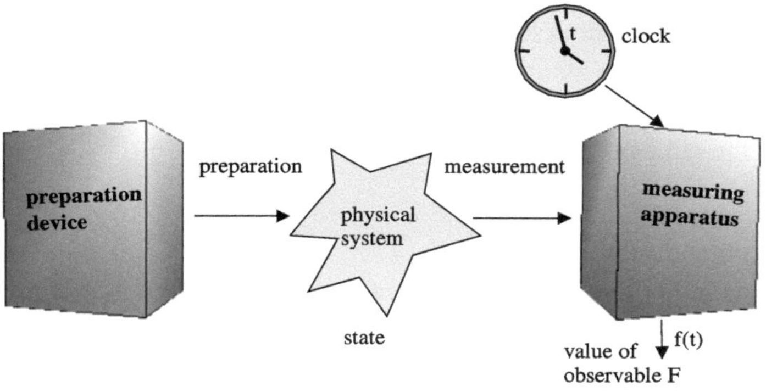

Figure 1: Schematic representation of the preparation/measurement act.

We get all information about the physical world through results of *measurements*, and the fundamental goal of theoretical physics is to describe and predict these results. Any act of measurement requires the presence of at least three objects (see Figure 1): the *preparation device*, the *physical system* and the *measuring apparatus*. The preparation device arranges the physical system in a specific *state*. This state has certain attributes or properties. If the state's attribute can be associated with a numerical value, it will be called a physical quantity or *observable* F. Observables are measured by bringing the system into contact with the measuring apparatus. The result of the measurement is a numerical value of the observable, i. e., a real number f. We assume that each measurement of the observable F always produces *some* result f, i. e., the measuring apparatus never misfires.

This is just a short list of important concepts. Let us now dwell on each of them in more detail.

https://doi.org/10.1515/9783110492132-206

Physical systems

We will call a physical system any object that is capable of causing a response (= measurement) in the measuring apparatus. Physical systems can be either *elementary* (they are also called *particles*) or *composite*, i. e., consisting of two or more particles.

In this book, we confine ourselves to examining *isolated systems*, which do not interact with the outside world or with any external potential. By insisting on this, we exclude from consideration interesting physical systems, such as an atom in external electric and magnetic fields. However, in fact, this does not limit the generality of our approach. Indeed, one can always combine the atom and the field creating device into a single unified system that now becomes isolated.

States

Each physical system can take different states. The book can be on the table or in the library, open or closed, at rest or moving with great speed. It is not always easy to understand whether we are dealing with different systems or with different states of the same system. For example, a pair of separated particles (an electron and a proton) does not at all look like a hydrogen atom. So, it is easy to make a mistake and decide that $e^- + p^+$ and ^1H are two different physical systems, although in reality these are two different states of the same two-particle system.

Preparation and measuring devices

In general, devices used to prepare different states of physical systems and perform measurements on them can be very complex. It would be hopeless to try to include in our theory a detailed description of these devices and their interactions with physical systems. Instead, we will use an idealized representation of the acts of preparation and measurement. In particular, we will assume that the measuring device is a black box that somehow "interacts" with the system and produces one real number f – the value of the measured observable. We are not interested in what is the inner working of this device and what is the mechanism of the device–system interaction. Such an abstraction is necessary to avoid a logical vicious circle. After all, if we begin to study components of the measuring apparatus, then we will have to operate with their observables (positions, velocities, etc.). Should we then introduce in our theory also outside instruments, which measure the measuring apparatus? No.

Time and clocks

In physics, there are numerical quantities which are not associated with any physical system and, therefore, they are not observables. These include, for example, the number of spatial dimensions (3) and the Planck constant. The most important physical quantity, not included in the class of observables, is *time*. We consider clock as an integral part of any laboratory and not as a physical system.[7] The role of the clock is to give a time stamp (= numerical parameter) to each measurement of true observables. This label simply records when the measurement was made, and the value of the label does not depend on the state of the observed system. Thus, we cannot say that time is an attribute or property of a physical system, because the "measurement" of time (= looking at the positions of the clock hands) does not involve interaction with any physical system. Time can be "measured," even if there is no physical system in our laboratory. The special places occupied by the clock and time in the acts of measurement are noted in Figure 1.

Observables

Theoretical physics seeks to study the simplest physical systems and their most fundamental observable properties: mass, velocity, spin, etc.

Without doubt, there are devices capable of measuring only one observable.[8] The result of such a measurement is one real number (= the value of the observable). Classical physics makes a stronger assumption, that one can also measure several observables simultaneously. In quantum physics, we will be more cautious and let the theory itself tell us which pairs of observables are simultaneously measurable and which pairs are not. The observables that are measurable simultaneously will be called *compatible*. Examples of compatible observables are x-, y- and z-components of one particle's position. Examples of *incompatible* observables are the x-components of the position and momentum of the same particle.

We will also see that in the quantum world, one act of preparation/measurement is insufficient to completely characterize the state of the studied system. For a complete picture, it is necessary to prepare many copies of the same system under the same conditions[9] and make independent measurements in each copy. An unexpected and still mysterious property of nature is that for such repeated measurements there is no guarantee of obtaining the same result, even if the preparation conditions are controlled in the most rigorous way. We will conclude that this scatter of values is purely random, and that a system's description can be only probabilistic. This idea is the starting point of *quantum theory*, which we discuss in Chapter 1.

---

7 Of course, the experimenter can decide that the laboratory clock is a physical system worthy of research, and undertake its experimental study. In particular, he can explore the quantum uncertainty of the positions of the clock's hands. However, such a clock ceases to be an effective device for recording time. Some other device should be used as the laboratory clock in this case.

8 We will assume that in each specific act of measurement the given observable can be measured absolutely precisely. Of course, real measuring devices are not ideal, but we will assume that with some effort it is always possible to achieve more and more perfect measurements, so that the precision is, in principle, unlimited.

9 This set of copies is called the *ensemble* of physical systems.

Spectra of observables

Some observables can take any value on the entire real axis $\mathbb{R}$. The Cartesian components of the position R_x, R_y and R_z are the simplest examples of such (unbounded and continuous) observables. However, there are also examples of observables whose values can occupy only a certain subset of the real axis. Such a subset is called the *spectrum* of the observable. For example, it is known (see Chapter 5) that each component of the velocity of a particle (V_x, V_y or V_z) cannot exceed the speed of light c, so the spectrum of these observables is within the interval $[-c, c]$. The position and velocity have *continuous* spectra. There are also quantities with *discrete* spectra. For example, the "number of particles" can take only integer values $n = 0, 1, 2, \ldots$. Later we will also meet observables whose spectrum is a combination of discrete and continuous parts, such as the energy spectrum of the hydrogen atom; see Figure 2.

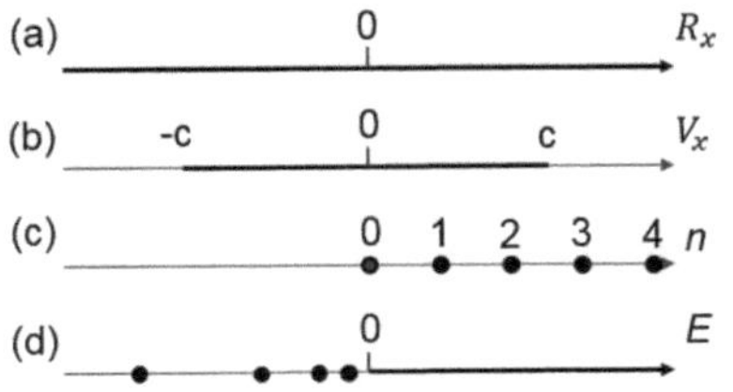

Figure 2: Examples of spectra of observables: (a) x-component of position R_x; (b) x-component of velocity V_x; (c) number of particles n; (d) energy E of a particle in an attractive potential.

Observers

We will call the *observer* O a set of measuring devices that are adapted to measure all possible observables. The minimum set of instruments includes a ruler for measuring distances, a clock for recording time, a fixed origin and three mutually perpendicular axes erected at this point. *Laboratory* is the complete experimental setup, i. e., a preparation device plus an observer O with all its measuring instruments.

In this book, we only consider *inertial observers* (= reference frames, = laboratories). These are observers/laboratories that are either at rest or moving uniformly without acceleration and rotation; their velocities and the directions of their coordinate axes do not change with time. The importance of choosing inertial observers will become clear in Section 2.1, where we will see that measurements made by these observers satisfy an important *relativity principle*.

Inertial transformations

In addition to physical systems, an observer can also measure properties of her fellow observers. With the set of devices listed above, each observer O can characterize another observer O' by ten parameters $\{\boldsymbol{v}, \boldsymbol{\varphi}, \boldsymbol{r}, t\}$. These parameters include (i) the time shift t between O and O'; (ii) the radius-vector $\boldsymbol{r}$ connecting the origins of O and O';

(iii) the rotation angle[10] $\boldsymbol{\varphi}$ between orientations of the coordinate axes in O' and O and (iv) the speed $\mathbf{v}$ of O' with respect to O.

It is convenient to introduce the notion of *inertial transformation* of the observer or laboratory. Such transformations include:
- rotations,
- spatial translations,
- changes of the speed or *boosts*,
- time translations.

There are three independent rotations (about the axes x, y and z), three independent shifts and three independent boosts. So, together with time translations, we have 10 basic types of inertial transformations. The general inertial transformation (g) can be obtained by the successive application of two (or more) basic transformations. We postulate that for any pair of inertial observers O and O' there is always an inertial transformation g that takes O to O'. This fact will be denoted by $O' = gO$. Conversely, applying any inertial transformation g to a given observer O will transform it to another valid inertial observer $O' = gO$. In Chapter 2 we will make an important observation that transformations g form a *Lie group*.

One of the most important tasks of physics is to establish a connection between results of measurements made by two different observers on the same physical system. In particular, if the values of observables measured in the reference frame O are known, then the theory should be able to calculate the values of the same observables in the frame of reference $O' = gO$. The most important and demanded is the solution of this problem in the case of *dynamics* or the time evolution, i. e., when g is a time translation.

Tasks of physics

Let us summarize our reflections by pointing out the five basic tasks of theoretical physics that we will discuss in this book:
- classify physical systems;
- for each physical system, list its observables and their spectra;
- for each physical system, list its possible states;
- for each state of the system, find results of measurements of all relevant observables;
- if a description of the system is given from the point of view of one observer, then predict how the same system looks to other inertial observers.

10 Vector parametrization of rotation angles is explained in Appendix D.5.

1 Quantum logic

The nature of light is a subject of no material importance to the concerns of life or to the practice of the arts, but it is in many other respects extremely interesting.
Thomas Young

In this chapter, we will continue our discussion of the connections between preparation devices, physical systems and measuring instruments, started in the Introduction. In particular, we will try to understand what is actually measured by the instruments, and how these results depend on the state of the observed system.

Until the end of the 19th century such questions could only raise eyebrows. In classical mechanics and in all pre-quantum physics, it was tacitly assumed that in each state the physical system possesses a set of quantities (position, momentum, mass, etc.). These quantities simply show up in measurements. They can be measured simultaneously, accurately and reproducibly. Yes, of course, every measurement is limited by a certain imprecision, but this is only a technical difficulty that can and should be neglected in a fundamental theory. All this was considered so obvious that it was not even mentioned in textbooks.

However, since the end of the 19th century, these traditional ideas began to disintegrate under the onset of new discoveries such as the radiation spectrum of heated bodies, the discrete spectrum of atoms and the photoelectric effect. Solutions of all these problems have been found within the framework of *quantum mechanics* – a completely new approach to physics that emerged in the first third of the 20th century as a result of joint efforts and passionate debates of such outstanding scientists as Bohr, Born, de Broglie, Dirac, Einstein, Fermi, Fock, Heisenberg, Pauli, Planck, Schrödinger, Wigner and many others. Out of all these studies, a completely unexpected and paradoxical picture of the physical world has emerged that was completely unlike the orderly and transparent classical picture. In spite of their counterintuitive strangeness, predictions of quantum mechanics are extraordinarily accurate: they are checked daily in countless physical and chemical laboratories around the world, and have never been refuted. This makes quantum mechanics the most successful physical theory of all time.

There are dozens of good textbooks explaining the laws of quantum mechanics and how to use them to analyze systems and predict observations in each particular case. We assume that the reader is fairly familiar with these laws. We will be more interested in the deeper meaning and interpretation of the quantum formalism, which still generates bitter controversies. Why does nature behave in a random way? Or is this randomness only apparent, but in fact there is a deeper level of reality, where quantum uncertainty gives way to some new laws? How can different states (alive and dead Schrödinger's cats) exist in a superposition? Is it possible to change the rules of quantum mechanics (for example, by adding some nonlinearity to the Hilbert space) without being in contradiction with experiments? People are increasingly asking such

https://doi.org/10.1515/9783110492132-001

questions recently, when the search for the quantum theory of gravitation has intensified, and one popular trend is to look for alternative formulations of quantum mechanics in order to "harmonize" it with the general theory of relativity [47].

In this chapter we will present a rather old, but not well-known point of view on the origin of quantum laws. This approach seeks explanation of the quantum behavior in the fundamental logical structure of physics. In particular, this approach asserts that the true logical relationships between results of measurements are different from the classical laws of Aristotle and Boole. The usual classical logic needs to be replaced by the so-called *quantum logic*.

Since ancient Greece, logic has been considered the queen of sciences, perhaps not even a science as such, but something even more fundamental: a metascience, a framework for our perception of the world and for the construction of all other sciences. Therefore, it is difficult to imagine anything more revolutionary and provocative than an encroachment on the laws of logic. Nevertheless, there are enough convincing reasons to make just this step.

> In introductory quantum physics classes (especially in the United States), students are informed *ex cathedra* that the state of a physical system is represented by a complex-valued wave function ψ, that observables correspond to self-adjoint operators, that the temporal evolution of the system is governed by a Schrödinger equation and so on. Students are expected to accept all this uncritically, as their professors probably did before them. Any question of why is dismissed with an appeal to authority and an injunction to wait and see how well it all works. Those students whose curiosity precludes blind compliance with the gospel according to Dirac and von Neumann are told that they have no feeling for physics and that they would be better off studying mathematics or philosophy. A happy alternative to teaching by dogma is provided by basic quantum logic, which furnishes a sound and intellectually satisfying background for the introduction of the standard notions of elementary quantum mechanics – D. J. Foulis [30].

The idea that the most fundamental difference between classical and quantum mechanics lies in their different logical structures belongs to Birkhoff and von Neumann. In this chapter, we briefly outline their ideas of *quantum logic* [10] as well as later contributions made especially by Mackey [50] and Piron [66, 67]; see also [19].

We will argue that the formalism of quantum mechanics (including the algebras of state vectors and Hermitian operators in the Hilbert space) follows almost inevitably from the simplest properties of measurements and logical relationships between them. These properties and relationships are so simple and fundamental that it seems impossible to modify them, and therefore it would be almost impossible to change quantum laws without violating their internal consistency and agreement with experiment. The practical conclusion is that the unification of quantum mechanics and relativity will not be achieved by changing or modifying quantum laws.[1]

---

1 In Volume 3 we will explain how one should change the formalism of special relativity to make it compatible with quantum mechanics.

In Section 1.1 we will examine limitations of classical approaches by analyzing the two-holes (two-slits) interference experiment from the points of view of the wave and corpuscular theories of light.

The logical structure of classical physics will be presented in Sections 1.2 and 1.3. In particular, we will discuss the close relationship between the classical Boolean logic and the phase space formalism. In Section 1.4 we will note the remarkable fact that the only difference between classical and quantum logics (and, therefore, between classical and quantum physics in general) lies in two inconspicuous axioms of *distributivity*. This postulate of classical logic should be replaced by the *orthomodular* postulate of quantum logic. In Section 1.5 this will lead us (via Piron's theorem) to the standard formalism of quantum mechanics with its Hilbert spaces, Hermitian operators, wave functions, etc. In Section 1.6 we will add some thoughts to the endless philosophical debate about interpretations of quantum mechanics.

1.1 Why do we need quantum mechanics?

The inadequacy of the classical concepts becomes clear if we analyze the dispute between corpuscular and wave theories of light. Let us illustrate the essence of this, without exaggeration, centuries-old debate by the example of a thought experiment with the *camera obscura*.

1.1.1 Corpuscular theory of light

You may have seen or heard about a simple optical device called *camera obscura* or pinhole camera. It is easy to make this device yourself. Take a lightproof box, make a small hole in one of its walls and place a photographic plate at the opposite wall, as shown in Figure 1.1. The light entering the inside of the box through the hole will create a clear inverted image of the outside world on the photographic plate.

You can achieve even greater clarity by reducing the size of the hole. But this, of course, will reduce the brightness of the image. This behavior of light has been known for centuries. The first scientific explanation for this and many other properties of light

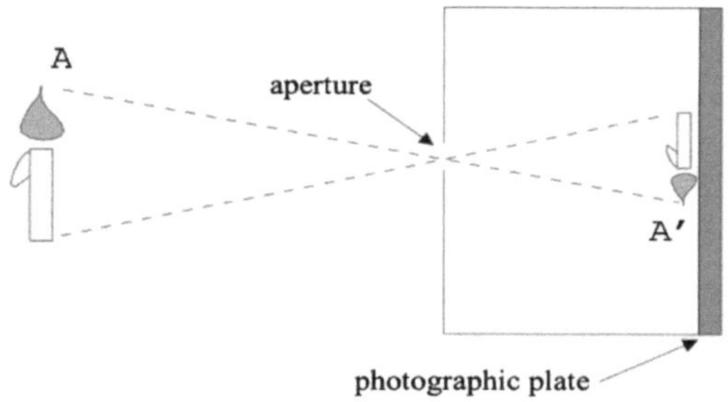

Figure 1.1: The image in the pinhole camera is created by rectilinear beams (rays) of light.

(reflection, refraction, etc.) was suggested by Newton. In a slightly modernized language, his *corpuscular theory* explained the formation of the image as follows:

> ***Corpuscular theory:*** Light is a stream of tiny particles (photons) flying along straight classical trajectories (light rays). [For example, the ray that lands at the point A' in Figure 1.1 was emitted from the point A and passed right through the hole.] Each such particle carries a certain amount of energy. When the particle collides with the photographic plate, this energy is released within one grain of the emulsion and creates a single image point. Bright light contains so many photons that their individual spots flood the photographic plate. All these points merge into one continuous image, and the density of the image is proportional to the number of particles hitting the plate during the time of exposure.

Let us continue our experiment with the pinhole camera, making the hole size smaller and smaller. Corpuscular theory asserts that shrinking holes will produce a clearer, but dimmer image. However, the experiment shows something completely different! At some point, as the size of the hole is reduced, the image will begin to blur; and in the limit of a very small hole all the details will disappear, and the picture will turn into one circular diffuse spot, as in Figure 1.2 (a). The shape and size of this blur are no longer dependent on the light source outside the camera. It would seem that the light rays, passing through the small hole, are randomly scattered in all directions. This effect was discovered by Grimaldi in the middle of the 17th century and was subsequently dubbed *diffraction*.

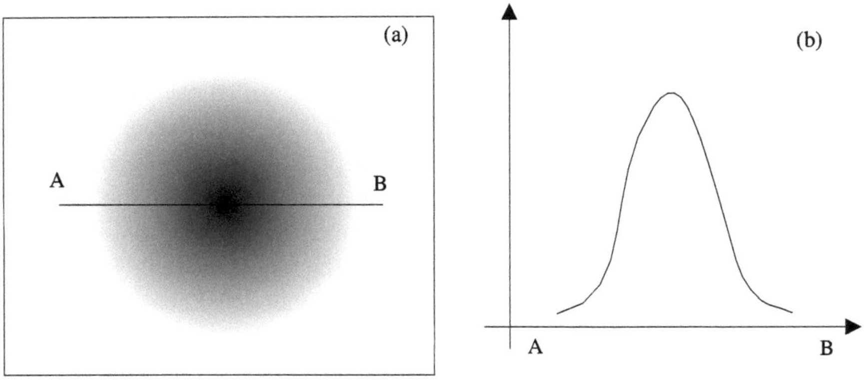

Figure 1.2: (a) The image in the pinhole camera with a very small hole. (b) Image density along the line *AB*.

Diffraction does not fit in the corpuscular theory. Why on earth do light corpuscles deviate from straight-line trajectories? Maybe this is due to their interaction with the material of the walls surrounding the hole? However, this explanation should be rejected, if only because the diffraction pattern does not depend on the material – paper or steel – from which the walls of the box are made.

The most striking evidence of the fallacy of the naïve corpuscular theory of light is the *interference* effect, discovered by Young in 1802. To see the interference, we can slightly modify our pinhole camera: instead of one hole, make two holes that are close to each other so that their diffraction blurs on the photographic plate overlap. We already know that if we leave open the left hole and close the right one, then we get a diffuse blur L (the left dashed line in Figure 1.3 (a)). If, on the contrary, we close the left hole and open the right one, we get another diffuse blur R. Let us now try to predict what would happen if both holes are opened.

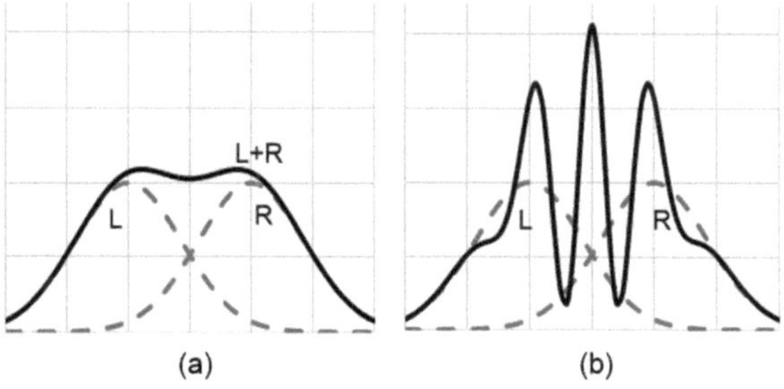

Figure 1.3: Image density in a two-hole camera. (a) In the naïve corpuscular theory. (b) In reality.

Following the logic of the corpuscular theory, we could conclude that the photons reaching the photographic plate are of two kinds: those that have passed through the left and the right hole, respectively. If the two holes are open simultaneously, then the density of "left" photons should add with the density of "right" photons, and the resulting image $L + R$ must be a superposition of the two images (solid line in Figure 1.3 (a)). Right? No, wrong! This seemingly logical reasoning is at odds with the experiment. The actual image on the photographic plate has an additional structure (brighter and darker areas shown by solid line in Figure 1.3 (b)), called the *interference pattern*. There are regions where the image density is higher than $L + R$ (constructive interference) and regions with the density lower than $L + R$ (destructive interference).

How can corpuscular theory explain this strange interference pattern? For example, we could assume that there is some interaction between light corpuscles, so that the passage of particles through the left and right holes are not independent events, and the law of addition of probabilities is not applicable to them. However, this idea should be rejected, because the interference pattern does not disappear even if we release the photons one by one, so that their interaction is excluded.

For example, in a two-hole interference experiment performed by Taylor in 1909 [89], the light intensity was so low that no more than one photon was present in the camera at any given time. This removed any possibility of interaction between photons and any effect of this interaction on the interference pattern. Does this mean that a single photon can interfere with itself? Maybe the photon somehow splits apart, passes through both holes and then reconnects before colliding with the photographic plate? This explanation also does not stand up to criticism, because one photon can blacken only one grain of the emulsion. Nobody has ever seen a "half-photon".

Perhaps, a particle passing through the right hole somehow knows whether the left hole is open or closed, and adjusts its trajectory, accordingly? This just does not make sense, and we have to admit that our simple corpuscular theory does not have a logical explanation for all these observations.

1.1.2 Wave theory of light

The inability to explain such fundamental properties of light as diffraction and interference was a heavy blow to the Newtonian corpuscular theory. These effects, like other light properties known in the pre-quantum era (reflection, refraction, polarization, etc.) were brilliantly explained by the *wave theory* of light developed by Huygens, Young, Fresnel and others. During the 19th century, the wave theory gradually supplanted the Newtonian corpuscles. The idea that light is a wave process received its strongest support from the Maxwell theory, which combined optics with electromagnetic phenomena. This theory explained that light is made of oscillating electric $\boldsymbol{E}(t, \boldsymbol{r})$ and magnetic $\boldsymbol{B}(t, \boldsymbol{r})$ fields – sinusoidal waves propagating with the speed of light c. According to Maxwell, the energy of this wave and, accordingly, the intensity of light I is proportional to the square of the amplitude of the field vectors: $I \propto \boldsymbol{E}^2$. Then, from the point of view of the wave theory, the formation of the photographic image can be explained as follows.

> ***Wave theory:*** Light is a continuous oscillating wave or field propagating through space. When a light wave meets with molecules of the photographic emulsion, charged parts of the molecules begin to oscillate under the action of the electric and magnetic vectors in the light field. In those places where the amplitude of the electromagnetic oscillations is maximal, the charges of the molecules are subjected to the strongest force, and the density of the photographic image is the highest.

This model explains both diffraction and interference in a fairly natural way: diffraction simply means that light waves are capable of going around obstacles, just like other types of waves (sea waves, sound waves, etc.) do.[2] To explain the interference at two holes, it is sufficient to note that when two parts of a monochromatic wave pass through different apertures and meet on a photographic plate, their electric (and magnetic) vectors add up. However, the wave intensities are proportional to the squares of the vectors and, therefore, are not additive: $I \propto (\boldsymbol{E}_1 + \boldsymbol{E}_2)^2 = \boldsymbol{E}_1^2 + 2\boldsymbol{E}_1 \cdot \boldsymbol{E}_2 + \boldsymbol{E}_2^2 \neq \boldsymbol{E}_1^2 + \boldsymbol{E}_2^2 \propto I_1 + I_2$. From simple geometric considerations it follows that there are places

---

2 The wavelength of the visible light varies between 0.4 micron for violet light and 0.7 microns for red light. So, for large obstacles or holes, the effect of diffraction is very small, and the corpuscular theory of light works quite well.

where these two waves always come with the same phase ($\boldsymbol{E}_1 \uparrow\uparrow \boldsymbol{E}_2$ and $\boldsymbol{E}_1 \cdot \boldsymbol{E}_2 > 0$), which means *constructive interference*, and there are other places where the waves come out of phase ($\boldsymbol{E}_1 \uparrow\downarrow \boldsymbol{E}_2$ and $\boldsymbol{E}_1 \cdot \boldsymbol{E}_2 < 0$), i. e., *destructive interference*.

1.1.3 Light of low intensity and other experiments

In the 19th century physics, the particle–wave dispute was resolved in favor of the wave theory of light. However, further experiments showed that the victory was declared prematurely. To understand the problems of the wave theory, let us continue our thought experiment with the interference pattern. This time, we will gradually reduce the intensity of the light source. At first we will not notice anything unusual: quite predictably the image density on the photographic plate will decrease. However, from a certain point we will notice that the image ceases to be uniform and continuous, as before. We will see that it consists of separate dots, as if light were incident on some grains of the emulsion and did not touch others. This observation is difficult to explain from the point of view of the wave theory. How can a continuous wave create this dotted image? But the corpuscular theory copes easily: obviously, these dots are created by separate particles (photons), which bombard the surface of the photographic plate.

In the late 19th and early 20th centuries, other experiments appeared that challenged the wave theory of light. The most famous of them was the photoelectric effect: it was found out that when light falls on a piece of metal, it can knock electrons out of the metal into the vacuum. In itself, this discovery was not surprising. However, it was surprising how the number of knocked-out electrons depended on the frequency of light and its intensity. It was found out that only light with a frequency above a certain threshold ω_0 could knock out electrons from the metal. Light of a lower frequency was unable to do this, even if its intensity was very high. Why was this observation so surprising? From the point of view of the wave theory, it could be assumed that the electrons are emitted from the metal by the forces originated from the electric $\boldsymbol{E}$ and magnetic $\boldsymbol{B}$ fields in the light wave. The higher intensity of light (= the larger magnitudes of the vectors $\boldsymbol{E}$ and $\boldsymbol{B}$) naturally means a higher force acting on the electrons and a greater probability of the electron emission. So why could not intensive low-frequency light cope with this work?

In 1905, Einstein explained the photoelectric effect by returning to the long-forgotten Newtonian corpuscles in the form of light quanta, later called *photons*. Einstein described light as "... consisting of finite number of energy quanta which are localized at points in space, which move without dividing and which can only be produced and absorbed as complete units." [4]. According to Einstein, each photon carries the energy $\hbar\omega$, where ω is the light frequency[3] and $\hbar$ is the *Planck constant*.

3 ω is the so-called *cyclic frequency* (measured in radians per second), which is related to the ordinary frequency ν (measured in cycles per second) by formula $\omega = 2\pi\nu$.

Each photon has a chance to encounter only one electron in the metal and transfer its energy to it. Only high-energy photons (i. e., photons present in high-frequency light) can transmit enough energy to the electron to overcome the energy barrier E_b between the volume of the metal and the vacuum. Low-frequency light has low-energy photons $\hbar\omega < E_b \approx \hbar\omega_0$. Hence, regardless of the intensity (= the number of photons) of such light, its photons are simply too weak and unable to kick the electrons strong enough to overcome the barrier.[4]

In the Compton experiments (1923), the interaction of X-ray radiation with free electrons was studied in much detail, and indeed, this interaction was more like a collision of two particles than a shaking of the electron by a periodic electromagnetic wave.

These observations should confirm our conclusion that light is a stream of corpuscles, as Newton said. But how about interference? We have already established that corpuscular theory is unable to give a logical explanation for this effect!

So, the young quantum theory faced the seemingly impossible task of reconciling two classes of experiments with light. Some experiments (diffraction, interference) were easily explained within the framework of the wave theory of light, but did not agree with the corpuscles – photons. Other experiments (photoelectric effect, Compton scattering) contradicted the wave properties and clearly indicated that light consists of particles. To all this confusion, in 1924 de Broglie added the hypothesis that the *particle–wave dualism* is characteristic not only of photons. He argued that all material particles – for example, electrons – have wave properties. This "crazy" idea was soon confirmed by Davisson and Germer, who observed the interference of electron beams in 1927.

Without a doubt, in the first quarter of the 20th century, physics approached the greatest crisis in its history. Heisenberg described this situation as follows.

> I remember discussions with Bohr which went through many hours till very late at night and ended almost in despair; and when at the end of the discussion I went alone for a walk in the neighboring park I repeated to myself again and again the question: Can nature possibly be as absurd as it seemed to us in those atomic experiments? – W. Heisenberg [36].

1.2 Classical logic

In order to advance in our understanding of the paradoxes mentioned above, we need to go beyond the framework of classical physics. Therefore, to begin with, we are going to outline this framework, i. e., to look at classical mechanics. For simplicity, we consider the classical description of a single particle in a one-dimensional space.

---

4 In fact, low-frequency light can lead to the electron emission when two low-energy photons collide simultaneously with the same electron. But such events are unlikely and become noticeable only at very high light intensities.

1.2.1 Phase space of one classical particle

The *state* ϕ of a classical particle is completely and uniquely determined by specifying the particle's position x and momentum p.[5] Such states will be called *pure classical*.

Thus, all possible states of a one-particle system are labeled by a pair of numbers (x, p) and can be represented by points on a plane, which will be called the *phase space* of the system and denoted $\mathcal{S}$ (see Figure 1.4). Then particle dynamics is represented by lines (= trajectories) in the phase space $\mathcal{S}$.

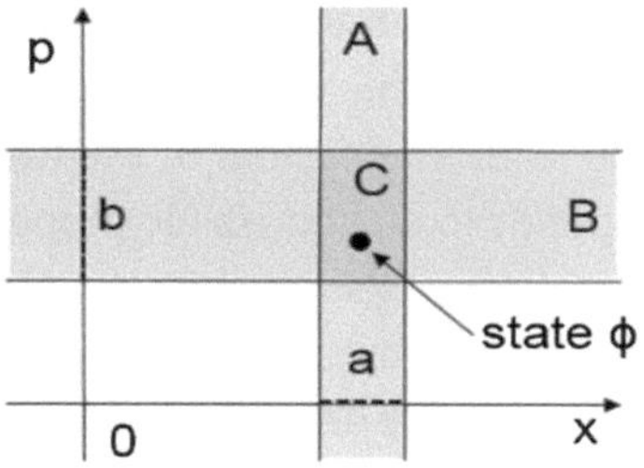

Figure 1.4: Phase space of a particle in one spatial dimension.

1.2.2 Propositions in phase space

In order to make it easier to switch to the quantum description in the future, let us introduce the concept of experimental (or logical) *proposition*, sometimes also called *yes–no question*. The most obvious are propositions about individual observables. For example, the proposition $\mathcal{A}$ = "the particle position is in the interval a" is meaningful. In each pure state of the system, this proposition can be either true or false (the question answered "yes" or "no").

Experimentally, such a proposition can be realized using a one-dimensional "Geiger counter", which occupies the region a of space. The counter clicks (= the proposition is true) if the particle passes through the counter's discharge chamber and does not click (= the proposition is false) if the particle is outside the region a.

Propositions can be represented in the phase space. For example, the above proposition $\mathcal{A}$ is associated with the strip A in Figure 1.4. The proposition is true if the point representing the state ϕ is inside the strip A. Otherwise, the proposition is false.

Similarly, propositions about the momentum are represented by strips parallel to the x axis. For example, the strip B in Figure 1.4 corresponds to the logical proposition $\mathcal{B}$ = "the particle momentum belongs to the interval b."

---

5 It might seem more natural to work with the particle's velocity v instead of its momentum. However, we will see later that our choice of primary observables has a special meaning, because x and p are "canonically conjugated" variables.

We will denote by $\mathcal{L}$ the set of all propositions about the physical system.[6] In the rest of this section we will study the structure of this set and establish its connection with the classical *Boolean logic*. The set of all possible states of the system will be denoted $\mathfrak{S}$. In this chapter, our goal is to study the mathematical relationships between the elements $\mathcal{X} \in \mathcal{L}$ and $\phi \in \mathfrak{S}$ in these two sets.

1.2.3 Operations with propositions

In classical theory, in addition to the above propositions $\mathcal{A}$ and $\mathcal{B}$ about single observables, we can also associate a proposition with each region of the phase space. For example, in Figure 1.4 we showed a rectangle C, which is an intersection of the two strips $C = A \cap B$. Apparently, this rectangle also corresponds to an admissible proposition $\mathcal{C} = \mathcal{A} \wedge \mathcal{B}$ = 'the particle position is in the interval a *and* its momentum is in the interval b." In other words, this proposition is obtained by applying the logical operation "AND" to the two elementary propositions $\mathcal{A}$ and $\mathcal{B}$. We denote this logical operation (*meet*) by the symbol $\mathcal{A} \wedge \mathcal{B}$.[7]

The four other logical operations listed in Table 1.1 are also naturally defined in the language of propositions–regions. For example, the rectangle $C = A \cap B$ lies en-

Table 1.1: Five operations and two special elements in the theory of subsets of the phase space $\mathcal{S}$, in the classical logic and in the lattice theory.

Symbol for subsets in $\mathcal{S}$	Name in logic	Meaning in classical logic	Name in lattice theory	Symbol in lattice theory
Operations with subsets/propositions				
$X \subseteq Y$	implication	$\mathcal{X}$ IMPLIES $\mathcal{Y}$	less or equal	$\mathcal{X} \leq \mathcal{Y}$
$X \subseteq Y, X \neq Y$	implication	$\mathcal{X}$ IMPLIES $\mathcal{Y}$	less	$\mathcal{X} < \mathcal{Y}$
$X \cap Y$	conjunction	$\mathcal{X}$ AND $\mathcal{Y}$	meet	$\mathcal{X} \wedge \mathcal{Y}$
$X \cup Y$	disjunction	$\mathcal{X}$ OR $\mathcal{Y}$	join	$\mathcal{X} \vee \mathcal{Y}$
$\mathcal{S} \setminus X$	negation	NOT $\mathcal{X}$	orthocomplement	$\mathcal{X}^{\perp}$
Special subsets/propositions				
$\mathcal{S}$	tautology	always true	maximal element	$\mathcal{I}$
$\emptyset_{\mathcal{S}}$	absurdity	always false	minimal element	$\emptyset$

6 $\mathcal{L}$ is also called the *propositional system* or *logic*.

7 This symbol differs from the symbol $A \cap B$ for the intersection of two regions in the phase space, thereby emphasizing that we are dealing with the logical operation "AND", which relates specifically to propositions. In classical logic, there is an equivalence between propositions and regions in the phase space $\mathcal{S}$, so having two different notations may seem superfluous. However, in the quantum case, such an equivalence is lost, the idea of the phase space is not applicable and only the logical notation $\mathcal{A} \wedge \mathcal{B}$ makes sense.

tirely inside the strip A. From the point of view of logic, we can say that proposition $\mathcal{C}$ "IMPLIES" proposition $\mathcal{A}$. Indeed, in any state where $\mathcal{C}$ is true, the proposition $\mathcal{A}$ is also true. This logical connection will be denoted by $\mathcal{C} \leq \mathcal{A}$.[8]

The proposition $\mathcal{A}$ "OR" $\mathcal{B}$ corresponds to the union ($A \cup B$) of two regions in the phase space. This proposition will be written as $\mathcal{A} \vee \mathcal{B}$. If either $\mathcal{A}$ or $\mathcal{B}$ is true, then the *join* $\mathcal{A} \vee \mathcal{B}$ is definitely true.

The last operation is the complement of a phase-space region A.[9] Obviously its logical equivalent is the negation of the proposition $\mathcal{A}$, which we denote by $\mathcal{A}^{\perp}$ (= "NOT"$\mathcal{A}$, *orthocomplement*).

In addition to these four operations, we will need two special propositions, listed in Table 1.1.

Maximal proposition (or *tautology*) $\mathcal{I} \in \mathcal{L}$ corresponds to the whole phase space, i. e., the maximal subset of $\mathcal{S}$. This proposition can be expressed in different verbal forms. For example: $\mathcal{I}$ = "particle position is somewhere on the real axis" or $\mathcal{I}$ = "particle momentum is somewhere on the real axis." Both these propositions are always true for any state.[10]

Propositions like "the value of the observable is not on the real axis" or "the value of the observable lies in the empty subset of the real axis" are always false and equal to the single *minimal* (or *absurd*) proposition $\emptyset$ in the set $\mathcal{L}$.

1.2.4 Axioms of logic

Five operations and two special propositions, presented above, define a rich mathematical structure. To work with these objects, it is necessary to establish their mutual relations, i. e., laws (or axioms) of logic.

The easiest way to establish these laws is to use the equivalence between logical propositions and subsets of the phase space. This means that the properties of logical operations ("IMPLIES," "AND," "OR," "NOT") coincide with the properties of operations on subsets ("inclusion," "intersection," "union," "complement"). From this analogy, it is not difficult to obtain the laws of classical logic listed in lines 1 through 19 of Table 1.2.[11]

---

8 If $\mathcal{C}$ "IMPLIES" $\mathcal{A}$ and definitely $\mathcal{A} \neq \mathcal{C}$, then we will use the symbol $\mathcal{C} < \mathcal{A}$.

9 That is, the region consisting of phase-space points not belonging to A.

10 Measurements of observables always yield *some* result, because we agreed in the Introduction that an ideal measuring device never misfires.

11 Actually, the choice of the axioms of logic is rather arbitrary. There are different approaches to the axiomatization of logic, and our approach is not the most economical. We tried to select our axioms so that they had the most transparent meaning.

Table 1.2: Basic axioms of classical and quantum logics.

	Name	Formula
Axioms of orthocomplemented lattices		
1	Reflectivity	$\mathcal{X} \le \mathcal{X}$
2	Symmetry	$(\mathcal{X} \le \mathcal{Y})\ \&\ (\mathcal{Y} \le \mathcal{X}) \Rightarrow \mathcal{X} = \mathcal{Y}$
3	Transitivity	$(\mathcal{X} \le \mathcal{Y})\ \&\ (\mathcal{Y} \le \mathcal{Z}) \Rightarrow \mathcal{X} \le \mathcal{Z}$
4	Definition of $\mathcal{I}$	$\mathcal{X} \le \mathcal{I}$
5	Definition of $\emptyset$	$\emptyset \le \mathcal{X}$
6	Definition of $\wedge$	$\mathcal{X} \wedge \mathcal{Y} \le \mathcal{X}$
7	Definition of $\wedge$	$(\mathcal{Z} \le \mathcal{X})\ \&\ (\mathcal{Z} \le \mathcal{Y}) \Rightarrow \mathcal{Z} \le (\mathcal{X} \wedge \mathcal{Y})$
8	Definition of $\vee$	$\mathcal{X} \le \mathcal{X} \vee \mathcal{Y}$
9	Definition of $\vee$	$(\mathcal{X} \le \mathcal{Z})\ \&\ (\mathcal{Y} \le \mathcal{Z}) \Rightarrow (\mathcal{X} \vee \mathcal{Y}) \le \mathcal{Z}$
10	Commutativity	$\mathcal{X} \vee \mathcal{Y} = \mathcal{Y} \vee \mathcal{X}$
11	Commutativity	$\mathcal{X} \wedge \mathcal{Y} = \mathcal{Y} \wedge \mathcal{X}$
12	Associativity	$(\mathcal{X} \vee \mathcal{Y}) \vee \mathcal{Z} = \mathcal{X} \vee (\mathcal{Y} \vee \mathcal{Z})$
13	Associativity	$(\mathcal{X} \wedge \mathcal{Y}) \wedge \mathcal{Z} = \mathcal{X} \wedge (\mathcal{Y} \wedge \mathcal{Z})$
14	Noncontradiction	$\mathcal{X} \wedge \mathcal{X}^\perp = \emptyset$
15	Noncontradiction	$\mathcal{X} \vee \mathcal{X}^\perp = \mathcal{I}$
16	Double negation	$(\mathcal{X}^\perp)^\perp = \mathcal{X}$
17	Contraposition	$\mathcal{X} \le \mathcal{Y} \Rightarrow \mathcal{Y}^\perp \le \mathcal{X}^\perp$
Additional assertions of classical logic		
18	Distributivity	$\mathcal{X} \vee (\mathcal{Y} \wedge \mathcal{Z}) = (\mathcal{X} \vee \mathcal{Y}) \wedge (\mathcal{X} \vee \mathcal{Z})$
19	Distributivity	$\mathcal{X} \wedge (\mathcal{Y} \vee \mathcal{Z}) = (\mathcal{X} \wedge \mathcal{Y}) \vee (\mathcal{X} \wedge \mathcal{Z})$
Additional postulate of quantum logic		
20	Orthomodularity	$\mathcal{X} \le \mathcal{Y} \Rightarrow \mathcal{X} \leftrightarrow \mathcal{Y}$

For example, the transitivity property 3 from Table 1.2 allows us to build syllogisms, such as the one analyzed by Aristotle:

If all humans are mortal,
and all Greeks are humans,
then all Greeks are mortal.

Indeed, we have three propositions: $\mathcal{X}$ = "this is a Greek," $\mathcal{Y}$ = "this is a human being" and $\mathcal{Z}$ = "this is mortal." We know that $\mathcal{X}$ implies $\mathcal{Y}$ (i. e., $\mathcal{X} \le \mathcal{Y}$). We also know that $\mathcal{Y}$ implies $\mathcal{Z}$ ($\mathcal{Y} \le \mathcal{Z}$). Then the transitivity property tells us that $\mathcal{X}$ implies $\mathcal{Z}$ ($\mathcal{X} \le \mathcal{Z}$, i. e., "every Greek is mortal").

Property 14 says that a proposition $\mathcal{X}$ and its negation $\mathcal{X}^\perp$ cannot be true at the same time, i. e., their meet $\mathcal{X} \wedge \mathcal{X}^\perp$ is equal to the absurd proposition $\emptyset$. Property 15 is the famous *tertium non datur* law of logic: either the proposition $\mathcal{X}$ or its negation $\mathcal{X}^\perp$ is true, and *the third is not given.*

A set of objects with operations and special elements from Table 1.1, subject to properties 1–17 from Table 1.2, is referred to as the *orthocomplemented lattice* by mathematicians.

Many useful logical relationships can be derived from the axioms of orthocomplemented lattices. Some of them are formulated in the form of lemmas and theorems in Appendix B.2. However, these axioms 1–17 are still not enough to describe the classical logic of propositions unequivocally. Subsets of the phase space and propositions of classical logic are subject to additional *distributive laws* 18 and 19 in Table 1.2.

Like other properties in the upper portion of Table 1.2, the distributive laws are easily derived from our analogy "proposition" ↔ "region of the phase space." Nevertheless, we put these laws in a separate category. As we shall see in Section 1.4, it is these laws that determine the difference between classical and quantum logics. In Table 1.2 we call them "assertions," because we do not consider them to be true in fundamental quantum theory.[12]

1.2.5 Phase space from axioms of classical logic

Thus, we have shown that in the phase space of classical mechanics the set of all propositions $\mathcal{L}$ is an orthocomplemented lattice with distributive laws 18 and 19. Such lattices will be called *Boolean algebras* or *classical logics*.[13]

For us, it is very important that one can prove the converse statement, which is the following.

Theorem 1.1 (representation of classical logic). *For each classical logic $\mathcal{L}$ defined by properties* 1–19 *from Table* 1.2*, there exist a set $\mathcal{S}$*[14] *and an isomorphism $h(\mathcal{X})$ between logical propositions $\mathcal{X} \in \mathcal{L}$ and subsets of $\mathcal{S}$, such that logical operations in $\mathcal{L}$ match with set-theoretical operations in $\mathcal{S}$, as follows:*

$$\begin{aligned}
\mathcal{X} \le \mathcal{Y} &\Leftrightarrow h(\mathcal{X}) \subseteq h(\mathcal{Y}), \\
h(\mathcal{X} \wedge \mathcal{Y}) &= h(\mathcal{X}) \cap h(\mathcal{Y}), \\
h(\mathcal{X} \vee \mathcal{Y}) &= h(\mathcal{X}) \cup h(\mathcal{Y}), \\
h(\mathcal{X}^{\perp}) &= \mathcal{S} \setminus h(\mathcal{X}), \\
h(\mathcal{I}) &= \mathcal{S}, \\
h(\emptyset) &= \emptyset_{\mathcal{S}};
\end{aligned}$$

see Table 1.1.

12 In our book we distinguish *postulates*, *statements* and *assertions*. Postulates form the basis of our theory. In many cases, they undoubtedly follow from experiments, and we do not question their validity. Statements follow logically from the Postulates, and we consider them to be correct. Assertions refer to claims that are made in other theories, but do not have place in our approach (RQD).

13 Strictly speaking, the definition of classical logic involves also a technical condition of the lattice *atomicity*. In our case this means the existence of "minimal nonzero" propositions – *atoms*, which correspond to points in the phase space.

14 In classical mechanics, the set $\mathcal{S}$ is called the *phase space*.

The importance of this theorem lies in the possibility to derive foundations of classical physics (e. g., the structure of the phase space) from axioms of logic. Starting with the Boolean logic, we come to the idea of the phase space, where states are represented by points. From here it is not far to other elements of classical mechanics, such as, for example, the description of dynamics by trajectories.

1.2.6 Classical observables

In classical mechanics, an observable (= physical quantity) F is represented by a real function $f : \mathcal{S} \to \mathbb{R}$ on the phase space. To each point (= state) of the phase space the function f associates a single number – the value of the observable in this state. Three examples of such observables/functions are shown in Figure 1.5. They are the position x, the momentum p and the energy H of a one-dimensional oscillator (a pendulum) with a quadratic Hamiltonian. The values taken by the corresponding functions f are from the spectra of the observables. In the case of x and p, the spectrum is the entire real axis $\mathbb{R} = (-\infty, +\infty)$, and the spectrum of H is the set of nonnegative numbers $[0, +\infty)$.

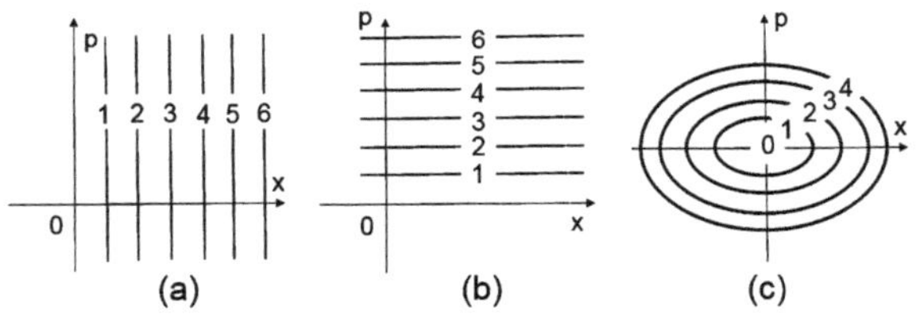

Figure 1.5: Observables in the language of propositions in the phase space: (a) position x, (b) momentum p, (c) energy of the harmonic oscillator $H(x,p) = p^2/(2m) + \alpha x^2$.

Each such function-observable f defines a set of constant-value lines $x, p, H = \dots, 1, 2, 3, 4, \dots$ in $\mathcal{S}$ (shown in Figure 1.5), which in turn can be interpreted as subsets $\mathcal{S}_f$ or logical propositions in the phase space. The proposition $\mathcal{S}_f \in \mathcal{L}$ is pronounced "the observable F has the value f." Thus, each observable can be equivalently described as a map $\mathcal{F}$ from the spectrum of the observable into the set of all propositions $\mathcal{L}$. This map[15] has the following properties:

(1) The function $\mathcal{F}$ associates to each point f of the spectrum of the observable F one and only one logical proposition $\mathcal{S}_f \in \mathcal{L}$.
(2) Propositions corresponding to different points ($f \neq f'$) of the spectrum are disjoint.[16] On the phase plane, such disjoint propositions correspond to nonintersecting regions, otherwise we would have absurd states possessing two different values of the same observable simultaneously.

15 It is also called the *proposition-valued measure*.
16 Two propositions $\mathcal{X}$ and $\mathcal{Y}$ are called *disjoint*, if $\mathcal{X} \leq \mathcal{Y}^\perp$ (or, equivalently, $\mathcal{Y} \leq \mathcal{X}^\perp$).

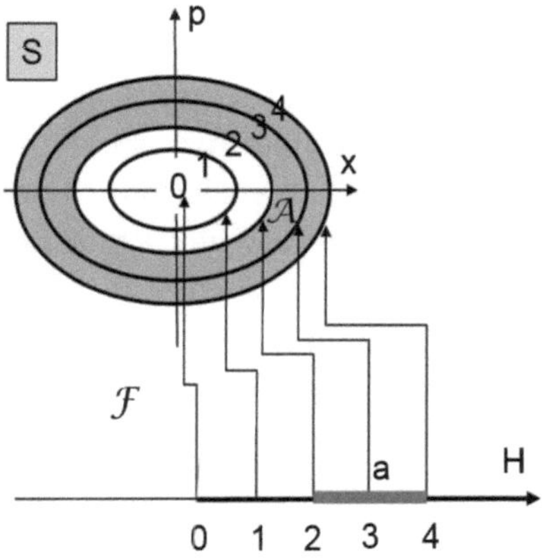

Figure 1.6: Observable H as a mapping $\mathcal{F}$ from the spectrum of H into the set of propositions in the phase space S. The function $\mathcal{F}$ maps the spectrum interval $a = [2, 4]$ to the subset-proposition $\mathcal{A}$.

(3) The join (union) of the propositions $\mathcal{S}_f$ over all spectrum points is equal to the maximal (trivial) proposition ($\vee_f \mathcal{S}_f = \mathcal{I}$), which is equivalent to the entire phase space. This condition indicates that in each state it is possible to measure some value of the observable. There are no states (= points in the phase space) where the observable is not measurable.

So, with each observable F and with each subset a of the real axis $\mathbb{R}$ we associate an experimental *proposition* $\mathcal{A}$ = "the value of the observable F is inside the subset $a \subseteq \mathbb{R}$." Obviously, $\mathcal{A}$ is equal to the join of elementary propositions $\mathcal{S}_f$ over all points of the spectrum lying inside the interval a. The mapping subset→proposition is illustrated in Figure 1.6.

1.3 Measurements and probabilities

In the previous section, we developed the classical logic of strictly deterministic states in which the answer to any yes–no question could be either definite "yes" or definite "no." However, such states are rarely found in real experiments. As a rule, measurements are associated with randomness, uncertainties, errors, etc. To describe such unpredictable outcomes we need the concepts of an ensemble and a probability measure.

1.3.1 Ensembles and measurements

We will call *experiment* a procedure for preparing an *ensemble*[17] and measuring the same observable in each member of the ensemble.[18]

17 Ensemble is a set of identical copies of the physical system, made in – as much as possible – the same conditions.

18 It is important to note that in this book we do not consider repeated measurements performed on the same copy of the physical system. We will assume that after the measurement has been made, the

So, let us prepare many copies of the system, all in one state ϕ (= ensemble) and perform measurements of the same proposition $\mathcal{X}$ in all these copies. As we already know, there is no guarantee that the outcomes of these measurements will be the same. Hence, for some members of the ensemble the proposition $\mathcal{X}$ will be found true, and for other members it will be false. Using these data, we can introduce a function $(\phi|\mathcal{X})$, which we call *probability measure* and which associates to each state ϕ and each proposition $\mathcal{X}$ the probability that $\mathcal{X}$ is true in the state (ensemble) ϕ. The value of this function (a real number in the interval between 0 and 1) is obtained as a result of the following steps:

(i) prepare an instance of the system in the state ϕ;
(ii) make a measurement and determine whether the proposition $\mathcal{X}$ is true or false;
(iii) repeat steps (i) and (ii) N times and calculate the probability by the formula

$$(\phi|\mathcal{X}) = \lim_{N\to\infty} \frac{M}{N},$$

where M is the number of times the proposition $\mathcal{X}$ was found true.

In order to obtain the most complete description of the physical system, it is necessary to perform such experiments with all possible propositions $\mathcal{X} \in \mathcal{L}$ for all possible ensembles (= states) $\phi \in \mathfrak{S}$.

1.3.2 States as probability measures

If we are not too lazy to complete all such measurements, we will notice that the probability measure $(\phi|\mathcal{X})$ has the following properties:

- The probability corresponding to the maximal (trivial) proposition is 1 in all states, so

$$(\phi|\mathcal{I}) = 1. \tag{1.1}$$

- The probability corresponding to the minimal (absurd) proposition is 0 in all states, so

$$(\phi|\emptyset) = 0. \tag{1.2}$$

- The probability corresponding to the join of disjoint propositions is the sum of individual probabilities, so

$$(\phi|\mathcal{X} \vee \mathcal{Y}) = (\phi|\mathcal{X}) + (\phi|\mathcal{Y}), \quad \text{if } \mathcal{X} \leq \mathcal{Y}^{\perp}. \tag{1.3}$$

used copy of the system is discarded. A fresh copy is required for each new measurement. In particular, this means that we are not interested in the state of the system after the measurement. The description of successive measurements in one instance of a physical system is an interesting task, but it is beyond the scope of our book.

The first two statements follow directly from definitions of special logic elements $\mathcal{I}$ and $\emptyset$. The third statement is known as the *third Kolmogorov probability axiom*: the probability of observing either one of the two (or several) mutually exclusive events is equal to the sum of event probabilities.

1.3.3 Probability distributions and statistical mechanics

In classical physics, the description of random events is handled by *statistical mechanics*. In this discipline, states that have an element of randomness are called *mixed classical* states. Mathematically, they are represented by *probability distributions*, which are functions $\rho(x,p)$ on the phase space that

(1) are nonnegative: $\rho(x,p) \geq 0$;
(2) normalized (their integral over the entire phase space is equal to 1),

$$\int_{-\infty}^{+\infty} dx \int_{-\infty}^{+\infty} dp\rho(x,p) = 1;$$

(3) express the probability of the answer *yes* to the question $\mathcal{X}$ by the formula

$$(\phi|\mathcal{X}) = \int_X dxdp\rho(x,p), \tag{1.4}$$

where X is the region of the phase space corresponding to the question (= proposition) $\mathcal{X}$.

By combining the notion of probability distribution with the laws of logic 1–19 from Table 1.2, we can arrive at the classical theory of probability. But we will not dwell on it here. In the next two sections, we will be more interested in quantum logic and quantum probability theory. Here, we will finish our discussion of classical probabilities with a few remarks about determinism.

The randomness present in mixed classical states is usually associated with our inability to provide identical preparation conditions for all members in the ensemble. For example, when we throw a die, it falls in an accidental, unpredictable manner. However, we believe that this unpredictability is simply due to our inability to strictly control the movement of our hand. Thus, classical randomness and probabilities are technical in nature rather than fundamental.

Therefore, classical physics is based on one tacitly assumed axiom, which we formulate here as an assertion.

Assertion 1.2 (full determinism). *It is possible to prepare such ensembles (states) of the physical system where measurements of all observables produce the same result every time. In other words, we assume the existence of pure classical states representable by points in the phase space.*

Pure states are also representable in the language of probability distributions. They correspond to delta-like functions $\rho(x,p) = \delta(x-x_0)\delta(p-p_0)$ on the phase space. Then formula (1.4) confirms that in such states all experimental results are deterministic. A proposition $\mathcal{X}$ is either true ($(\phi|\mathcal{X}) = 1$), if the point (x_0, p_0) of the phase space belongs to the subset X, or false ($(\phi|\mathcal{X}) = 0$) otherwise, without any intermediate possibility.

1.4 Logic of quantum mechanics

To clarify the basic ideas of quantum mechanics, let us return to the experiment with photons passing through one hole (see Subection 1.1.3). We found out that in the low intensity regime, when the photons are emitted one by one, the image on the screen (or photographic plate) consists of individual dots, which are randomly distributed sites of particle (photon) hits. This means that results of measuring the photon position are not reproducible, even if the state preparation conditions are controlled in the most careful way!

From this we conclude that the behavior of photons involves some *random* element. This is the most fundamental statement of quantum mechanics.[19]

Statement 1.3 (fundamental randomness). *Measurements in the microworld have an element of randomness. This randomness is fundamental and cannot be explained or reduced (as we did in the classical case) to some inaccuracies in the preparation of initial states or experimental errors.*

By adopting this Statement, we conclude that classical Assertion 1.2 (complete determinism) is incorrect. In the pinhole camera setup, it is impossible to prepare such an ensemble of photons, in which all of them hit the same point on the screen. What is the reason for this scatter? Honestly, no one knows. This is one of the greatest mysteries of nature. Quantum theory does not even attempt to explain the physical causes of such a random behavior of microsystems. This theory takes randomness as a given and simply tries to find its mathematical description. To proceed, we have to move beyond the simple declaration of randomness and introduce more precise statements and definitions.

1.4.1 Partial determinism of quantum mechanics

We begin our construction of the formalism from the following postulate.

19 In Subection 1.5.2, we will see that this statement is, in fact, a consequence of the even more fundamental Postulate 1.6.

Postulate 1.4 (connection between states and propositions). For each yes–no question there is an ensemble in which the answer "yes" is found with certainty (the probability = 1).

Indeed, it makes no sense to talk about an experimental proposition, if there is not a single ensemble in which this proposition can be unambiguously measured.

In Subsection 1.2.6, we identified observables with mappings from $\mathbb{R}$ into the set of yes–no questions. Thus, each value f of the observable F maps to a proposition S_f. According to Postulate 1.4, we can always prepare an ensemble in which this proposition is 100 % true.

Statement 1.5 (partial determinism). *For each observable F and each value f from its spectrum, an ensemble can be prepared in which measurements of this observable are reproducible, i. e., repeatedly yield the same value f.*

Postulate 1.4 and Statement 1.5 are weakened versions of the classical Assertion 1.2. Instead of requiring the reproducibility of measurements for all observables and propositions at once, we limit this property to single observables.[20] Hence the quantum postulate is a softer requirement, and quantum mechanics is a more general theory than classical mechanics. Moreover, we expect the quantum theory to include classical mechanics as a special case.

So, in quantum mechanics we do not question the existence of propositions about one observable. This means that propositions represented by the strips A and B in Figure 1.4 continue to have well-defined meanings.[21] However, the quantum and classical approaches diverge when it comes to propositions involving more than one observable. For example, in quantum mechanics we cannot guarantee that the proposition corresponding to the rectangle $C = A \cap B$ in Figure 1.4[22] exists and can be realized in the form of an instrumental setup.

Heisenberg was the first to question the simultaneous measurability of certain pairs of observables. He gave the following heuristic arguments. Imagine that we want to accurately measure both the position and the momentum of an electron. For this, we have to look through the microscope. To see the electron, we have to illuminate it. For a more accurate determination of the position we should use light with a short wavelength. However, photons of this light have high energy (momentum). Colliding

20 Notice also that Statement 1.5 does not forbid the existence of certain groups of (*compatible*) observables, whose measurements can be reproducible within the same ensemble. For example, in Chapter 4, we will see that three components (p_x, p_y, p_z) of the particle momentum are compatible observables. The same is true for three components (r_x, r_y, r_z) of the particle position. However, the pairs (p_x, x), (p_y, y) and (p_z, z) are incompatible.

21 That is, there are ensemble states in which these propositions are true.

22 This is a proposition about simultaneous measurement of both the position and the momentum of one particle.

with the electron under study, such photons will inevitably give it a kick, making it impossible to accurately determine the electron's velocity or momentum.

So, we suspect that for sufficiently narrow strips A and B in Figure 1.4 their intersection $C = A \cap B$ may be just "too small" to correspond to any real experimental proposition. Thus, there are no experimental propositions about points in the phase space. In other words, in nature there is no device that could realize the proposition "the particle's position is x_0 and the particle's momentum is p_0." This also means that the true lattice of propositions cannot coincide with the Boolean lattice of subsets in the phase space. What can we do? What lattice should we take to build the logic of questions in quantum physics?

Our plan for constructing quantum theory is as follows:

- First, we will establish the logic of propositions in our theory. We will call it *quantum logic*. As we saw above, we expect it to differ from the classical Boolean logic (= orthocomplemented distributive lattice).
- Next, we will formulate the representation theorem of Piron, which asserts that propositions of quantum logic can be represented by subspaces in a Hilbert space.
- From this result it will be easy to derive all basic properties of the quantum formalism: the superposition principle, the probability interpretation of wave functions, observables as Hermitian operators, etc.

1.4.2 Axioms of quantum logic from probability measures

Note that in our derivation of the axioms of classical logic we used the equivalence between logical propositions and subsets of the phase space. In the quantum case, this equivalence does not work, and we have to look for other ways. To implement the first point of our plan, we will show that many axioms from Table 1.2 can be derived even without reference to the phase space.[23] For these derivations we will need only the simplest properties of probability measures $(\phi|\mathcal{X})$.

Suppose that we have prepared two state ensembles ϕ and ψ of our physical system and measured values of the probability measures $(\phi|\mathcal{X})$ and $(\psi|\mathcal{X})$ by going over all possible experimental propositions $\mathcal{X}$. If, as a result of this gigantic work, we find that $(\phi|\mathcal{X}) = (\psi|\mathcal{X})$ for all $\mathcal{X}$, then the states ϕ and ψ will be regarded as equal ($\phi = \psi$). Indeed, there is no physical difference between these two states, where measurements give the same results (= probabilities).

For similar reasons, we will say that two propositions $\mathcal{X}$ and $\mathcal{Y}$ are equal ($\mathcal{X} = \mathcal{Y}$) if for all states ϕ

$$(\phi|\mathcal{X}) = (\phi|\mathcal{Y}). \tag{1.5}$$

23 These axioms will be transferred without changes from the classical logic to the quantum one.

It then follows that the probability measure $(\phi|\mathcal{X})$, considered as a function on the set of all states $\mathfrak{S}$, is a unique representative of the proposition $\mathcal{X}$.[24] Hence, we can study properties of propositions by analyzing properties of probability measures $(\phi|\mathcal{X})$. For this, there is no need to deal with regions of the phase space, which is exactly what we want.

For example, we will say that $\mathcal{X} \leq \mathcal{Y}$ if measurements for all states $\phi \in \mathfrak{S}$ show that $(\phi|\mathcal{X}) \leq (\phi|\mathcal{Y})$. The relation $\mathcal{X} \leq \mathcal{Y}$ defines the *partial ordering* on the propositional system $\mathcal{L}$.

After the partial ordering $\leq$ is established on the entire set $\mathcal{L}$, it is not difficult to define the *meet* operation $\mathcal{X} \wedge \mathcal{Y}$ for all pairs $\mathcal{X}$, $\mathcal{Y}$. For example, if $\mathcal{X}$ and $\mathcal{Y}$ are given, we should be able to find a set of all propositions $\mathcal{Z}'$ that are "less than or equal to" both $\mathcal{X}$ and $\mathcal{Y}$,[25] $\mathcal{Z}' \leq \mathcal{X}$ and $\mathcal{Z}' \leq \mathcal{Y}$. It is reasonable to assume that there is a single maximal proposition $\mathcal{Z}$ in this set. We shall call it the *meet* of $\mathcal{X}$ and $\mathcal{Y}$: $\mathcal{Z} = \mathcal{X} \wedge \mathcal{Y}$.

The *join* $\mathcal{X} \vee \mathcal{Y}$ for all pairs $\mathcal{X}$, $\mathcal{Y}$ is defined in a similar way: it is the unique smallest proposition that is greater than or equal to both $\mathcal{X}$ and $\mathcal{Y}$. These definitions are formalized as properties 6–9 in Table 1.2. Properties 10–13 follow naturally from these definitions.

Further, suppose that for some pair of propositions $\mathcal{X}$, $\mathcal{Y}$ we notice that for all states $(\phi|\mathcal{X}) = 1 - (\phi|\mathcal{Y})$. Then we will say that the two propositions are orthocomplemented: $\mathcal{X} = \mathcal{Y}^{\perp}$ or, equivalently, $\mathcal{Y} = \mathcal{X}^{\perp}$.

Reasoning in this way, it is not difficult to derive (see Appendix B) all axioms 1–17 of classical logic from Table 1.2, except for the axioms of distributivity 18 and 19. The latter two axioms cannot be justified using our approach with probability measures $(\phi|\mathcal{X})$. For this reason, we regard distributive laws as less valid and call them simply assertions.

So, in quantum mechanics, we are not allowed to use the distributive laws of logic. However, in order to obtain a nontrivial theory, it is necessary to find some kind of substitute for these two laws. On the one hand, this new postulate must be sufficiently specific, so that it can be used to develop a nontrivial logical and physical theory. On the other hand, it must be general enough and include distributive laws as a special case, because we want to have the classical theory as a limiting case of the quantum one.

So, how should we formulate this new quantum axiom?

1.4.3 Compatibility of propositions

To answer this question, let us turn to the important concept of compatibility. We will say that two propositions $\mathcal{X}$ and $\mathcal{Y}$ are *compatible* (denoted $\mathcal{X} \leftrightarrow \mathcal{Y}$) if

$$\mathcal{X} = (\mathcal{X} \wedge \mathcal{Y}) \vee (\mathcal{X} \wedge \mathcal{Y}^{\perp}) \tag{1.6}$$

24 That is, different propositions define different functions $(\phi|\mathcal{X})$ on the set of states $\mathfrak{S}$.

25 At least one such proposition $\emptyset$ always exists.

and

$$\mathcal{Y} = (\mathcal{X} \wedge \mathcal{Y}) \vee (\mathcal{X}^{\perp} \wedge \mathcal{Y}). \tag{1.7}$$

In Subection 1.5.3 we will see that two experimental propositions can be measured simultaneously if and only if they are compatible. Therefore, we should not be surprised by Theorem B.14, which states that the compatibility (= simultaneous measurability) of all propositions is a characteristic property of classical Boolean lattices.

1.4.4 Logic of quantum mechanics

From the Heisenberg microscope example it should be clear that, unlike in the classical case, in quantum mechanics not all propositions are measurable simultaneously (= compatible). Then, as the basic statement of quantum logic, we postulate that two propositions are definitely compatible if one follows from the other, and we leave it to mathematics to decide on the compatibility of other pairs.[26]

Postulate 1.6 (orthomodularity). Propositions about physical systems obey the *orthomodular law*: If $\mathcal{B}$ follows from $\mathcal{A}$, then $\mathcal{A}$ and $\mathcal{B}$ are compatible, i. e.,

$$\mathcal{A} \leq \mathcal{B} \quad \Rightarrow \quad \mathcal{A} \leftrightarrow \mathcal{B}. \tag{1.8}$$

Orthocomplemented lattices[27] with the additional orthomodular Postulate 1.6 (property 20 in Table 1.2) are called *orthomodular lattices*. With the addition of technical conditions of atomicity and irreducibility, these lattices become the so-called *quantum logics*. The relationships between different types of lattices and logics are shown in Figure 1.7.

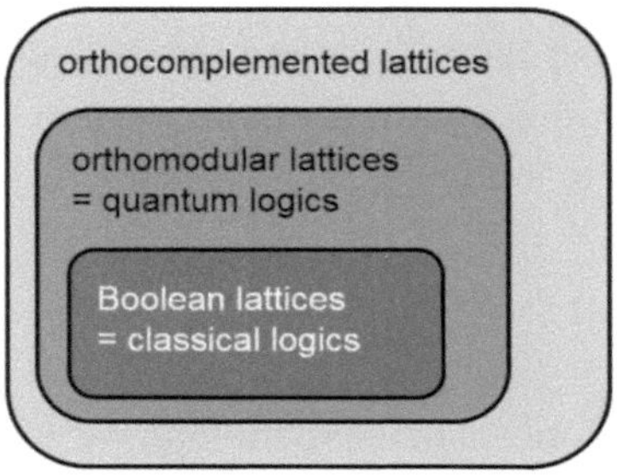

Figure 1.7: Relationships between different types of lattices and logics.

26 The author does not know any deeper justification for this postulate. The strongest argument is that this postulate really works, i. e., it leads to the well-known mathematical structure of quantum mechanics, which has been tested extensively in experiments.
27 That is, described by properties 1–17 in Table 1.2.

1.4.5 Quantum logic and Hilbert space

We have seen that in classical mechanics we do not have to use the exotic lattice theory. Instead, we can apply Theorem 1.1 and go over to the physically transparent language of the phase space. Is there a similar equivalence theorem in the quantum case? The answer is "yes." It is not difficult to notice close analogies between the quantum system of propositions described above and the algebra of projections on closed subspaces in a complex Hilbert space $\mathscr{H}$ (see Appendices F and H). In particular, if operations between projections (or subspaces) in the Hilbert space are translated into the language of lattice operations in accordance with Table 1.3,[28] then all axioms of quantum logic are easily verified.

Table 1.3: Translations of symbols between equivalent languages: (i) subspaces in the Hilbert space $\mathscr{H}$, (ii) projections on these subspaces and (iii) propositions in quantum logic $\mathcal{L}$.

Subspaces in $\mathscr{H}$	Projections in $\mathscr{H}$	Propositions in $\mathcal{L}$
$\mathscr{X} \subseteq \mathscr{Y}$	$P_{\mathscr{X}} P_{\mathscr{Y}} = P_{\mathscr{Y}} P_{\mathscr{X}} = P_{\mathscr{X}}$	$\mathcal{X} \leq \mathcal{Y}$
$\mathscr{X} \cap \mathscr{Y}$	$P_{\mathscr{X} \cap \mathscr{Y}}$	$\mathcal{X} \wedge \mathcal{Y}$
$\mathscr{X} \uplus \mathscr{Y}$	$P_{\mathscr{X} \uplus \mathscr{Y}}$	$\mathcal{X} \vee \mathcal{Y}$
$\mathscr{X}'$	$1 - P_{\mathscr{X}}$	$\mathcal{X}^{\perp}$
$\mathscr{X}$ and $\mathscr{Y}$ compatible	$[P_{\mathscr{X}}, P_{\mathscr{Y}}] = 0$	$\mathcal{X} \leftrightarrow \mathcal{Y}$
$\mathscr{X} \perp \mathscr{Y}$	$P_{\mathscr{X}} P_{\mathscr{Y}} = P_{\mathscr{Y}} P_{\mathscr{X}} = 0$	$\mathcal{X} \leq \mathcal{Y}^{\perp}$
$\emptyset$	0	$\emptyset$
$\mathscr{H}$	1	$\mathcal{I}$
ray x	$\vert x\rangle\langle x\vert$ (1D projection)	x is an atom

For example, the violation of distributive laws follows from the fact that in $\mathscr{H}$ there are pairs of incompatible subspaces (see Appendix H). The validity of the orthomodular law in $\mathscr{H}$ is proved in Theorem H.6.

1.4.6 Piron's theorem

Thus, we have established that the set of subspaces in a complex Hilbert space $\mathscr{H}$ is indeed a representative of some quantum logic. Can we claim also the opposite, i. e., that for each quantum logic one can construct a representation by subspaces in some Hilbert space? The (positive) answer to this question is given by the famous *Piron*

28 We have denoted by $\mathscr{X} \uplus \mathscr{Y}$ the *linear span* of two subspaces $\mathscr{X}$ and $\mathscr{Y}$ (see Appendix C.2). Here $\mathscr{X} \cap \mathscr{Y}$ denotes the intersection of these subspaces and $\mathscr{X}'$ is the orthogonal complement of $\mathscr{X}$. In the Hilbert space *atoms* are one-dimensional subspaces. They are also referred to as *rays*.

theorem [66, 67], which forms the basis of the mathematical formalism of quantum mechanics and is a quantum analog of the classical representation Theorem 1.1.

Theorem 1.7 (Piron's theorem). *Each quantum logic $\mathcal{L}$ is isomorphic to the lattice of closed subspaces in a Hilbert space $\mathscr{H}$. The correspondences proposition $\leftrightarrow$ subspace are defined by the rules shown in Table* 1.3.

The proof of this theorem is beyond the scope of our book.[29]

1.4.7 Should we abandon classical logic?

So, we came to the paradoxical conclusion that classical logic and classical probability theory are not suitable for describing quantum microscopic systems. How could this be? After all, classical logic is the foundation of all mathematics and indeed of the whole scientific method! All proofs of mathematical theorems use the laws of Boolean logic, including the distributive laws that were discarded by us.[30] Even theorems of quantum mechanics are being proved in the framework of classical logic. Are we not entering into a contradiction when we claim that the true logic of experimental statements is not classical, but quantum [70]?

In everyday life, as in ordinary mathematics, we have the right to use inaccurate classical logic, because we usually deal with fixed objects that are not subject to quantum fluctuations. Theorems of Euclidean geometry speak of well-defined circles and triangles, not of statistical ensembles of figures with randomly distributed parameters. Therefore, in the proofs of such theorems, it is perfectly acceptable to use the laws of classical logic. However, when we go to the microworld, where results of measurements are subject to randomness and observables may be incompatible with each other, then we have to admit that classical distributive laws are no longer valid and that quantum logic should take over.

The theory of orthomodular lattices is well known to mathematicians. In principle, we could make all constructions and calculations in quantum theory, based on

29 Piron's theorem does not specify the nature of scalars in the Hilbert space. It leaves the possibility of choosing any *division ring with involutive antiautomorphism* as the collection of scalars in $\mathscr{H}$. We can substantially reduce this unwanted freedom if we recall the important role played by real numbers in physics (for example, the values of observables always lie in $\mathbb{R}$). Therefore, it makes sense to consider only those rings that include $\mathbb{R}$ as a subring. In 1877 Frobenius proved that there are only three such rings. These are *real numbers* $\mathbb{R}$, *complex numbers* $\mathbb{C}$ and *quaternions* $\mathbb{H}$. Although there is fairly extensive literature on the real and, especially, quaternionic quantum mechanics [88, 40, 58, 57], the significance of these exotic theories for physics remains unclear. Therefore, in our book we will adhere to the standard quantum mechanics in complex Hilbert spaces.

30 Note, however, attempts [22] to develop the so-called *quantum mathematics*, which is based on the laws of quantum logic.

this formalism. Such an approach would have certain advantages, since all its components would have a clear physical meaning: the elements $\mathcal{X}$ of the lattice are experimental propositions implementable in the laboratory and the probabilities $(\phi|\mathcal{X})$ can be measured directly in experiments. However, such a theory would encounter insurmountable difficulties, mainly because lattices are rather exotic mathematical objects; we lack experience and intuition to work with them. In addition, this approach would require us to abandon the familiar distributive laws of logic and thus would greatly complicate our reasoning.

Historically, the development of quantum theory took another route. Thanks to Piron's theorem,[31] the physically transparent but mathematically cumbersome lattices could be replaced by physically obscure but mathematically convenient Hilbert spaces, wave functions and Hermitian operators. In the next section we will briefly summarize this traditional formalism.

1.5 Physics in Hilbert space

In the previous section, we established a one-to-one correspondence between experimental propositions and subspaces of the Hilbert space. In this section, we will use this fact to construct the mathematical formalism of quantum mechanics. In particular, we will see that, in accordance with textbooks, observables are expressed by Hermitian operators in $\mathscr{H}$, and pure quantum states are unit length vectors in the same space.

1.5.1 Quantum observables

In Subsection 1.2.6, we saw that, in the language of logic, an observable F is a mapping $\mathcal{F}$ associating a proposition in $\mathcal{L}$ (= a subspace in $\mathscr{H}$) with each point of the spectrum of F. The points f in the spectrum of the observable F are called *eigenvalues* of this observable. The subspace $\mathscr{F}_f \subseteq \mathscr{H}$ corresponding to the eigenvalue f is called the *eigensubspace*, and the projection P_f on this subspace is called the *spectral projection*. Each vector in the eigensubspace $\mathscr{F}_f$ will be called an *eigenvector* of the observable.

Let us consider two distinct eigenvalues f and g of one observable F. According to the definition from Subsection 1.2.6, the corresponding propositions $\mathcal{F}_f$ and $\mathcal{F}_g$ are disjoint, and their eigensubspaces are orthogonal. The linear span of the subspaces $\mathscr{F}_f$, where f runs through the entire spectrum of the observable F, coincides with the entire Hilbert space $\mathscr{H}$. Consequently, spectral projections P_f of any observable form

31 Of course, the fathers of quantum mechanics did not know about this theorem, which was formulated only in the 1960s.

a *resolution of the identity* (see Appendix H.1). Thus, according to formula (H.2), we can associate an Hermitian operator

$$F = \sum_f f P_f \tag{1.9}$$

with each observable F. In the following, we often use the terms "observable" and "Hermitian operator" as synonyms.

1.5.2 States

As we know from Subsection 1.3.2, each state ϕ defines a probability measure $(\phi|\mathcal{X})$ on propositions in $\mathcal{L}$. In accordance with the quantum isomorphism "proposition" $\leftrightarrow$ "subspace," the state ϕ also defines a *probability measure* $(\phi|\mathscr{X})$ on subspaces $\mathscr{X}$ in the Hilbert space $\mathscr{H}$. This probability measure is a function that maps subspaces into the interval $[0,1] \subseteq \mathbb{R}$ and has the following properties:

- The probability corresponding to the entire space $\mathscr{H}$ is equal to 1 in all states,

$$(\phi|\mathscr{H}) = 1. \tag{1.10}$$

- The probability corresponding to the zero (empty) subspace is 0 in all states,

$$(\phi|\emptyset) = 0. \tag{1.11}$$

- The probability corresponding to the direct sum of orthogonal subspaces is the sum of probabilities for each subspace,[32]

$$(\phi|\mathscr{X} \oplus \mathscr{Y}) = (\phi|\mathscr{X}) + (\phi|\mathscr{Y}), \quad \text{if } \mathscr{X} \perp \mathscr{Y}. \tag{1.12}$$

The following important theorem [33] provides a classification of all such probability measures (= all possible states of a quantum system).

Theorem 1.8 (Gleason's theorem). *If $(\phi|\mathscr{X})$ is a probability measure on subspaces in the Hilbert space $\mathscr{H}$ with properties (1.10)–(1.12), then there is a nonnegative[33] Hermitian operator $\hat{\rho}$ in $\mathscr{H}$ such that*

$$\mathrm{Tr}(\hat{\rho}) = 1 \tag{1.13}$$

and for any subspace $\mathscr{X}$ and its projection $P_{\mathscr{X}}$, the value of the probability measure is[34]

$$(\phi|\mathscr{X}) = \mathrm{Tr}(P_{\mathscr{X}}\hat{\rho}). \tag{1.14}$$

32 This is equivalent to the third Kolmogorov probability axiom (1.3). By the symbol $\oplus$ (direct sum) we denote the linear span ($\uplus$) of two subspaces in the case where these subspaces are orthogonal.

33 A Hermitian operator is called *nonnegative* if all its eigenvalues are greater than or equal to zero.

34 Tr means the *trace* of the matrix of the operator $\hat{\rho}$, i. e., the sum of its diagonal elements; see Appendix G.3.

The operator $\hat{\rho}$ is usually called the *density operator* or *density matrix*.

Proving Gleason's theorem is not easy, and we refer the curious reader to the original papers [33, 72]. Here we will only touch upon the physical interpretation of this result. First, in accordance with the spectral Theorem G.6, one can find an orthonormal basis $|e_i\rangle$ in $\mathscr{H}$ where the density operator $\hat{\rho}$ reduces to the diagonal form

$$\hat{\rho} = \sum_i \rho_i |e_i\rangle\langle e_i|, \tag{1.15}$$

where the eigenvalues ρ_i have the properties

$$0 \le \rho_i \le 1, \tag{1.16}$$

$$\sum_i \rho_i = 1. \tag{1.17}$$

Among all states with properties (1.16)–(1.17), one can select those in which only one coefficient ρ_i is nonzero and $\rho_j = 0$ for all other indices $j \neq i$. In this case the density operator reduces to the projection onto a one-dimensional subspace

$$\hat{\rho} = |e_i\rangle\langle e_i|. \tag{1.18}$$

Such states will be referred to as *pure quantum* states. For pure states, the formula (1.14) for calculating probabilities is simplified. Formally, using Lemma G.2 and Theorem H.1, we find that the probability for the proposition $\mathcal{X}$ to be true in the state (1.18) is equal to the square of the modulus of the projection of $|e_i\rangle$ onto the subspace $\mathscr{X}$, i. e.,

$$(\phi|\mathcal{X}) = \mathrm{Tr}(P_{\mathscr{X}}|e_i\rangle\langle e_i|) = \mathrm{Tr}(\langle e_i|P_{\mathscr{X}}|e_i\rangle) = \langle e_i|P_{\mathscr{X}}P_{\mathscr{X}}|e_i\rangle = \|P_{\mathscr{X}}|e_i\rangle\|^2. \tag{1.19}$$

Therefore, it is customary to describe a pure state by a vector $|e_i\rangle$ of unit length chosen arbitrarily from the corresponding one-dimensional subspace (= ray).[35]

In Subsection 1.5.1 we introduced the notion of an eigenvector of the observable F. Pure states corresponding to such eigenvectors will be called *eigenstates* of the observable F. Obviously, observables have definite values (= eigenvalues) in their eigenstates. This means that the eigenstates are precisely those states (= ensembles) whose existence was guaranteed by Statement 1.5.

Importantly, there are no quantum probability measures (= states) that give definite answers to all experimental questions. Thus, by assuming the orthomodularity of

---

35 Obviously, the vector $|e_i\rangle$ is defined only up to a *phase factor* $e^{i\alpha}$, which has a unit modulus ($|e^{i\alpha}| = 1$, where $\alpha \in \mathbb{R}$). Indeed, being substituted in (1.19), the vector $e^{i\alpha}|e_i\rangle$ leads to the same probability value,

$$\left\|P_{\mathscr{X}}\left(e^{i\alpha}|e_i\rangle\right)\right\|^2 = |e^{i\alpha}|^2\|P_{\mathscr{X}}|e_i\rangle\|^2 = \|P_{\mathscr{X}}|e_i\rangle\|^2,$$

so that both vectors $|e_i\rangle$ and $e^{i\alpha}|e_i\rangle$ are legitimate representatives of the state ϕ.

the propositional lattice (Postulate 1.6), we automatically explained the probabilistic nature of quantum states (Statement 1.3).

Mixed quantum states are expressed as mixtures (1.15) of pure states. The coefficients ρ_i in this formula reflect the probabilities of the pure states in the mixture. Thus, in quantum mechanics there are two types of uncertainties. The first type is present in mixed states. This is the same uncertainty familiar to us from classical (statistical) physics. It appears in situations where the experimenter does not have complete control over the preparation of the system, for example, when he throws a die. The second type of uncertainty is present even in pure quantum states (1.18). It has no analog in classical physics, it cannot be gotten rid of by improved control of the initial conditions. This uncertainty reflects the unavoidable presence of chance in microscopic phenomena.

We will not discuss mixed quantum states in this book. Therefore, we will only deal with uncertainties of the second fundamental type. Hence, speaking of a quantum state ϕ, we always have in mind a certain state vector $|\phi\rangle$, determined up to a phase factor $e^{i\alpha}$. In the following, we will use the terms "quantum state" and "state vector" as synonyms.

1.5.3 Complete sets of commuting observables

In Subsection 1.4.3 we defined the idea of compatible propositions, and in Lemma H.5 we showed that the compatibility of propositions is equivalent to the commutativity of the corresponding projections. For physics, these properties are important because for a pair of compatible propositions (= projections, = subspaces), there are states in which both these propositions have certain values, i. e., they are simultaneously measurable without any statistical randomness. Similar claims can be made about two compatible (= commuting) Hermitian operators (= observables). In accordance with Theorem H.9, a pair of such operators has a common basis of eigenvectors (= eigenstates). In these eigenstates both observables have definite (eigen)values.

We assume that for every physical system one can always find at least one minimal and complete set of mutually commuting observables $K, L, M, \ldots$.[36] Then we can construct an orthonormal basis $|e_i\rangle$ of common eigenvectors of these operators so that each such eigenvector is uniquely marked by eigenvalues $k_i, l_i, m_i, \ldots$ of the operators $K, L, M, \ldots$. That is, if $|e_i\rangle$ and $|e_j\rangle$ are two different basis vectors, then their sets of eigenvalues $\{k_i, l_i, m_i, \ldots\}$ and $\{k_j, l_j, m_j, \ldots\}$ are not the same.

36 A set $K, L, M, \ldots$ is called *minimal* if not one observable from this set can be expressed as a function of other observables in the set. The set is *complete* if no new observable can be added to it without destroying the minimality property. An example of a complete set of mutually commuting observables for one massive particle is $\{M, P_x, P_y, P_z, S_z\}$, where M, $\boldsymbol{P}$ and $\boldsymbol{S}$ are the operators of mass, momentum and spin, respectively (see Section 5.1).

1.5.4 Wave functions

Each state vector $|\phi\rangle$ can be represented as a linear combination of the basis vectors constructed in the previous subsection,

$$|\phi\rangle = \sum_i \phi_i |e_i\rangle, \tag{1.20}$$

where in the bra-ket notation (see Appendix F.3) the coefficients are expressed as

$$\phi_i = \langle e_i|\phi\rangle. \tag{1.21}$$

The set of complex numbers ϕ_i can be considered as a function $\phi(k, l, m, \ldots)$ on the common spectrum of the observables $K, L, M, \ldots$. This is called the *wave function* of the state $|\phi\rangle$ in the *representation* defined by the observables $K, L, M, \ldots$. We will discuss examples of one-particle wave functions in Sections 5.2–5.3.

1.5.5 Expectation values

Formula (1.9) defines a spectral resolution of the observable F, where index f runs through all eigenvalues of F. The spectral projections P_f can be expanded through basis eigenvectors, so we have

$$P_f \equiv \sum_{i=1}^{m} |e_i^f\rangle\langle e_i^f|. \tag{1.22}$$

Here $|e_i^f\rangle$ are orthogonal eigenvectors of the operator F that are inside the eigensubspace $\mathscr{F}_f$, and m is the dimension of this subspace.[37] Then from (1.19) one can find the probability for measuring f in each pure state ϕ,

$$(\phi|P_f) = \left\| \sum_{i=1}^{m} |e_i^f\rangle\langle e_i^f|\phi\rangle \right\|^2 = \sum_{i=1}^{m} |\langle e_i^f|\phi\rangle|^2. \tag{1.23}$$

Sometimes we need to know the weighted average, or the *expectation value*, $\langle F\rangle$ of the observable F in the state $|\phi\rangle$,

$$\langle F\rangle \equiv \sum_f (\phi|P_f) f.$$

Substituting here equation (1.23), we obtain

$$\langle F\rangle = \sum_{j=1}^{n} |\langle e_j|\phi\rangle|^2 f_j \equiv \sum_{j=1}^{n} |\phi_j|^2 f_j,$$

37 If $m > 1$, then the eigenvalue f is called *degenerate*.

where the summation is carried out over the entire basis of eigenvectors $|e_j\rangle$. From expansions (1.20), (1.9) and (1.22) it follows that the combination $\langle\phi|F|\phi\rangle$ is a more compact notation for the expectation value $\langle F\rangle$. Indeed

$$\begin{aligned}\langle\phi|F|\phi\rangle &= \Big(\sum_i \phi_i^* \langle e_i|\Big)\Big(\sum_j |e_j\rangle f_j \langle e_j|\Big)\Big(\sum_k \phi_k |e_k\rangle\Big)\\ &= \sum_{ijk} \phi_i^* f_j \phi_k \langle e_i|e_j\rangle\langle e_j|e_k\rangle = \sum_{ijk} \phi_i^* f_j \phi_k \delta_{ij}\delta_{jk}\\ &= \sum_j |\phi_j|^2 f_j = \langle F\rangle. \end{aligned} \tag{1.24}$$

1.5.6 Basic rules of classical and quantum mechanics

The results obtained in this chapter can be summarized as follows. If the physical system is prepared in a pure state ϕ and we want to calculate the probability ω to measure the observable F within the interval $E \subseteq \mathbb{R}$, then we need to perform the following steps.

In classical mechanics:
(1) Determine the phase space $\mathcal{S}$ of the physical system.
(2) Find the real function $f : \mathcal{S} \to \mathbb{R}$ corresponding to our observable F.
(3) Find the subset $U \subseteq \mathcal{S}$ corresponding to the spectral interval E, where U is the collection of all points $s \in \mathcal{S}$ such that $f(s) \in E$ (see Figure 1.6).
(4) Find the point $s_\phi \in \mathcal{S}$ representing the pure classical state ϕ.
(5) The probability ω is 1 if $s_\phi \in U$ and $\omega = 0$ otherwise.

In quantum mechanics:
(1) Determine the Hilbert space $\mathscr{H}$ of the physical system.
(2) Find the Hermitian operator F corresponding to our observable in $\mathscr{H}$.
(3) Find the eigenvalues and eigenvectors of F.
(4) Find the spectral projection P_E corresponding to the spectral interval E.
(5) Find the unit vector $|\phi\rangle$ (defined up to a phase factor) representing the state ϕ in the Hilbert space $\mathscr{H}$.
(6) Substitute all these ingredients in the probaility formula $\omega = \langle\phi|P_E|\phi\rangle$.

At the moment, the classical and quantum recipes seem completely unrelated to each other. Nevertheless, we are sure that such a connection must exist, because we know that both these theories are variants of the probability formalism on orthomodular lattices. In Section 6.6, we will see that in the macroscopic world with massive objects and poor resolution of measuring devices, the classical recipe appears as a reasonable approximation to the quantum one.

1.6 Interpretations of quantum mechanics

So far in this chapter, we were occupied with the mathematical formalism of quantum mechanics. Many details of this formalism (wave functions, superpositions of states, Hermitian operators, nonstandard logic, etc.) seem very abstract and detached from reality. This situation has generated a lot of debates about the physical meaning and interpretation of quantum laws. In this section, we will suggest our point of view on these controversies.

1.6.1 Quantum nonpredictability

Experiments with quantum microsystems revealed one simple but nonetheless mysterious fact: if we prepare N absolutely identical physical systems under the same conditions and measure the value of the same physical quantity, we can obtain N different results.

Let us illustrate this statement with two examples. From experience we know that each photon passing through the aperture of the pinhole camera will hit some point on the photographic plate. However, the next photon, most likely, will hit another point. And, in general, the locations of hits are randomly distributed over the surface. Quantum mechanics does not even try to predict the fate of each individual photon. It only knows how to calculate the probability density for the points of impact, but the behavior of each individual photon remains completely random and unpredictable.

Another example of this – obviously random – behavior is the decay of radioactive nuclei. The ^{232}Th nucleus has a half-life of 14 billion years. This means that in any sample containing thorium, approximately half of all ^{232}Th nuclei will decay during this period. In principle, quantum physicists can calculate the decay probability of a nucleus by solving the corresponding Schrödinger equation.[38] However, they cannot even approximately guess when the given nucleus decays. It can happen today or in 100 billion years.

It would be wrong to think that the probabilistic nature of microscopic systems has little effect on our macroscopic world. Very often the effects of random quantum processes can be amplified and lead to macroscopic phenomena, which are equally random. One well-known example of such amplification is the thought experiment with "Schrödinger's cat" [77].

---

38 Although our current knowledge of the nature of nuclear forces is completely inadequate to perform this kind of calculation for thorium.

1.6.2 Collapse of wave function

In the *orthodox interpretation* of quantum mechanics, the behavior described above is called the "collapse of the quantum probability distribution" and is often surrounded with a certain aura of mystery.[39] In this interpretation, the most controversial point of quantum mechanics is its different attitude to the physical system and the measuring device. The system is regarded as a quantum object that can exist in strange superpositions,[40] while the measuring device is a classical object whose state (readout) is fully unambiguous. It is believed that at the time of measurement, an uncontrolled interaction between the system and the measuring device occurs, so that the superposition collapses into one of its components, which is recorded by the instrument. Inside the theory, this difference of attitudes is expressed in the fact that the system is described by a wave function, but the measuring device is described by an Hermitian operator. This leads to a number of unpleasant questions.

Indeed, the measuring device consists of the same atoms as the physical system and the rest of the universe. Therefore, it is rather strange when such devices are put into a separate category of objects. But if we decided to combine the device and the system into one wave function, when would it collapse? Maybe this collapse would require the participation of a conscious observer? Does this mean that by making observations, we control the course of physical processes?

Sometimes a mystery is seen in the fact that the quantum-mechanical probability distribution (= wave function) has two mutually exclusive laws of evolution. While we are not looking at the system, this distribution develops smoothly and predictably (in accordance with the Schrödinger equation), and at the time of measurement it experiences a random unpredictable collapse.

1.6.3 Collapse of classical probability distribution

By itself, the collapse of the probability distribution is not something strange. A similar collapse occurs in the classical world as well. For example, when shooting from a rifle at a target, it is almost impossible to predict the hit location of each specific bullet. Therefore, the state of the bullet before it hits the target (= before the measurement) is conveniently described by the probability distribution. At the moment of the

---

39 To emphasize the analogy with the classical case, here we specifically talk about the collapse of the "probability distribution," and not about the collapse of the "wave function," as in other works. It is precisely the probability distribution that is subject to experimental observation, and the wave function is a purely theoretical concept.

40 The electron in the previous example is allegedly in a superposition of states smeared over the surface of the photographic plate, and the thorium nucleus is in a superposition of the decayed and undecayed states.

hit, the bullet punches the target at a specific place, and the probability is immediately replaced by certainty. The measurement leads to the "collapse of the probability distribution," exactly as in the formalism of quantum mechanics.

The probability density for the bullet changes smoothly (spreads out) from the moment of the shot and up to the time of impact. The unpredictable collapse of this probability distribution occurs instantaneously in the entire space. These behaviors are completely analogous to the two (allegedly contradictory) variants of quantum evolution, but the classical collapse does not raise any controversy among theorists and philosophers.

We rightly believe that the collapse of classical probability is the natural behavior of any probability distribution. Then, why does the collapse of quantum probability still trouble theoreticians?

The fact is that in the case of the bullet and the target, we are sure that the bullet was *somewhere* at each time instant, even when we did not see it. In all these moments the bullet had a definite position, momentum, rotation speed about its axis (spin) and other properties. Our description of the bullet had some element of randomness only because of our laziness, unwillingness or inability to completely control the act of shooting. By describing the state of the bullet by a probability distribution, we simply admitted the level of our ignorance. When we looked at the pierced target and thus "collapsed" the probability distribution, we had absolutely no influence on the state of the bullet, but simply improved (updated) our knowledge about it. The probability distribution and its collapse are things that occur exclusively in our heads and do not have actual physical existence.

1.6.4 Hidden variables

Einstein believed that the same logic should be applied to measurements in the microworld. He wrote:

> I think that a particle must have a separate reality independent of the measurements. That is an electron has spin, location and so forth even when it is not being measured. I like to think that the moon is there even if I am not looking at it.

If we follow this logic blindly, we must admit that even at the microscopic level, nature must be regular and deterministic. Then the observed randomness of quantum processes should be explained by some yet unknown "hidden" variables that cannot be observed and controlled *yet*. If we exaggerate somewhat, the theory of hidden variables reduces to the idea that each electron has a navigation system that directs it to the designated point on the photographic plate. Each nucleus has an alarm clock inside it, and the nucleus decays at the call of this alarm clock. The behavior of quantum systems only seems random to us, since we have not yet penetrated the designs of these "navigation systems" and "alarm clocks".

According to the theory of "hidden variables," the randomness in the microworld has no special quantum-mechanical nature. This is the same classic pseudo-randomness that we see when shooting at a target or throwing dice. Then we have to admit that modern quantum mechanics is not the last word. Future theory will teach us how to fully describe the properties of individual systems and to predict events without reference to the quantum chance. Of course, such faith cannot be refuted, but so far no one has succeeded in constructing a convincing theory of hidden variables predicting (at least approximately, but beyond the limits of quantum probabilities) the results of microscopic measurements.

1.6.5 Quantum-logical interpretation

The most famous thought experiment in quantum mechanics is the two-hole interference, which demonstrates the limits of classical probability theory. Recall that in this experiment (see Section 1.1 and Subsection 6.6.7) we did not have the right to add the probabilities for passing through alternative holes. Instead, quantum mechanics recommended adding the so-called probability amplitudes and then squaring the resulting sum [28].

This observation leads to the suspicion that the usual postulates of probability (and logic) do not operate in microsystems. Thus, we naturally approach the idea of quantum logic as the basis of quantum mechanics. It turns out that both fundamental features of quantum measurements – the randomness of outcomes and the addition of probability amplitudes for alternative events – find a simple and concise explanation in quantum logic (see Section 1.4). Both these laws of quantum mechanics follow directly from the orthomodular logic of experimental propositions. As we know from Piron's theorem, such logic is realized by a system of projections in the Hilbert space, and by Gleason's theorem any state (= probability measure) on such a system must be stochastic, random.

As we saw in Section 1.4 (see Figure 1.7), the Boolean deterministic logic of classical mechanics is only a particular case of the orthomodular quantum logic with its probabilities. Thus, even in formal reasoning, it is the *particular* classical theory that needs a special explanation and interpretation, and not the *general* quantum mechanics.

> ... classical mechanics is loaded with metaphysical hypotheses which clearly exceed our everyday experience. Since quantum mechanics is based on strongly relaxed hypotheses of this kind, classical mechanics is less intuitive and less plausible than quantum mechanics. Hence classical mechanics, its language and its logic cannot be the basis of an adequate interpretation of quantum mechanics – P. Mittelstaedt [55].

1.6.6 Quantum randomness and limits of knowledge

So, we came to the conclusion that quantum probability, its collapse and the existence of superpositions of states are inevitable consequences of the special orthomodular nature of the logic of experimental propositions. The laws of probability, built on this logic, differ from the classical laws of probability that are familiar to us. In particular, any state (= a probability measure on logical propositions) must be stochastic, i. e., it is impossible to get rid of the element of chance in measurements. This also means that there is no mystery in the collapse of the wave function, and there is no need to introduce an artificial boundary between the physical system and the measuring apparatus.

The imaginary paradox of the quantum formalism is connected, on the one hand, with the weirdness of quantum logic, and on the other hand with unrealistic expectations about the power of science. Theoretical physicists experience an internal protest when faced with real physically measurable effects,[41] which they are powerless to control and/or predict. These are facts without explanations, effects without causes. It seems that microparticles are subject to some annoying mysterious random force. But in our view, instead of grieving, physicists should have celebrated their success.

To us, the idea of the fundamental, irreducible and fundamentally inexplicable nature of quantum probabilities seems very attractive, because it may signal the fulfillment of the centuries-old dream of scientists searching for deep laws of nature. Perhaps, in such an elegant way, nature has evaded the need to answer our endless questions "why?" Indeed, if at the fundamental level nature were deterministic, then we would face a terrifying prospect of unraveling the endless sequences of cause–effect relationships. Each phenomenon would have its own cause, which, in turn, would have a deeper reason, and so on, *ad infinitum*. Quantum mechanics breaks this chain and at some point gives us the full right to answer: "I don't know. It's just an accident." And if some phenomenon is truly random, then there is no need to seek an explanation for it. The chain of questions "why?" breaks. The quest for understanding ends in a logical, natural and satisfying way.

So, perhaps, the apparent "incompleteness" of quantum theory is not a problem to be solved, but an accurate reflection of the fundamental essence of nature, in particular, its inherent unpredictability? In this connection, the following quote from Einstein seems suitable:

> I now imagine a quantum theoretician who may even admit that the quantum-theoretical description refers to ensembles of systems and not to individual systems, but who, nevertheless, clings to the idea that the type of description of the statistical quantum theory will, in its essential features, be retained in the future. He may argue as follows: True, I admit that the quantum-theoretical description is an incomplete description of the individual system. I even admit that

---

41 Such as random hits of electrons on the screen or decay of nuclei.

> a complete theoretical description is, in principle, thinkable. But I consider it proven that the search for such a complete description would be aimless. For the lawfulness of nature is thus constructed that the laws can be completely and suitably formulated within the framework of our incomplete description. To this I can only reply as follows: Your point of view – taken as theoretical possibility – is incontestable – A. Einstein [26].

The most important philosophical lesson of quantum mechanics is the call to abandon speculations about unobservable things and their use in the foundations of the theory. Quantum mechanics does not know whether the moon is there or not. Quantum mechanics says that the moon will be there when we look.

2 Poincaré group

This subject has been thoroughly worked out and is now understood. A thesis on this topic, even a correct one, will not get you a job.
Ray F. Streater

In the previous chapter, we learned that each physical system can be described mathematically in a Hilbert space. Rays in this space are in one-to-one correspondence with (pure) states of the system. The observables are expressed by Hermitian operators. These formal statements are still insufficient to create a working theory. We still lack a classification of possible physical systems; we do not yet know which operators correspond to basic physical quantities, such as position, momentum, mass, energy, spin, etc., and how these operators are related to each other; we cannot yet say how the states and observables develop in time. Our theory is incomplete.

It turns out that many of the mentioned gaps become filled if we combine quantum mechanics with the principle of relativity – one of the most profound ideas in physics. This principle has universal applicability. It works regardless of what physical system, observable or state we are considering. In its essence, this principle says that there is no distinguished inertial frame of reference. All frames are equivalent if they are at rest or move uniformly without acceleration and rotation. In addition, this principle establishes the group properties of inertial transformations between frames of reference. Our main goal in this chapter is to explain that the group of transformations between inertial observers is the famous Poincaré group. Throughout the book, we will have many opportunities to appreciate the fundamental importance of this idea for physics.

2.1 Inertial observers

2.1.1 Principle of relativity

As we said in the Introduction, in our book we consider only inertial laboratories. What is so special about them? The answer is that they are subject to the so-called *relativity principle*. The essence of this principle was best explained by Galilei more than 370 years ago, when he poetically described the movement of insects and drops of water in the cabin of a moving ship [32]. Galilei realized that inertial laboratories cannot be distinguished from each other by performing experiments inside these laboratories. Each experiment performed in one laboratory will produce the same result as an identical experiment in any other inertial laboratory. The results will remain the same no matter how far away these labs are from each other and what are their relative orientations and speeds. Moreover, we can repeat the experiment tomorrow or many

https://doi.org/10.1515/9783110492132-002

years later – the results will not change. These observations allow us to formulate one of the most important postulates of physics.

Postulate 2.1 (principle of relativity). In all inertial laboratories, the laws of physics are the same: they do not change with time and they do not depend on the position, the orientation of the laboratory in space and its speed. In other words, the laws of nature are invariant with respect to inertial transformations.

2.1.2 Inertial transformations

Our next task is to study inertial transformations between laboratories in more detail. Imagine a world populated by observers/laboratories. In such a world, observers can measure parameters $\{\boldsymbol{v}; \boldsymbol{\varphi}; \boldsymbol{r}; t\}$ (speed, angle,[1] distance and time shift) of their fellow observers.

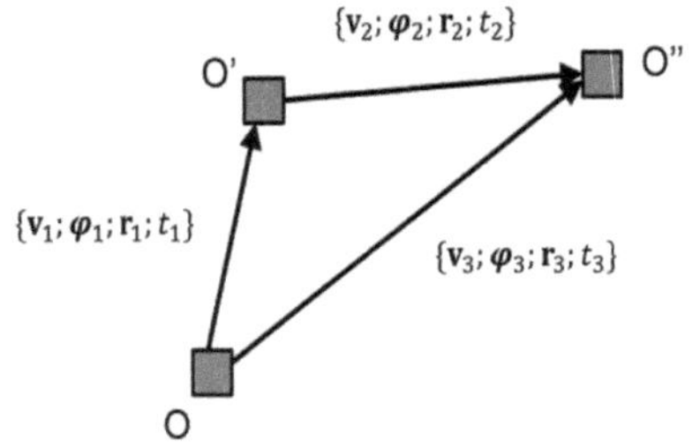

Figure 2.1: Inertial transformations between three reference frames, O, O', O''.

Suppose now that there are three different inertial observers, O, O' and O'', as shown in Figure 2.1. There is an inertial transformation $\{\boldsymbol{v}_1; \boldsymbol{\varphi}_1; \boldsymbol{r}_1; t_1\}$ connecting O and O', written

$$O' = \{\boldsymbol{v}_1; \boldsymbol{\varphi}_1; \boldsymbol{r}_1; t_1\}O, \tag{2.1}$$

where parameters $\boldsymbol{v}_1$, $\boldsymbol{\varphi}_1$, $\boldsymbol{r}_1$ and t_1 are measured by clocks and rulers belonging to the reference frame O with respect to its basis vectors **i**, **j**, **k**. Similarly, there exists an inertial transformation connecting O' and O'', written

$$O'' = \{\boldsymbol{v}_2; \boldsymbol{\varphi}_2; \boldsymbol{r}_2; t_2\}O', \tag{2.2}$$

where parameters $\boldsymbol{v}_2$, $\boldsymbol{\varphi}_2$, $\boldsymbol{r}_2$ and t_2 are defined with respect to basis vectors, rulers and clocks of the observer O'. Finally, there is also a transformation between O and O'', written

$$O'' = \{\boldsymbol{v}_3; \boldsymbol{\varphi}_3; \boldsymbol{r}_3; t_3\}O, \tag{2.3}$$

1 In Appendix D.5 we explain how to parameterize rotations by 3-vectors $\boldsymbol{\varphi}$.

whose parameters refer to O. We can regard transformation (2.3) as a *composition* or a *product* of transformations (2.1) and (2.2),

$$\{\boldsymbol{v}_3;\boldsymbol{\varphi}_3;\boldsymbol{r}_3;t_3\} = \{\boldsymbol{v}_2;\boldsymbol{\varphi}_2;\boldsymbol{r}_2;t_2\}\{\boldsymbol{v}_1;\boldsymbol{\varphi}_1;\boldsymbol{r}_1;t_1\}. \tag{2.4}$$

Obviously, this product has the associativity property (C.1). Also, there is a *trivial* transformation $\{\mathbf{0};\mathbf{0};\mathbf{0};0\}$, which leaves all observers unchanged; and for each transformation $\{\boldsymbol{v};\boldsymbol{\varphi};\boldsymbol{r};t\}$ there is an inverse transformation $\{\boldsymbol{v};\boldsymbol{\varphi};\boldsymbol{r};t\}^{-1}$ such that their product is equal to the trivial one,

$$\{\boldsymbol{v};\boldsymbol{\varphi};\boldsymbol{r};t\}\{\boldsymbol{v};\boldsymbol{\varphi};\boldsymbol{r};t\}^{-1} = \{\boldsymbol{v};\boldsymbol{\varphi};\boldsymbol{r};t\}^{-1}\{\boldsymbol{v};\boldsymbol{\varphi};\boldsymbol{r};t\} = \{\mathbf{0};\mathbf{0};\mathbf{0};0\}. \tag{2.5}$$

In other words, we have just shown that the set of all inertial transformations forms a mathematical group (see Appendix C.1). Moreover, since the inertial transformations depend smoothly on their parameters, this group is a Lie group (see Appendix E.1). The main purpose of this chapter is to study in more detail the properties of this group. In particular, we need explicit formulas for Lie bracket relationships in the group's Lie algebra.

To begin the solution of this problem, let us fix an arbitrary frame of reference O as our basic observer, and consider other examples of observers:

(i) observer $\{\mathbf{0};\mathbf{0};\mathbf{0};t_1\}O$, displaced in time with respect to O by the amount t_1;
(ii) observer $\{\mathbf{0};\mathbf{0};\boldsymbol{r}_1;0\}O$, shifted in space by vector $\boldsymbol{r}_1$;
(iii) observer $\{\mathbf{0};\boldsymbol{\varphi}_1;\mathbf{0};0\}O$, rotated by vector $\boldsymbol{\varphi}_1$;
(iv) observer $\{\boldsymbol{v}_1;\mathbf{0};\mathbf{0};0\}O$, moving with velocity $\boldsymbol{v}_1$.

We assume that the general inertial transformation $\{\boldsymbol{v};\boldsymbol{\varphi};\boldsymbol{r};t\}$ can be always expressed as a product of elementary transformations (i)–(iv). Since these elementary transformations usually do not commute, we need to agree on the canonical order in this product. For our purposes, we found convenient the following choice:

$$\{\boldsymbol{v};\boldsymbol{\varphi};\boldsymbol{r};t\}O = \{\boldsymbol{v};\mathbf{0};\mathbf{0};0\}\{\mathbf{0};\boldsymbol{\varphi};\mathbf{0};0\}\{\mathbf{0};\mathbf{0};\boldsymbol{r};0\}\{\mathbf{0};\mathbf{0};\mathbf{0};t\}O, \tag{2.6}$$

meaning that in order to obtain the observer $O' = \{\boldsymbol{v};\boldsymbol{\varphi};\boldsymbol{r};t\}O$, we first move O in time by the amount t, then shift the resulting observer in space by vector $\boldsymbol{r}$, then rotate it by angle $\boldsymbol{\varphi}$ and, finally, give it velocity $\boldsymbol{v}$.

2.2 Galilei group

In this section, we begin our study of the group of inertial transformations in the nonrelativistic world, where observers move with low velocities $v \ll c$. This is a relatively simple task, because we can use our daily experience and common sense for its completion. We will get to the relativistic group of transformations in Section 2.3 by generalizing results obtained here.

2.2.1 Composition law in Galilei group

Let us first consider four examples of products (2.4), in which $\{\boldsymbol{v}_1; \boldsymbol{\varphi}_1; \boldsymbol{r}_1; t_1\}$ is an inertial transformation of general form and $\{\boldsymbol{v}_2; \boldsymbol{\varphi}_2; \boldsymbol{r}_2; t_2\}$ is one of the elementary transformations from the list (i)–(iv). Applying a time translation to the general frame of reference $\{\boldsymbol{v}_1; \boldsymbol{\varphi}_1; \boldsymbol{r}_1; t_1\}O$, we simply shift its timestamp and change its position in space according to the equality

$$\{\mathbf{0}; \mathbf{0}; \mathbf{0}; t_2\}\{\boldsymbol{v}_1; \boldsymbol{\varphi}_1; \boldsymbol{r}_1; t_1\}O = \{\boldsymbol{v}_1; \boldsymbol{\varphi}_1; \boldsymbol{r}_1 + \boldsymbol{v}_1 t_2; t_1 + t_2\}O. \tag{2.7}$$

Space translation changes the frame's position, as in

$$\{\mathbf{0}; \mathbf{0}; \boldsymbol{r}_2; 0\}\{\boldsymbol{v}_1; \boldsymbol{\varphi}_1; \boldsymbol{r}_1; t_1\}O = \{\boldsymbol{\varphi}_1; \boldsymbol{v}_1; \boldsymbol{r}_1 + \boldsymbol{r}_2; t_1\}O. \tag{2.8}$$

Rotations affect all vector parameters, so we have[2]

$$\{\mathbf{0}; \boldsymbol{\varphi}_2; \mathbf{0}; 0\}\{\boldsymbol{v}_1; \boldsymbol{\varphi}_1; \boldsymbol{r}_1; t_1\}O = \{\boldsymbol{\varphi}_2\boldsymbol{v}_1; \boldsymbol{\varphi}_2 \circ \boldsymbol{\varphi}_1; \boldsymbol{\varphi}_2\boldsymbol{r}_1; t_1\}O. \tag{2.9}$$

Boosts change the velocity, so

$$\{\boldsymbol{v}_2; \mathbf{0}; \mathbf{0}; 0\}\{\boldsymbol{v}_1; \boldsymbol{\varphi}_1; \boldsymbol{r}_1; t_1\}O = \{\boldsymbol{v}_1 + \boldsymbol{v}_2; \boldsymbol{\varphi}_1; \boldsymbol{r}_1; t_1\}O. \tag{2.10}$$

Now, using (2.6)–(2.10), we are ready to calculate the product of two arbitrary inertial transformations in (2.4). We obtain[3]

$$\begin{aligned}
&\{\boldsymbol{v}_2; \boldsymbol{\varphi}_2; \boldsymbol{r}_2; t_2\}\{\boldsymbol{v}_1; \boldsymbol{\varphi}_1; \boldsymbol{r}_1; t_1\}\\
&\quad = \{\boldsymbol{v}_2; \mathbf{0}; \mathbf{0}; 0\}\{\mathbf{0}; \boldsymbol{\varphi}_2; \mathbf{0}; 0\}\{\mathbf{0}; \mathbf{0}; \boldsymbol{r}_2; 0\}\{\mathbf{0}; \mathbf{0}; \mathbf{0}; t_2\}\{\boldsymbol{v}_1; \boldsymbol{\varphi}_1; \boldsymbol{r}_1; t_1\}\\
&\quad = \{\boldsymbol{v}_2; \mathbf{0}; \mathbf{0}; 0\}\{\mathbf{0}; \boldsymbol{\varphi}_2; \mathbf{0}; 0\}\{\mathbf{0}; \mathbf{0}; \boldsymbol{r}_2; 0\}\{\boldsymbol{v}_1; \boldsymbol{\varphi}_1; \boldsymbol{r}_1 + \boldsymbol{v}_1 t_2; t_1 + t_2\}\\
&\quad = \{\boldsymbol{v}_2; \mathbf{0}; \mathbf{0}; 0\}\{\mathbf{0}; \boldsymbol{\varphi}_2; \mathbf{0}; 0\}\{\boldsymbol{v}_1; \boldsymbol{\varphi}_1; \boldsymbol{r}_1 + \boldsymbol{v}_1 t_2 + \boldsymbol{r}_2; t_1 + t_2\}\\
&\quad = \{\boldsymbol{v}_2; \mathbf{0}; \mathbf{0}; 0\}\{\boldsymbol{\varphi}_2\boldsymbol{v}_1; \boldsymbol{\varphi}_2 \circ \boldsymbol{\varphi}_1; \boldsymbol{\varphi}_2\boldsymbol{r}_1 + \boldsymbol{\varphi}_2\boldsymbol{v}_1 t_2 + \boldsymbol{\varphi}_2\boldsymbol{r}_2; t_1 + t_2\}\\
&\quad = \{\boldsymbol{\varphi}_2\boldsymbol{v}_1 + \boldsymbol{v}_2; \boldsymbol{\varphi}_2 \circ \boldsymbol{\varphi}_1; \boldsymbol{\varphi}_2\boldsymbol{r}_1 + \boldsymbol{\varphi}_2\boldsymbol{v}_1 t_2 + \boldsymbol{\varphi}_2\boldsymbol{r}_2; t_1 + t_2\}.
\end{aligned} \tag{2.11}$$

By direct substitution in equation (2.5), it is easy to verify that the inverse element is given by the formula

$$\{\boldsymbol{v}; \boldsymbol{\varphi}; \boldsymbol{r}; t\}^{-1} = \{-\boldsymbol{\varphi}^{-1}\boldsymbol{v}; \boldsymbol{\varphi}^{-1}; -\boldsymbol{r} + \boldsymbol{v}t; -t\}. \tag{2.12}$$

Indeed, according to (2.11),

---

2 Products like $\boldsymbol{\varphi}_2\boldsymbol{v}_1$ and $\boldsymbol{\varphi}_2 \circ \boldsymbol{\varphi}_1$ are defined in Appendices D.5–D.6.

3 Sometimes the product of transformations in the Galilei group is written in other forms; see, for example, Section 3.2 in [7], where the choice of the canonical order of factors is different from our (2.6).

$$
\begin{aligned}
&\{-\boldsymbol{\varphi}^{-1}\boldsymbol{v};\boldsymbol{\varphi}^{-1};-\boldsymbol{r}+\boldsymbol{v}t;-t\}\{\boldsymbol{v};\boldsymbol{\varphi};\boldsymbol{r};t\}\\
&\quad=\{\boldsymbol{\varphi}^{-1}\boldsymbol{v}-\boldsymbol{\varphi}^{-1}\boldsymbol{v};\boldsymbol{\varphi}^{-1}\circ\boldsymbol{\varphi};\boldsymbol{\varphi}^{-1}\boldsymbol{r}-\boldsymbol{\varphi}^{-1}\boldsymbol{v}t-\boldsymbol{\varphi}^{-1}\boldsymbol{r}+\boldsymbol{\varphi}^{-1}\boldsymbol{v}t;t-t\}\\
&\quad=\{\boldsymbol{0};\boldsymbol{0};\boldsymbol{0};0\}.
\end{aligned}
$$

Equations (2.11) and (2.12) define the composition and inversion laws that completely determine the structure of the Lie group of inertial transformations in nonrelativistic physics. It is called the *Galilei group*.

2.2.2 Lie algebra of Galilei group

In physical applications, the Lie algebra of the group of inertial transformations plays an even larger role than the group itself. According to Appendix E, we can get the basis $\{\mathcal{H},\mathcal{P},\mathcal{K},\mathcal{J}\}$ of the Galilei Lie algebra by taking derivatives with respect to the parameters of one-parameter subgroups. For example, the *time translation generator* is formally represented as

$$\mathcal{H}=\lim_{t\to 0}\frac{d}{dt}\{\boldsymbol{0};\boldsymbol{0};\boldsymbol{0};t\}.$$

For generators of space translations and boosts along the x-axis we obtain

$$
\begin{aligned}
\mathcal{P}_x&=\lim_{x\to 0}\frac{d}{dx}\{\boldsymbol{0};\boldsymbol{0};x,0,0;0\},\\
\mathcal{K}_x&=\lim_{v\to 0}\frac{d}{dv}\{v,0,0;\boldsymbol{0};\boldsymbol{0};0\}.
\end{aligned}
$$

The generator of rotations about the x-axis is defined as

$$\mathcal{J}_x=\lim_{\varphi\to 0}\frac{d}{d\varphi}\{\boldsymbol{0};\varphi,0,0;\boldsymbol{0};0\}.$$

Similar formulas are valid for the y- and z-components. According to (E.2), finite transformations are expressed as exponents of the generators

$$\{\boldsymbol{0};\boldsymbol{0};\boldsymbol{0};t\}=e^{\mathcal{H}t}\approx 1+\mathcal{H}t, \tag{2.13}$$

$$\{\boldsymbol{0};\boldsymbol{0};\boldsymbol{r};0\}=e^{\mathcal{P}\cdot\boldsymbol{r}}\approx 1+\mathcal{P}\cdot\boldsymbol{r}, \tag{2.14}$$

$$\{\boldsymbol{0};\boldsymbol{\varphi};\boldsymbol{0};0\}=e^{\mathcal{J}\cdot\boldsymbol{\varphi}}\approx 1+\mathcal{J}\cdot\boldsymbol{\varphi},$$

$$\{\boldsymbol{v};\boldsymbol{0};\boldsymbol{0};0\}=e^{\mathcal{K}\cdot\boldsymbol{v}}\approx 1+\mathcal{K}\cdot\boldsymbol{v}. \tag{2.15}$$

Hence, any element of the group is a product of exponentials in the canonical order (2.6), so we have

$$\{\boldsymbol{v};\boldsymbol{\varphi};\boldsymbol{r};t\}\equiv\{\boldsymbol{v};\boldsymbol{0};\boldsymbol{0};0\}\{\boldsymbol{0};\boldsymbol{\varphi};\boldsymbol{0};0\}\{\boldsymbol{0};\boldsymbol{0};\boldsymbol{r};0\}\{\boldsymbol{0};\boldsymbol{0};\boldsymbol{0};t\}=e^{\mathcal{K}\cdot\boldsymbol{v}}e^{\mathcal{J}\cdot\boldsymbol{\varphi}}e^{\mathcal{P}\cdot\boldsymbol{r}}e^{\mathcal{H}t}. \tag{2.16}$$

Let us now find commutators between the generators, i. e., structure constants of the Galilei Lie algebra. Consider, for example, translations in time and space. Using (2.11), we obtain

$$\{\mathbf{0};\mathbf{0};\mathbf{0};t\}\{\mathbf{0};\mathbf{0};x,0,0;0\} = \{\mathbf{0};\mathbf{0};x,0,0;0\}\{\mathbf{0};\mathbf{0};\mathbf{0};t\}.$$

This implies

$$e^{\mathcal{H}t}e^{\mathcal{P}_x x} = e^{\mathcal{P}_x x}e^{\mathcal{H}t},$$
$$1 = e^{\mathcal{P}_x x}e^{\mathcal{H}t}e^{-\mathcal{P}_x x}e^{-\mathcal{H}t}.$$

From equations (2.13) and (2.14) we obtain in the first order in x and t

$$\begin{aligned}1 &\approx (1+\mathcal{P}_x x)(1+\mathcal{H}t)(1-\mathcal{P}_x x)(1-\mathcal{H}t)\\ &= 1+\mathcal{P}_x\mathcal{H}xt-\mathcal{P}_x\mathcal{H}xt-\mathcal{H}\mathcal{P}_x xt+\mathcal{P}_x\mathcal{H}xt+\cdots\\ &= 1-\mathcal{H}\mathcal{P}_x xt+\mathcal{P}_x\mathcal{H}xt+\cdots.\end{aligned}$$

Hence the Lie bracket of the generators of $\mathcal{P}_x$ and $\mathcal{H}$ vanishes:

$$[\mathcal{P}_x,\mathcal{H}]_L \equiv \mathcal{P}_x\mathcal{H}-\mathcal{H}\mathcal{P}_x = 0.$$

Similarly, one can find zero Lie brackets,

$$[\mathcal{H},\mathcal{P}_i]_L = [\mathcal{P}_i,\mathcal{P}_j]_L = [\mathcal{K}_i,\mathcal{K}_j]_L = [\mathcal{K}_i,\mathcal{P}_j]_L = 0,$$

for all components $i,j = x,y,z$ (or $i,j = 1,2,3$). The noncommuting pair of time translation and boost is more interesting. From the general formula (2.11) we obtain

$$\begin{aligned}e^{\mathcal{K}_x v}e^{\mathcal{H}t}e^{-\mathcal{K}_x v} &= \{v,0,0;\mathbf{0};\mathbf{0},0\}\{\mathbf{0};\mathbf{0};\mathbf{0};t\}\{-v,0,0;\mathbf{0};\mathbf{0};0\}\\ &= \{v,0,0;\mathbf{0};\mathbf{0};0\}\{-v,0,0;\mathbf{0};-vt,0,0;t\}\\ &= \{\mathbf{0};\mathbf{0};-vt,0,0;t\} = e^{\mathcal{H}t}e^{-\mathcal{P}_x vt} = e^{\mathcal{H}t-\mathcal{P}_x vt}.\end{aligned}$$

On the other hand, applying (E.14) to the left-hand side of this equality, we get

$$e^{\mathcal{K}_x v}e^{\mathcal{H}t}e^{-\mathcal{K}_x v} = e^{(\mathcal{H}+v[\mathcal{K}_x,\mathcal{H}]_L+\cdots)t}.$$

Therefore

$$\mathcal{H}t + [\mathcal{K}_x,\mathcal{H}]_L vt = \mathcal{H}t - \mathcal{P}_x vt,$$
$$[\mathcal{K}_x,\mathcal{H}]_L = -\mathcal{P}_x.$$

Proceeding in a similar way with other pairs of transformations, we obtain the complete set of Lie brackets of the Galilei Lie algebra,

$$[\mathcal{J}_i,\mathcal{P}_j]_L = \sum_{k=1}^{3}\epsilon_{ijk}\mathcal{P}_k, \tag{2.17}$$

$$[\mathcal{J}_i,\mathcal{J}_j]_L = \sum_{k=1}^{3}\epsilon_{ijk}\mathcal{J}_k, \tag{2.18}$$

$$[\mathcal{J}_i, \mathcal{K}_j]_L = \sum_{k=1}^{3} \epsilon_{ijk} \mathcal{K}_k, \quad (2.19)$$

$$[\mathcal{J}_i, \mathcal{H}]_L = 0, \quad (2.20)$$

$$[\mathcal{P}_i, \mathcal{P}_j]_L = [\mathcal{P}_i, \mathcal{H}]_L = 0, \quad (2.21)$$

$$[\mathcal{K}_i, \mathcal{K}_j]_L = 0, \quad (2.22)$$

$$[\mathcal{K}_i, \mathcal{P}_j]_L = 0, \quad (2.23)$$

$$[\mathcal{K}_i, \mathcal{H}]_L = -\mathcal{P}_i. \quad (2.24)$$

Analyzing these relations, we can distinguish several important subalgebras and, consequently, subgroups of the Galilei group. In particular, there is the Abelian subgroup of spatial and time translations (with generators $\boldsymbol{\mathcal{P}}$ and $\mathcal{H}$, respectively), the subgroup of rotations (with generators $\boldsymbol{\mathcal{J}}$) and the Abelian subgroup of boosts (with generators $\boldsymbol{\mathcal{K}}$).

2.2.3 Rotations applied to generators

Let us consider two reference frames O and O' related to each other by an element g of the Galilei group,

$$O' = gO.$$

Suppose that observer O applies an (active[4]) inertial transformation h to a physical object S. Let, for example, h be a translation along the x-axis, as in Figure 2.2. We want to find the transformation h' that refers to the observer O' in the same way as h refers to O (that is, h' is a translation along the x'-axis belonging to the observer O'). As can be seen from the example in Figure 2.2, the transformation h' of the object S can be performed in three steps: first go from the frame O' to the frame O, perform the translation h there, and then go back to the frame O', so

$$h' = ghg^{-1}.$$

Similarly, if $\mathcal{G}$ is a generator of a one-dimensional subgroup of inertial transformations in the frame O, then

$$\mathcal{G}' = g\mathcal{G}g^{-1} \quad (2.25)$$

is "the same" generator in the reference frame $O' = gO$.

4 So far we have discussed how inertial transformations are applied to observers; such transformations are called "passive." But inertial transformations can be also applied to the physical system (or the state preparation device). They are called "active." We will discuss this difference in more detail in Subsection 5.3.3.

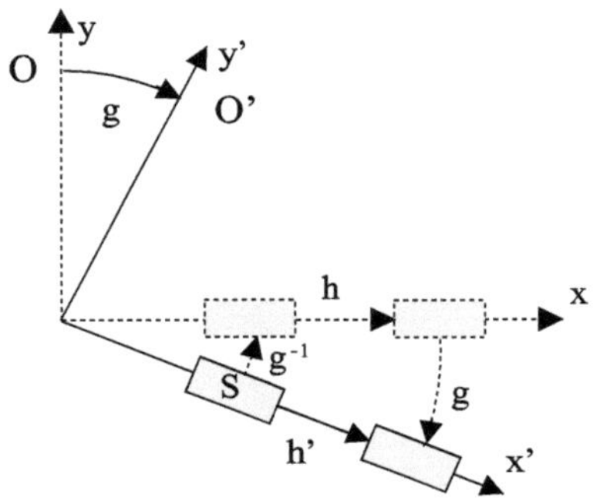

Figure 2.2: Connection between similar transformations h and h' in different reference frames. The transformation $g = \exp(J_z\varphi)$ is a rotation about the z-axis, which is perpendicular to the book's page.

As an example, let us consider the effect of a rotation about the z-axis on generators of the Galilei group. We can write

$$\begin{aligned}
\mathcal{G}'_x &\equiv \mathcal{G}_x(\varphi) = e^{\mathcal{J}_z\varphi}\mathcal{G}_x e^{-\mathcal{J}_z\varphi}, \\
\mathcal{G}'_y &\equiv \mathcal{G}_y(\varphi) = e^{\mathcal{J}_z\varphi}\mathcal{G}_y e^{-\mathcal{J}_z\varphi}, \\
\mathcal{G}'_z &\equiv \mathcal{G}_z(\varphi) = e^{\mathcal{J}_z\varphi}\mathcal{G}_z e^{-\mathcal{J}_z\varphi},
\end{aligned}$$

where $\boldsymbol{\mathcal{G}}$ is one of the vector generators ($\boldsymbol{\mathcal{P}}$, $\boldsymbol{\mathcal{J}}$ or $\boldsymbol{\mathcal{K}}$) in the frame O. From formulas (2.17)–(2.19) we obtain

$$\frac{d}{d\varphi}\mathcal{G}_x(\varphi) = e^{\mathcal{J}_z\varphi}(\mathcal{J}_z\mathcal{G}_x - \mathcal{G}_x\mathcal{J}_z)e^{-\mathcal{J}_z\varphi} = e^{\mathcal{J}_z\varphi}\mathcal{G}_y e^{-\mathcal{J}_z\varphi} = \mathcal{G}_y(\varphi), \tag{2.26}$$

$$\frac{d}{d\varphi}\mathcal{G}_y(\varphi) = e^{\mathcal{J}_z\varphi}(\mathcal{J}_z\mathcal{G}_y - \mathcal{G}_y\mathcal{J}_z)e^{-\mathcal{J}_z\varphi} = -e^{\mathcal{J}_z\varphi}\mathcal{G}_x e^{-\mathcal{J}_z\varphi} = -\mathcal{G}_x(\varphi),$$

$$\frac{d}{d\varphi}\mathcal{G}_z(\varphi) = e^{\mathcal{J}_z\varphi}(\mathcal{J}_z\mathcal{G}_z - \mathcal{G}_z\mathcal{J}_z)e^{-\mathcal{J}_z\varphi} = 0. \tag{2.27}$$

Taking derivatives with respect to φ from both sides of (2.26), we obtain a second-order differential equation,

$$\frac{d^2}{d\varphi^2}\mathcal{G}_x(\varphi) = \frac{d}{d\varphi}\mathcal{G}_y(\varphi) = -\mathcal{G}_x(\varphi),$$

with the general solution

$$\mathcal{G}_x(\varphi) = \mathcal{B}\cos\varphi + \mathcal{D}\sin\varphi,$$

where $\mathcal{B}$ and $\mathcal{D}$ are arbitrary functions of generators. From the initial conditions it follows that

$$\begin{aligned}
\mathcal{B} &= \mathcal{G}_x(0) = \mathcal{G}_x, \\
\mathcal{D} &= \frac{d}{d\varphi}\mathcal{G}_x(\varphi)\Big|_{\varphi=0} = \mathcal{G}_y
\end{aligned}$$

and the full solution is

$$\mathcal{G}_x(\varphi) = \mathcal{G}_x \cos\varphi + \mathcal{G}_y \sin\varphi. \tag{2.28}$$

Similar calculations yield

$$\mathcal{G}_y(\varphi) = -\mathcal{G}_x \sin\varphi + \mathcal{G}_y \cos\varphi, \tag{2.29}$$
$$\mathcal{G}_z(\varphi) = \mathcal{G}_z. \tag{2.30}$$

Comparing (2.28)–(2.30) with equation (D.11), we see that

$$\mathcal{G}_i' \equiv e^{\mathcal{J}_z\varphi}\mathcal{G}_i e^{-\mathcal{J}_z\varphi} = \sum_{j=1}^{3}(R_z)_{ij}\mathcal{G}_j, \tag{2.31}$$

where R_z is the matrix of rotation about the z-axis. Using equation (D.22), we can generalize our result to the case of an arbitrary rotation $\{\mathbf{0}; \boldsymbol{\varphi}; \mathbf{0}; 0\}$,

$$\begin{aligned}\boldsymbol{\mathcal{G}}' &= e^{\boldsymbol{\mathcal{J}}\cdot\boldsymbol{\varphi}}\boldsymbol{\mathcal{G}}e^{-\boldsymbol{\mathcal{J}}\cdot\boldsymbol{\varphi}}\\ &= \boldsymbol{\mathcal{G}}\cos\varphi + \frac{\boldsymbol{\varphi}}{\varphi}\left(\boldsymbol{\mathcal{G}}\cdot\frac{\boldsymbol{\varphi}}{\varphi}\right)(1-\cos\varphi) + \left[\boldsymbol{\mathcal{G}}\times\frac{\boldsymbol{\varphi}}{\varphi}\right]\sin\varphi\\ &\equiv R_{\boldsymbol{\varphi}}\boldsymbol{\mathcal{G}} \equiv \boldsymbol{\varphi}\boldsymbol{\mathcal{G}}.\end{aligned} \tag{2.32}$$

This simply means that generators $\boldsymbol{\mathcal{G}} = \boldsymbol{\mathcal{P}}, \boldsymbol{\mathcal{J}}, \boldsymbol{\mathcal{K}}$ are 3-vectors. From the Lie bracket (2.20) it follows that $\mathcal{H}$ is not affected by rotations, i. e., this is a 3-scalar.

2.2.4 Space inversion

In this book, we will not discuss physical consequences of discrete transformations, such as spatial inversion or time reversal. It is physically impossible to prepare an exact mirror image or a time-reversed image of a laboratory, so the postulate of relativity does not tell us anything about such transformations. Indeed, it has been proven experimentally that such discrete symmetries are not exact. Nevertheless, it will be useful to find out how the Galilei generators behave with respect to space inversion.

Suppose that we have a classical physical system S and its image S' inverted with respect to the origin of coordinates $\mathbf{0}$ (see Figure 2.3). How will the image S' change if we apply an inertial transformation to S?

Obviously, if we shift S by the vector $\boldsymbol{x}$, then S' shifts by $-\boldsymbol{x}$. This can be interpreted as a change of the sign of the spatial translation generator $\boldsymbol{\mathcal{P}}$ under inversion. The same is true for boosts: the inverted image of S' gets the additional speed of $-\boldsymbol{v}$ if the original was boosted by $\boldsymbol{v}$. Hence, the inversion also changes the sign of the boost generator, so

$$\boldsymbol{\mathcal{K}} \to -\boldsymbol{\mathcal{K}}. \tag{2.33}$$

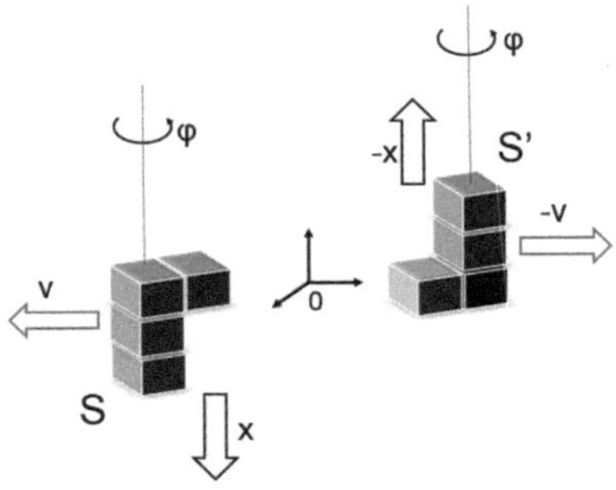

Figure 2.3: Transformations of generators under space inversion.

Such vectors as $\mathcal{P}$ and $\mathcal{K}$, changing their signs under inversion, are called *true vectors*. However, the rotation generator $\mathcal{J}$ is not a true vector. Indeed, if we rotate S by the angle $\boldsymbol{\varphi}$, then the image S' rotates by the same angle (see Figure 2.3). Hence, $\mathcal{J}$ does not change its sign after inversion. Such vectors are called *pseudo-vectors*. Similarly, we can introduce the concepts of true scalars/pseudo-scalars and true tensors/pseudo-tensors. It is customary to define their properties as opposite to the properties of true vectors/pseudo-vectors. In particular, true scalars and true tensors (of rank 2) do not change their signs under inversion. For example, $\mathcal{H}$ is a true scalar. Pseudo-scalars and pseudo-tensors of the second rank change their sign under inversion.

2.3 Poincaré group

It turns out that the Galilei group described in the previous section is suitable only for observers moving at low speeds. In a more general case, we have to use a different law of composition of inertial transformations. The resulting group is called the *Poincaré group*. This is a very important result that follows from the theory of relativity developed in the early 20th century by Einstein, Poincaré, Minkowski and others.

The derivation of the relativistic group of inertial transformations is not an easy task, because in everyday practice we do not have enough experience of observations from fast moving frames of reference. Therefore, we will use more formal mathematical arguments [15]. We will attempt to generalize commutation relations of the Galilei algebra (2.17)–(2.24) while requiring some simple physical conditions. As a result, we will see that the Lie algebra satisfying all our demands is practically unique. This is the Lie algebra of the Poincaré group.

2.3.1 Conditions on Poincaré generators

We cannot doubt the validity of the Galilean Lie brackets between the generators of space–time translations and rotations, because properties of these transformations were verified both in daily life and in physical experiments in a wide range of parameters (distances, times and angles). The situation with boosts is completely different. In everyday practice, we do not experience high speeds (comparable to the speed of light), and we lack the intuition that was so useful in deriving the commutators of the Galilei algebra. Therefore, we cannot be sure of the reasoning that led us to the Lie brackets (2.22)–(2.24) that involve boost generators; we should be open to the possibility that these brackets are just approximations valid only for slow observers. So, in our derivation of the exact relativistic group of inertial transformations, we apply the following ideas:

(I) Just as in the nonrelativistic case, we assume that the set of inertial transformations remains a 10-parameter Lie group.

(II) We also assume that it is necessary to revise only the brackets involving the boost generators.

(III) In addition, we assume that relativistic generators of boosts $\mathcal{K}$ continue to be components of a true vector, so that equations (2.19) and (2.33) remain valid.

(IV) The Galilei group perfectly describes slow boosts, and the speed of light c is the only fundamental constant with the dimension of speed. Therefore, we assume that the exact Lie brackets include c as a parameter and tend to their Galilean values in the limit $c \to \infty$.[5]

Summarizing requirements (I)–(IV), we can write the following relativistic generalizations for the Lie brackets (2.22)–(2.24):

$$[\mathcal{K}_i, \mathcal{K}_j]_L = \mathcal{T}_{ij},$$

$$[\mathcal{K}_i, \mathcal{P}_j]_L = \mathcal{U}_{ij}, \tag{2.34}$$

$$[\mathcal{K}_i, \mathcal{H}]_L = -\mathcal{P}_i + \mathcal{V}_i, \tag{2.35}$$

where $\mathcal{T}_{ij}$, $\mathcal{U}_{ij}$ and $\mathcal{V}_{ij}$ are some (as yet unknown) linear combinations of generators. The coefficients of these linear combinations must be chosen in such a way as to preserve all the characteristic properties of the Lie algebra, in particular the Jacobi identity (E.11). Let us now try to fulfill all these requirements step by step.

2.3.2 Lie algebra of Poincaré group

First, we note that the Lie bracket $[\mathcal{K}_i, \mathcal{P}_j]_L$ is a 3-tensor. Indeed, using equation (2.32), we get the tensor law (D.14) for transformations with respect to rotations. We have

5 Note that here we do not assume that c is an invariant or some kind of a limiting speed. These facts will appear in Chapter 5 as a result of applying our theory.

$$e^{\mathcal{J}\cdot\boldsymbol{\varphi}}[\mathcal{K}_i, \mathcal{P}_j]_L e^{-\mathcal{J}\cdot\boldsymbol{\varphi}} = \left[\sum_{k=1}^{3} R_{ik}(\boldsymbol{\varphi})\mathcal{K}_k, \sum_{l=1}^{3} R_{jl}(\boldsymbol{\varphi})\mathcal{P}_l\right]_L$$
$$= \sum_{kl=1}^{3} R_{ik}(\boldsymbol{\varphi})R_{jl}(\boldsymbol{\varphi})[\mathcal{K}_k, \mathcal{P}_l]_L.$$

Since both $\boldsymbol{\mathcal{K}}$ and $\boldsymbol{\mathcal{P}}$ change their sign under inversion, their Lie bracket is a true tensor. Therefore, $\mathcal{U}_{ij}$ must also be a true tensor, expressed as a linear combination of generators, among which we have a true scalar $\mathcal{H}$, a pseudo-vector $\boldsymbol{\mathcal{J}}$ and two true vectors $\boldsymbol{\mathcal{P}}$ and $\boldsymbol{\mathcal{K}}$. In accordance with Appendix D.4, the most general way to make a true tensor from all these ingredients is to apply formulas from the first and third rows of Table D.1. Hence we arrive at the following expression for the Lie bracket (2.34):

$$[\mathcal{K}_i, \mathcal{P}_j]_L = -\beta\mathcal{H}\delta_{ij} + \gamma\sum_{k=1}^{3} \epsilon_{ijk}\mathcal{J}_k,$$

where β and γ are as yet undefined real constants.

It follows from similar arguments that $\mathcal{T}_{ij}$ is also a true tensor. Because of the relation

$$[\mathcal{K}_i, \mathcal{K}_j]_L = -[\mathcal{K}_j, \mathcal{K}_i]_L,$$

this tensor is antisymmetric, which excludes the presence of the term proportional to δ_{ij}. So

$$[\mathcal{K}_i, \mathcal{K}_j]_L = \alpha\sum_{k=1}^{3} \epsilon_{ijk}\mathcal{J}_k,$$

where α is yet another undefined constant.

The quantity $\mathcal{V}_i$ in equation (2.35) should be a true vector, so the most general form of this Lie bracket is

$$[\mathcal{K}_i, \mathcal{H}]_L = -(1+\sigma)\mathcal{P}_i + \kappa\mathcal{K}_i.$$

Thus, we simplified the problem of generalizing the Galilei Lie algebra to the search for five real (presumably small) parameters α, β, γ, κ and σ. To proceed, we note that the modified Lie algebra must satisfy the Jacobi identity. Therefore, we can write

$$0 = [\mathcal{P}_x, [\mathcal{K}_x, \mathcal{H}]_L]_L + [\mathcal{K}_x, [\mathcal{H}, \mathcal{P}_x]_L]_L + [\mathcal{H}, [\mathcal{P}_x, \mathcal{K}_x]_L]_L$$
$$= \kappa[\mathcal{P}_x, \mathcal{K}_x]_L = \beta\kappa\mathcal{H},$$

which implies

$$\beta\kappa = 0. \tag{2.36}$$

Similarly,

$$\begin{aligned}0 &= [\mathcal{K}_x, [\mathcal{K}_y, \mathcal{P}_y]_L]_L + [\mathcal{K}_y, [\mathcal{P}_y, \mathcal{K}_x]_L]_L + [\mathcal{P}_y, [\mathcal{K}_x, \mathcal{K}_y]_L]_L \\ &= -\beta[\mathcal{K}_x, \mathcal{H}]_L - \gamma[\mathcal{K}_y, \mathcal{J}_z]_L + \alpha[\mathcal{P}_y, \mathcal{J}_z]_L = \beta(1+\sigma)\mathcal{P}_x - \beta\kappa\mathcal{K}_x - \gamma\mathcal{K}_x + \alpha\mathcal{P}_x \\ &= (\alpha + \beta + \beta\sigma)\mathcal{P}_x - (\beta\kappa + \gamma)\mathcal{K}_x = (\alpha + \beta + \beta\sigma)\mathcal{P}_x - \gamma\mathcal{K}_x\end{aligned}$$

implies

$$\alpha = -\beta(1+\sigma), \tag{2.37}$$

$$\gamma = 0. \tag{2.38}$$

The system of equations (2.36)–(2.37) has two possible solutions (in both cases the parameter σ remains undefined):
(i) If $\beta \neq 0$, then $\alpha = -\beta(1+\sigma)$ and $\kappa = 0$.
(ii) If $\beta = 0$, then $\alpha = 0$ and the parameter κ is arbitrary.

From condition (IV) in Subsection 2.3.1 we know that parameters α, β, σ, κ should depend on c and tend to zero in the limit $c \to \infty$, written

$$\lim_{c\to\infty} \alpha = \lim_{c\to\infty} \beta = \lim_{c\to\infty} \kappa = \lim_{c\to\infty} \sigma = 0. \tag{2.39}$$

In order to use this condition, let us analyze the physical dimensionalities of these parameters. For the argument of exponents in (2.16) to be dimensionless, we must assume the following dimensionalities (denoted by angle brackets) of the generators:

$$\begin{aligned}&<\mathcal{H}> = <\text{time}>^{-1}, \\ &<\mathcal{P}> = <\text{distance}>^{-1}, \\ &<\mathcal{K}> = <\text{speed}>^{-1}, \\ &<\mathcal{J}> = <\text{angle}>^{-1} = \text{dimensionless}.\end{aligned}$$

Then from the definitions of α, β, κ, σ it follows that

$$\begin{aligned}&<\alpha> = \frac{<\mathcal{K}>^2}{<\mathcal{J}>} = <\text{speed}>^{-2}, \\ &<\beta> = \frac{<\mathcal{K}><\mathcal{P}>}{<\mathcal{H}>} = <\text{speed}>^{-2}, \\ &<\kappa> = <\mathcal{H}> = <\text{time}>^{-1}, \\ &<\sigma> = \text{dimensionless}\end{aligned}$$

and we can satisfy condition (2.39) only by setting $\kappa = \sigma = 0$ (i. e., by realizing the choice (i) above) and $\beta = -\alpha \propto c^{-2}$. This approach does not allow us to establish the proportionality coefficient between β (and $-\alpha$) and c^{-2}. The agreement with experimental data requires that we choose this coefficient to be unity, so

$$\beta = -\alpha = \frac{1}{c^2}.$$

Thus, we obtain the following final expressions for the Lie brackets:

$$[\mathcal{J}_i, \mathcal{P}_j]_L = \sum_{k=1}^{3} \epsilon_{ijk}\mathcal{P}_k, \tag{2.40}$$

$$[\mathcal{J}_i, \mathcal{J}_j]_L = \sum_{k=1}^{3} \epsilon_{ijk}\mathcal{J}_k, \tag{2.41}$$

$$[\mathcal{J}_i, \mathcal{K}_j]_L = \sum_{k=1}^{3} \epsilon_{ijk}\mathcal{K}_k, \tag{2.42}$$

$$[\mathcal{J}_i, \mathcal{H}]_L = 0, \tag{2.43}$$

$$[\mathcal{P}_i, \mathcal{P}_j]_L = [\mathcal{P}_i, \mathcal{H}]_L = 0, \tag{2.44}$$

$$[\mathcal{K}_i, \mathcal{K}_j]_L = -\frac{1}{c^2}\sum_{k=1}^{3} \epsilon_{ijk}\mathcal{J}_k, \tag{2.45}$$

$$[\mathcal{K}_i, \mathcal{P}_j]_L = -\frac{1}{c^2}\mathcal{H}\delta_{ij}, \tag{2.46}$$

$$[\mathcal{K}_i, \mathcal{H}]_L = -\mathcal{P}_i. \tag{2.47}$$

These formulas express structure constants of the *Lie algebra* of the Poincaré group. They differ from structure constants of the Galilei Lie algebra (2.17)–(2.24) only by small terms on the right-hand sides of equations (2.45) and (2.46). The general element of the corresponding Poincaré group has the form

$$\{\boldsymbol{v}(\boldsymbol{\theta}); \boldsymbol{\varphi}; \boldsymbol{x}; t\} = e^{c\mathcal{K}\cdot\boldsymbol{\theta}} e^{\mathcal{J}\cdot\boldsymbol{\varphi}} e^{\mathcal{P}\cdot\boldsymbol{x}} e^{\mathcal{H}t}. \tag{2.48}$$

In contrast to elements of the Galilei group (2.16), where the boost parameter $\boldsymbol{v}$ coincides with the speed of the inertial system, in equation (2.48) we introduced a new boost parameter, denoted by $\boldsymbol{\theta}$ and called *rapidity*. The rapidity is related to the boost's speed by the following formulas:

$$\boldsymbol{v}(\boldsymbol{\theta}) = \frac{\boldsymbol{\theta}}{\theta} c \tanh\theta,$$
$$\cosh\theta = (1 - v^2/c^2)^{-1/2}.$$

The reason for introducing this new quantity is that rapidities of consecutive boosts in one direction add up, while velocities do not have this property.[6] Therefore, only

6 See the relativistic law of addition of velocities (4.7)–(4.9).

the rapidity can serve as a parameter of one-parameter subgroups of boosts and participate in the exponential notation (2.48).

Despite their simplicity, equations (2.40)–(2.47) are exceptionally important in physics, and they are confirmed by so many experimental data that one cannot doubt their validity. Therefore, we accept that the Poincaré group is the exact mathematical expression of relationships between different inertial laboratories.

Postulate 2.2 (Poincaré group). Transformations between inertial laboratories form the Poincaré group.

Even a superficial comparison of Lie brackets of the Poincaré (2.40)–(2.47) and Galilei (2.17)–(2.24) algebras reveals several important features of the relativistic theory. For example, a nonzero Lie bracket (2.45) shows that boosts, by themselves, no longer form a subgroup. However, together with rotations, the boosts form a six-dimensional subgroup of the Poincaré group, which is called the *Lorentz group*.

2.3.3 Boosts of translation generators

As an exercise, here we will apply the above Poincaré Lie brackets to derive transformations of the generators $\mathcal{P}$ and $\mathcal{H}$ under the action of boosts. Using equation (2.25) and Lie brackets (2.46)–(2.47), we find that if $\mathcal{P}_x$ and $\mathcal{H}$ are generators in the reference frame O, then in the frame O' moving along the x-axis they correspond to the following generators

$$
\begin{aligned}
\mathcal{H}(\theta) &= e^{c\mathcal{K}_x\theta}\mathcal{H}e^{-c\mathcal{K}_x\theta},\\
\mathcal{P}_x(\theta) &= e^{c\mathcal{K}_x\theta}\mathcal{P}_x e^{-c\mathcal{K}_x\theta}.
\end{aligned}
$$

Taking the derivatives of these equalities with respect to θ, we obtain

$$
\begin{aligned}
\frac{d}{d\theta}\mathcal{H}(\theta) &= ce^{c\mathcal{K}_x\theta}(\mathcal{K}_x\mathcal{H}-\mathcal{H}\mathcal{K}_x)e^{-c\mathcal{K}_x\theta} = -ce^{c\mathcal{K}_x\theta}\mathcal{P}_x e^{-c\mathcal{K}_x\theta} = -c\mathcal{P}_x(\theta),\\
\frac{d}{d\theta}\mathcal{P}_x(\theta) &= ce^{c\mathcal{K}_x\theta}(\mathcal{K}_x\mathcal{P}_x-\mathcal{P}_x\mathcal{K}_x)e^{-c\mathcal{K}_x\theta} = -\frac{1}{c}e^{c\mathcal{K}_x\theta}\mathcal{H}e^{-c\mathcal{K}_x\theta} = -\frac{1}{c}\mathcal{H}(\theta).
\end{aligned} \tag{2.49}
$$

Taking the derivatives of both sides of equation (2.49) again, we arrive at the differential equation

$$
\frac{d^2}{d\theta^2}\mathcal{P}_x(\theta) = -\frac{1}{c}\frac{d}{d\theta}\mathcal{H}(\theta) = \mathcal{P}_x(\theta)
$$

with the general solution

$$
\mathcal{P}_x(\theta) = \mathcal{B}\cosh\theta + \mathcal{D}\sinh\theta.
$$

From the initial conditions we obtain

$$\mathcal{B} = \mathcal{P}_x(0) = \mathcal{P}_x,$$
$$\mathcal{D} = \frac{d}{d\theta}\mathcal{P}_x(\theta)\Big|_{\theta=0} = -\frac{1}{c}\mathcal{H}$$

and finally

$$\mathcal{P}_x(\theta) = \mathcal{P}_x \cosh\theta - \frac{\mathcal{H}}{c}\sinh\theta. \tag{2.50}$$

Similar calculations lead to

$$\mathcal{H}(\theta) = \mathcal{H}\cosh\theta - c\mathcal{P}_x \sinh\theta, \tag{2.51}$$
$$\mathcal{P}_y(\theta) = \mathcal{P}_y,$$
$$\mathcal{P}_z(\theta) = \mathcal{P}_z.$$

Similar to our discussion of rotations in Appendix D.5, we can find transformations of $\boldsymbol{\mathcal{P}}$ and $\mathcal{H}$ with respect to a general boost $\boldsymbol{\theta}$ in a coordinate-independent form. First, we expand $\boldsymbol{\mathcal{P}}$ into the sum of two vectors $\boldsymbol{\mathcal{P}} = \boldsymbol{\mathcal{P}}_\| + \boldsymbol{\mathcal{P}}_\perp$, where the vector $\boldsymbol{\mathcal{P}}_\| = (\boldsymbol{\mathcal{P}} \cdot \frac{\boldsymbol{\theta}}{\theta})\frac{\boldsymbol{\theta}}{\theta}$ is parallel to the direction of the boost, and $\boldsymbol{\mathcal{P}}_\perp = \boldsymbol{\mathcal{P}} - \boldsymbol{\mathcal{P}}_\|$ is perpendicular to it (see Figure 2.4). The perpendicular part $\boldsymbol{\mathcal{P}}_\perp$ remains unchanged under the boost's action, and $\boldsymbol{\mathcal{P}}_\|$ is transformed according to (2.50), so we have

$$\boldsymbol{\mathcal{P}}'_\| = \exp(c\boldsymbol{\mathcal{K}} \cdot \boldsymbol{\theta})\boldsymbol{\mathcal{P}}_\| \exp(-c\boldsymbol{\mathcal{K}} \cdot \boldsymbol{\theta}) = \boldsymbol{\mathcal{P}}_\| \cosh\theta - \frac{\mathcal{H}\boldsymbol{\theta}}{c\theta}\sinh\theta.$$

Therefore

$$\boldsymbol{\mathcal{P}}' = e^{c\boldsymbol{\mathcal{K}}\cdot\boldsymbol{\theta}}\boldsymbol{\mathcal{P}}e^{-c\boldsymbol{\mathcal{K}}\cdot\boldsymbol{\theta}} = \boldsymbol{\mathcal{P}}'_\| + \boldsymbol{\mathcal{P}}_\perp = \boldsymbol{\mathcal{P}} + \frac{\boldsymbol{\theta}}{\theta}\left[\left(\boldsymbol{\mathcal{P}} \cdot \frac{\boldsymbol{\theta}}{\theta}\right)(\cosh\theta - 1) - \frac{\mathcal{H}}{c}\sinh\theta\right], \tag{2.52}$$
$$\mathcal{H}' = e^{c\boldsymbol{\mathcal{K}}\cdot\boldsymbol{\theta}}\mathcal{H}e^{-c\boldsymbol{\mathcal{K}}\cdot\boldsymbol{\theta}} = \mathcal{H}\cosh\theta - c\left(\boldsymbol{\mathcal{P}} \cdot \frac{\boldsymbol{\theta}}{\theta}\right)\sinh\theta. \tag{2.53}$$

It is clear from (2.52) and (2.53) that boosts induce linear transformations of the components $c\boldsymbol{\mathcal{P}}$ and $\mathcal{H}$. These transformations can be represented in a matrix form if we

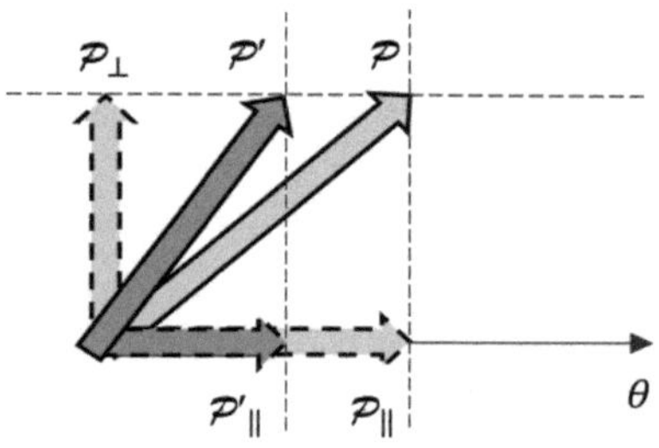

Figure 2.4: Transformation of the vector $\boldsymbol{\mathcal{P}} = \boldsymbol{\mathcal{P}}_\| + \boldsymbol{\mathcal{P}}_\perp$ to the vector $\boldsymbol{\mathcal{P}}' = \boldsymbol{\mathcal{P}}'_\| + \boldsymbol{\mathcal{P}}_\perp$ under the action of a passive boost $\boldsymbol{\theta}$.

collect four generators $(\mathcal{H}, c\boldsymbol{\mathcal{P}})$ into one 4-dimensional column vector,

$$\begin{bmatrix} \mathcal{H}' \\ c\mathcal{P}'_x \\ c\mathcal{P}'_y \\ c\mathcal{P}'_z \end{bmatrix} = \tilde{\boldsymbol{\theta}}^{-1} \begin{bmatrix} \mathcal{H} \\ c\mathcal{P}_x \\ c\mathcal{P}_y \\ c\mathcal{P}_z \end{bmatrix}. \tag{2.54}$$

The explicit form of the 4×4 matrix $\tilde{\boldsymbol{\theta}}$ is shown in equation (J.10).

3 Quantum mechanics and relativity

I am ashamed to tell you to how many figures I carried these computations, having no other business at the time.
Isaac Newton

The two previous chapters discussed the ideas of quantum mechanics and relativity in isolation from each other. Now it is time to combine them into one theory. The main contribution to such a unification was made by Wigner, who formulated and proved the famous theorem bearing his name and also developed the theory of unitary representations of the Poincaré group in Hilbert spaces. This theory is the mathematical basis of the relativistic quantum approach, which we develop in our book.

3.1 Inertial transformations in quantum mechanics

The postulate of relativity 2.1 tells us that each inertial laboratory L is physically equivalent to any other laboratory $L' = gL$ obtained from L by application of an inertial transformation g. This means that for equally organized experiments in these two laboratories, the corresponding quantum probability measures on subspaces in the Hilbert space $\mathscr{H}$ are the same. As shown in Figure 1, labs consist of two main parts: the preparation device P and the observer O. When applied to the whole laboratory, the inertial transformation g changes both these parts. The change of the preparation device can be interpreted as a change of the state of the system. Let us formally denote this transformation $\phi \to g\phi$. The change of the observer (or the measuring apparatus) can be considered as a change of experimental propositions $\mathcal{X} \to g\mathcal{X}$. Then, the principle of relativity in quantum mechanics can be expressed by a single formula,

$$(g\phi|g\mathcal{X}) = (\phi|\mathcal{X}), \tag{3.1}$$

which is valid for any g, ϕ and $\mathcal{X}$. In the remainder of this volume, we will develop a mathematical formalism for describing transformations $g\phi$ and $g\mathcal{X}$ in the Hilbert space. This is the formalism of unitary representations of the Poincaré group, which is the cornerstone of relativistic quantum physics.

3.1.1 Wigner's theorem

Let us first consider inertial transformations of experimental propositions $\mathcal{X} \to g\mathcal{X}$. As we know, propositions related to the observer O form an orthomodular lattice $\mathcal{L}(\mathscr{H})$, realized as a set of closed subspaces in the Hilbert space $\mathscr{H}$ of the given physical system. The observer $O' = gO$ can also represent her propositions in the form of subspaces of the same Hilbert space $\mathscr{H}$. Since these two observers are equivalent, we expect that

https://doi.org/10.1515/9783110492132-003

their propositional systems have the same mathematical structures, i. e., they are isomorphic. This means that there is a one-to-one mapping (= isomorphism) that translates propositions of O into propositions of O', and all the lattice relations remain unchanged by this mapping. In particular, we require that this mapping translates atoms into atoms and maps the minimum and maximum propositions of the observer O in the minimum and maximum propositions of the observer O', respectively. Thus, we postulate that to each inertial transformation g there corresponds a mapping $\mathbb{K}_g$ between subspaces in $\mathscr{H}$ that has the following properties:

$$\mathbb{K}_g(\mathscr{H}) = \mathscr{H}, \tag{3.2}$$

$$\mathbb{K}_g(\emptyset) = \emptyset, \tag{3.3}$$

$$\mathbb{K}_g(\mathscr{X} \cap \mathscr{Y}) = \mathbb{K}_g(\mathscr{X}) \cap \mathbb{K}_g(\mathscr{Y}), \tag{3.4}$$

$$\mathbb{K}_g(\mathscr{X} \uplus \mathscr{Y}) = \mathbb{K}_g(\mathscr{X}) \uplus \mathbb{K}_g(\mathscr{Y}), \tag{3.5}$$

$$\mathbb{K}_g(\mathscr{X}') = \mathbb{K}_g'(\mathscr{X}), \tag{3.6}$$

for all $\mathscr{X}, \mathscr{Y} \subseteq \mathscr{H}$.

We have already discussed in Section 1.4 that it is rather inconvenient to work with propositions/subspaces. It would be better to translate conditions (3.2)–(3.6) into the language of vectors in the Hilbert space. In other words, we would like to find a vector-to-vector transformation $\Bbbk_g : \mathscr{H} \to \mathscr{H}$, which *generates* the subspace-to-subspace transformation $\mathbb{K}_g$. More precisely, we require that if the subspaces $\mathscr{X}$ and $\mathscr{Y}$ are connected by our isomorphism (that is, $\mathbb{K}_g(\mathscr{X}) = \mathscr{Y}$), then the generator $\Bbbk_g$ should map all vectors from $\mathscr{X}$ to vectors from $\mathscr{Y}$, so that $\biguplus \Bbbk_g(x) = \mathscr{Y}$, where x runs over all vectors in $\mathscr{X}$.

The problem with generators $\Bbbk_g$ is that there are just too many choices of them. For example, if the ray $\mathscr{Z}$ transforms into the ray $\mathbb{K}_g(\mathscr{Z})$, then the generator $\Bbbk_g$ must map each vector $|z\rangle \in \mathscr{Z}$ somewhere inside the ray $\mathbb{K}_g(\mathscr{Z})$, but the exact image $\Bbbk_g|z\rangle$ is completely irrelevant. Indeed, we can multiply each vector image $\Bbbk_g|z\rangle$ by an arbitrary nonzero factor $\eta(|z\rangle)$ and still have a valid generator. The multipliers $\eta(|z\rangle)$ can be chosen independently for each $|z\rangle \in \mathscr{H}$. All this freedom is very inconvenient from the mathematical point of view.

The problem of the ambiguity of generators was solved by the famous Wigner theorem [99], which states that one can always choose the factors $\eta(|z\rangle)$ so that the vector-to-vector mapping $\eta(|z\rangle)\Bbbk_g$ is either unitary (linear) or antiunitary (antilinear).[1]

Theorem 3.1 (Wigner). *For each isomorphic mapping $\mathbb{K}_g$ of the lattice of subspaces of the Hilbert space $\mathscr{H}$ onto itself one can find either a unitary or an antiunitary transformation $\Bbbk_g$ of vectors in $\mathscr{H}$, which generates $\mathbb{K}_g$. This transformation is defined up to an arbitrary unimodular factor.*

1 For definitions of antilinear and antiunitary operators, see Appendix G.3.

In this formulation Wigner's theorem was proved in [91].[2] The importance of the theorem is due to the fact that mathematicians have a powerful mathematical apparatus for working with unitary and antiunitary operators $\Bbbk_g$, so that their properties (and, therefore, the properties of transformations $\mathbb{K}_g$) can be studied in detail by the familiar methods of linear algebra in Hilbert spaces.

From our analysis of inertial transformations in Chapter 2, we know that there is always a continuous path connecting the identity transformation $e = \{\mathbf{0}; \mathbf{0}; \mathbf{0}; 0\}$ with all other elements $g = \{\boldsymbol{v}; \boldsymbol{\varphi}; \boldsymbol{r}; t\}$ of the Poincaré group. The identity transformation e is naturally generated by the identity operator $\Bbbk_e = 1$ in $\mathscr{H}$, which, of course, is unitary. It also seems reasonable to require the maps $g \to \mathbb{K}_g$ and $g \to \Bbbk_g$ to be continuous. This means that a representative $\Bbbk_g$ cannot suddenly switch from unitary to antiunitary along the path connecting group elements e and g. Thus, we exclude antiunitary operators from representatives $\Bbbk_g$.[3] From now on, instead of $\Bbbk_g$ we will use the notation U_g to emphasize the unitary nature of vector transformations.

Although Wigner's theorem reduces the freedom of choice of generators, it does not remove this freedom altogether. If $\beta(g)$ is any unimodular complex number ($|\beta(g)| = 1$), then two unitary operators U_g and $\beta(g)U_g$ generate the same transformation of subspaces $\mathbb{K}_g$. Hence, for each g and $\mathbb{K}_g$ there is a set of generating unitary transformations U_g that differ from each other only by unimodular factors. Such a set is referred to as the *ray* of transformations $[U_g]$.

Results of this subsection can be summarized as follows. Each inertial transformation of observers can be represented by a unitary operator U_g in $\mathscr{H}$, defined up to an arbitrary unimodular factor: ket vectors are transformed as $|x\rangle \to U_g|x\rangle$ and bra vectors are transformed as $\langle x| \to \langle x|U_g^{-1}$. If $P_{\mathscr{X}} = \sum_i |e_i\rangle\langle e_i|$[4] is a projection (proposition) associated with the observer O, then the observer $O' = gO$ will represent the same proposition by the projection

$$P'_{\mathscr{X}} = \sum_i U_g|e_i\rangle\langle e_i|U_g^{-1} = U_g P_{\mathscr{X}} U_g^{-1}.$$

Similarly, if $F = \sum_i f_i|e_i\rangle\langle e_i|$ is the operator of an observable in the reference frame O, then

$$F' = \sum_i f_i U_g|e_i\rangle\langle e_i|U_g^{-1} = U_g F U_g^{-1} \tag{3.7}$$

is the same observable in the frame $O' = gO$.

2 See also [1].

3 Antiunitary operators can represent discrete transformations, such as time reversal, but we agreed not to touch these transformations in our book, because they are not exact symmetries.

4 Here $|e_i\rangle$ is an orthonormal basis in the subspace $\mathscr{X}$.

3.1.2 Inertial transformations of states

In the previous subsection we found out how the inertial transformation g changes observers, measuring devices, experimental propositions and observables. Now we turn to the action of g on preparation devices and states. We will try to answer the following question: if $|\Psi\rangle$ is a vector describing the state prepared by the device P, then which vector $|\Psi'\rangle$ describes the state prepared by the transformed device $P' = gP$?

To find the connection between $|\Psi\rangle$ and $|\Psi'\rangle$, we use the principle of relativity. According to equation (3.1), for each observable F its expectation value (1.24) should not change after inertial transformation of the entire laboratory (= preparation device + observer). In the bra-ket formalism this condition can be written as

$$\langle\Psi|F|\Psi\rangle = \langle\Psi'|F'|\Psi'\rangle = \langle\Psi'|(U_g F U_g^{-1})|\Psi'\rangle. \tag{3.8}$$

This equality must be satisfied for any observable F. Let us choose $F = |\Psi\rangle\langle\Psi|$, i. e., the projection on the ray of the vector $|\Psi\rangle$. Then equation (3.8) takes the form

$$\langle\Psi|\Psi\rangle\langle\Psi|\Psi\rangle = \langle\Psi'|U_g|\Psi\rangle\langle\Psi|U_g^{-1}|\Psi'\rangle = \langle\Psi'|U_g|\Psi\rangle\langle\Psi'|U_g|\Psi\rangle^* = |\langle\Psi'|U_g|\Psi\rangle|^2.$$

The left-hand side of this equation is 1. Hence, for each $|\Psi\rangle$ the transformed vector $|\Psi'\rangle$ is such that

$$|\langle\Psi'|U_g|\Psi\rangle|^2 = 1.$$

Since both $U_g|\Psi\rangle$ and $|\Psi'\rangle$ are unit vectors, we can write

$$|\Psi'\rangle = \beta(g)U_g|\Psi\rangle,$$

where $\beta(g)$ is some complex unimodular factor. The operator U_g is, anyway, defined only up to a unimodular factor; the state vector itself is defined only up to a factor. Hence, we are free to remove the $\beta(g)$ multiplier and finally write down the action of the inertial transformation g on state vectors,

$$|\Psi\rangle \to |\Psi'\rangle = U_g|\Psi\rangle. \tag{3.9}$$

Substituting this expression in (3.8), it is easy to verify the required invariance of the expectation value. We have

$$\langle F'\rangle = \langle\Psi'|F'|\Psi'\rangle = ((\langle\Psi|U_g^{-1})(U_g F U_g^{-1})(U_g|\Psi\rangle) = \langle\Psi|F|\Psi\rangle = \langle F\rangle. \tag{3.10}$$

Figure 3.1 illustrates results obtained so far in this section.

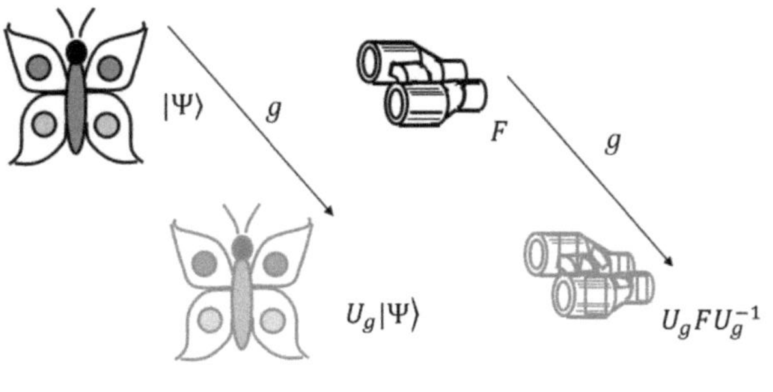

Figure 3.1: Action of the inertial transformation g on the physical system (state $|\Psi\rangle$) and the measuring apparatus (observable F).

3.1.3 Heisenberg and Schrödinger pictures

Let us return to inertial transformations of observers (= measuring devices). If we change the observer $O \to O' = gO$ and do not touch the preparation device, then operators of observables change according to (3.7), although the state vector remains the same $|\Psi\rangle$. As expected, this transformation changes the results of experiments. For example, the expectation value of the observable F is generally different for two observers O and $O' = gO$. We have

$$\langle F' \rangle = \langle \Psi | (U_g F U_g^{-1}) | \Psi \rangle \neq \langle \Psi | F | \Psi \rangle = \langle F \rangle. \tag{3.11}$$

On the other hand, if the inertial transformation is applied only to the preparation device, then the state of the system changes in accordance with (3.9), and the measurement results also change, so we have

$$\langle F'' \rangle = (\langle \Psi | U_g^{-1}) F (U_g | \Psi \rangle) \neq \langle \Psi | F | \Psi \rangle = \langle F \rangle. \tag{3.12}$$

Equations (3.11) and (3.12) play an important role, because many problems in physics can be formulated as questions about descriptions of systems subjected to inertial transformations. An important example is *dynamics*, i. e., the time evolution of the physical system. In this case, one considers time translation elements of the Poincaré group $g = \{\mathbf{0}; \mathbf{0}; \mathbf{0}; t\}$. Then equations (3.11) and (3.12) provide two equivalent approaches to the description of dynamics. Equation (3.11) describes dynamics in the so-called *Heisenberg picture*, where the state vector of the system remains fixed, while operators of observables change with time.

Equation (3.12) offers an alternative description of dynamics in the *Schrödinger picture*, where operators of observables are time-independent and state vectors change. These two pictures are equivalent, because, according to the relativity principle (3.1), the shift of the observer by g is equivalent to the shift of the preparation device by g^{-1}, written

$$(\phi | g\mathcal{X}) = (g^{-1}\phi | g^{-1}g\mathcal{X}) = (g^{-1}\phi | \mathcal{X}).$$

As follows from our derivation, the ideas of the Schrödinger and Heisenberg pictures can be generalized to nine other types of inertial transformations: space translations, rotations, boosts and any combinations thereof.

3.2 Unitary representations of Poincaré group

In the previous section, we discussed a representation of a particular inertial transformation g by the isomorphism $\mathbb{K}_g$ of the propositional lattice or by the ray of unitary operators $[U_g]$ acting on states (or observables) in the Hilbert space $\mathscr{H}$. From Chapter 2 we know that inertial transformations form the Poincaré group. Hence, the maps $\mathbb{K}_{g_1}, \mathbb{K}_{g_2}, \mathbb{K}_{g_3}, \ldots$ corresponding to different group elements $g_1, g_2, g_3, \ldots$ cannot be chosen arbitrarily. They must satisfy the conditions

$$\mathbb{K}_{g_2}\mathbb{K}_{g_1} = \mathbb{K}_{g_2 g_1}, \tag{3.13}$$

$$\mathbb{K}_{g^{-1}} = \mathbb{K}_g^{-1}, \tag{3.14}$$

$$\mathbb{K}_{g_3}(\mathbb{K}_{g_2}\mathbb{K}_{g_1}) = \mathbb{K}_{g_3}\mathbb{K}_{g_2 g_1} = \mathbb{K}_{g_3(g_2 g_1)} = \mathbb{K}_{(g_3 g_2)g_1} = (\mathbb{K}_{g_3}\mathbb{K}_{g_2})\mathbb{K}_{g_1}, \tag{3.15}$$

which reflect the group properties of inertial transformations g. In this section, our goal is to clarify the restrictions imposed by these conditions on the set of unitary representatives U_g.

3.2.1 Projective representations of groups

Suppose that for each element g of the Poincaré group we manage to construct a transformation $\mathbb{K}_g$ of subspaces in $\mathscr{H}$, so that conditions (3.13)–(3.15) are satisfied. As we know from Wigner's theorem, each $\mathbb{K}_g$ corresponds to a ray $[U_g]$ of unitary transformations. Let us now choose an arbitrary unitary representative $\mathfrak{U}_g$ in each ray $[U_g]$. Then, for any two elements of the group g_1 and g_2, we have representatives $\mathfrak{U}_{g_1} \in [U_{g_1}]$, $\mathfrak{U}_{g_2} \in [U_{g_2}]$ and $\mathfrak{U}_{g_2 g_1} \in [U_{g_2 g_1}]$. Obviously, the product $\mathfrak{U}_{g_2}\mathfrak{U}_{g_1}$ must generate the transformation $\mathbb{K}_{g_2 g_1}$, so it may differ from our chosen representative $\mathfrak{U}_{g_2 g_1}$, at most, by a unimodular factor $\alpha(g_2, g_1)$. Therefore, for any two transformations g_1 and g_2 we write

$$\mathfrak{U}_{g_2}\mathfrak{U}_{g_1} = \alpha(g_2, g_1)\mathfrak{U}_{g_2 g_1}. \tag{3.16}$$

The factors α have three important properties. First, as we mentioned, they are unimodular, so

$$|\alpha(g_2, g_1)| = 1. \tag{3.17}$$

Second, from the property (C.2) of the identity element, we get

$$\mathfrak{U}_g\mathfrak{U}_e = \alpha(g, e)\mathfrak{U}_g = \mathfrak{U}_g,$$

$$\mathfrak{U}_e\mathfrak{U}_g = \alpha(e, g)\mathfrak{U}_g = \mathfrak{U}_g,$$

which implies

$$\alpha(g, e) = \alpha(e, g) = 1 \tag{3.18}$$

for any g. Third, from the associative law (3.15), the following chain of equations follows:

$$\begin{aligned}
\mathfrak{U}_{g_3}(\mathfrak{U}_{g_2}\mathfrak{U}_{g_1}) &= (\mathfrak{U}_{g_3}\mathfrak{U}_{g_2})\mathfrak{U}_{g_1},\\
\mathfrak{U}_{g_3}(\alpha(g_2,g_1)\mathfrak{U}_{g_2g_1}) &= (\alpha(g_3,g_2)\mathfrak{U}_{g_3g_2})\mathfrak{U}_{g_1},\\
\alpha(g_2,g_1)\alpha(g_3,g_2g_1)\mathfrak{U}_{g_3g_2g_1} &= \alpha(g_3g_2,g_1)\alpha(g_3,g_2)\mathfrak{U}_{g_3g_2g_1},\\
\alpha(g_2,g_1)\alpha(g_3,g_2g_1) &= \alpha(g_3,g_2)\alpha(g_3g_2,g_1).
\end{aligned} \tag{3.19}$$

The mapping $\mathfrak{U}_g$ from group elements to unitary operators acting in the Hilbert space $\mathscr{H}$ is called the *projective representation* of the group if it satisfies conditions (3.16)–(3.19).

3.2.2 Generators of projective representation

In principle, we could fix our previously selected unitary representatives $\mathfrak{U}_{g_1} \in [U_{g_1}]$, $\mathfrak{U}_{g_2} \in [U_{g_2}], \ldots$ for all inertial transformations and work with such a projective representation of the Poincaré group. But this would lead to a rather cumbersome mathematical formalism. We could simplify our theory considerably if it were possible to use the freedom of choice of the representatives $\mathfrak{U}_g$ and to select them in such a way that the multipliers $\alpha(g_2, g_1)$ in the product law (3.16) are completely eliminated. Then, instead of the projective representation, we would deal with a simpler *linear* unitary representation of the group (see Appendix I). In Subsection 3.2.3, we will show that in any projective representation of the Poincaré group such an elimination of the factors $\alpha(g_2, g_1)$ is, indeed, possible [20, 87].

This proof is significantly simplified if the conditions (3.17)–(3.19) are rewritten in terms of the Lie algebra. Near the group identity element, we can use vectors $\boldsymbol{\zeta}$ from the Poincaré Lie algebra to identify all group elements by formula (E.2), so

$$g = e^{\zeta} = \exp\left(\sum_{a=1}^{10} \zeta^a t_a\right),$$

where t_a is the basis $\{\mathcal{H}, \mathcal{P}, \mathcal{J}, \mathcal{K}\}$ of the Poincaré Lie algebra, defined in Subsection 2.3.2. Then unitary representatives $\mathfrak{U}_g$ of inertial transformations can be also written in the exponential form[5]

$$\mathfrak{U}_{\zeta} = \exp\left(-\frac{i}{\hbar}\sum_{a=1}^{10} \zeta^a \mathfrak{T}_a\right), \tag{3.20}$$

5 Here we used Stone's Theorem I.1.

where $\hbar$ is a real constant, which we leave undefined here,[6] and $\mathfrak{T}_a$ are ten Hermitian operators,

$$\{-\mathfrak{H}, \mathfrak{P}, \mathfrak{J}, \mathfrak{K}\}, \tag{3.21}$$

in the Hilbert space $\mathscr{H}$. They are called *generators* of the unitary projective representation $\mathfrak{U}_g$. These operators generate time translations, space translations, rotations and boosts, respectively.

Then we can rewrite formula (3.16) in the form

$$\mathfrak{U}_{\zeta}\mathfrak{U}_{\xi} = \alpha(\boldsymbol{\zeta}, \boldsymbol{\xi})\mathfrak{U}_{\zeta\xi}. \tag{3.22}$$

Since α is a unimodular factor, it is appropriate to set $\alpha(\boldsymbol{\zeta}, \boldsymbol{\xi}) \equiv \exp[i\kappa(\boldsymbol{\zeta}, \boldsymbol{\xi})]$, where $\kappa(\boldsymbol{\zeta}, \boldsymbol{\xi})$ is a real function. Conditions (3.18) and (3.19) can then be rewritten in terms of the function κ, so we have

$$\kappa(\boldsymbol{\zeta}, \mathbf{0}) = \kappa(\mathbf{0}, \boldsymbol{\xi}) = 0, \tag{3.23}$$

$$\kappa(\boldsymbol{\xi}, \boldsymbol{\zeta}) + \kappa(\boldsymbol{\chi}, \boldsymbol{\xi}\boldsymbol{\zeta}) = \kappa(\boldsymbol{\chi}, \boldsymbol{\xi}) + \kappa(\boldsymbol{\chi}\boldsymbol{\xi}, \boldsymbol{\zeta}). \tag{3.24}$$

Note that the Taylor expansion of this function has the form

$$\kappa(\boldsymbol{\zeta}, \boldsymbol{\xi}) = \sum_{ab=1}^{10} h_{ab}\zeta^a\xi^b + \cdots. \tag{3.25}$$

The constant term, the terms linear in ζ^a and ξ^b and the terms proportional to the products $\zeta^a\zeta^b$ and $\xi^a\xi^b$ are missing on the right-hand side of (3.25) as a consequence of the condition (3.23).

Applying the same arguments as in our derivation of equation (E.7), we can expand all factors in (3.22) into Taylor series near $\boldsymbol{\zeta} = \boldsymbol{\xi} = \mathbf{0}$, so we have

$$\begin{aligned}
&\left(1 - \frac{i}{\hbar}\sum_{a=1}^{10}\xi^a\mathfrak{T}_a - \frac{1}{2\hbar^2}\sum_{bc=1}^{10}\xi^b\xi^c\mathfrak{T}_{bc} + \cdots\right) \\
&\quad\times\left(1 - \frac{i}{\hbar}\sum_{a=1}^{10}\zeta^a\mathfrak{T}_a - \frac{1}{2\hbar^2}\sum_{bc=1}^{10}\zeta^b\zeta^c\mathfrak{T}_{bc} + \cdots\right) \\
&= \left(1 + i\sum_{ab=1}^{10} h_{ab}\zeta^a\xi^b + \cdots\right) \\
&\quad\times\left[1 - \frac{i}{\hbar}\sum_{a=1}^{10}\left(\zeta^a + \xi^a + \sum_{bc=1}^{10} f^a_{bc}\xi^b\zeta^c + \cdots\right)\mathfrak{T}_a\right.
\end{aligned}$$

6 We will identify $\hbar$ with the *Planck constant* in Subsection 4.1.1.

$$-\frac{1}{2\hbar^2}\sum_{ab=1}^{10}(\zeta^a+\xi^a+\cdots)(\zeta^b+\xi^b+\cdots)\mathfrak{T}_{ab}+\cdots\Bigg].$$

Equating coefficients of the products $\xi^a\zeta^b$ on both sides, we obtain

$$-\frac{1}{2\hbar^2}(\mathfrak{T}_{ab}+\mathfrak{T}_{ba})=-\frac{1}{\hbar^2}\mathfrak{T}_a\mathfrak{T}_b-ih_{ba}+\frac{i}{\hbar}\sum_{c=1}^{10}f_{ab}^c\mathfrak{T}_c.$$

The left-hand side of this equality is symmetric with respect to the permutation of indices $a \leftrightarrow b$. The same must be true for the right-hand side. Hence we obtain commutators of the generators $\mathfrak{T}$,

$$\mathfrak{T}_a\mathfrak{T}_b-\mathfrak{T}_b\mathfrak{T}_a=i\hbar\sum_{c=1}^{10}C_{ab}^c\mathfrak{T}_c+i\mathfrak{Q}_{ab}, \tag{3.26}$$

where $C_{ab}^c \equiv f_{ab}^c - f_{ba}^c$ are the familiar structure constants of the Poincaré Lie algebra (2.40)–(2.47) and $\mathfrak{Q}_{ab} = \hbar^2(h_{ab}-h_{ba})$ are real constants that depend on our original choice of representatives $\mathfrak{U}_g$ in the rays $[U_g]$. These constants are called *central charges*. Our main task in the next two subsections is to prove that a new set of representatives

$$U_g=\beta(g)\mathfrak{U}_g \tag{3.27}$$

can be chosen in such a way that $\mathfrak{Q}_{ab} = 0$ for all a, b, i.e., that central charges are eliminated.

3.2.3 Commutators of projective generators

Here we start with writing the commutators (3.26) explicitly, using formulas (2.40)–(2.47). We have

$$[\mathfrak{J}_i,\mathfrak{P}_j]=i\hbar\sum_{k=1}^{3}\epsilon_{ijk}\mathfrak{P}_k+i\mathfrak{Q}_{ij}^{(1)}, \tag{3.28}$$

$$[\mathfrak{J}_i,\mathfrak{J}_j]=i\hbar\sum_{k=1}^{3}\epsilon_{ijk}(\mathfrak{J}_k-\mathfrak{Q}_k^{(2)}), \tag{3.29}$$

$$[\mathfrak{J}_i,\mathfrak{K}_j]=i\hbar\sum_{k=1}^{3}\epsilon_{ijk}\mathfrak{K}_k+i\mathfrak{Q}_{ij}^{(3)}, \tag{3.30}$$

$$[\mathfrak{P}_i,\mathfrak{P}_j]=i\mathfrak{Q}_{ij}^{(4)}, \tag{3.31}$$

$$[\mathfrak{J}_i,\mathfrak{H}]=i\mathfrak{Q}_i^{(5)}, \tag{3.32}$$

$$[\mathfrak{P}_i,\mathfrak{H}]=i\mathfrak{Q}_i^{(6)}, \tag{3.33}$$

$$[\mathfrak{K}_i, \mathfrak{K}_j] = -i\frac{\hbar}{c^2}\sum_{k=1}^{3}\epsilon_{ijk}(\mathfrak{J}_k - \mathfrak{Q}_k^{(7)}), \tag{3.34}$$

$$[\mathfrak{K}_i, \mathfrak{P}_j] = -i\frac{\hbar}{c^2}\mathfrak{H}\delta_{ij} + i\mathfrak{Q}_{ij}^{(8)}, \tag{3.35}$$

$$[\mathfrak{K}_i, \mathfrak{H}] = -i\hbar\mathfrak{P}_i + i\mathfrak{Q}_i^{(9)}, \tag{3.36}$$

where we combined $\mathfrak{Q}_{ab}$ into nine sets of central charges $\mathfrak{Q}^{(1)} \dots \mathfrak{Q}^{(9)}$. In equations (3.29) and (3.34), we also took into account the fact that their left-hand sides are antisymmetric tensors. This means that central charges also form antisymmetric tensors, i. e., $\mathfrak{Q}_k^{(2)}$ and $\mathfrak{Q}_k^{(7)}$ are 3-vectors.

Next, we use the requirement that commutators (3.28)–(3.36) must satisfy the Jacobi identity (E.11). This allows us to make many simplifications. For example, from (3.28) it follows that $\mathfrak{P}_3 = -\frac{i}{\hbar}[\mathfrak{J}_1, \mathfrak{P}_2] - \frac{1}{\hbar}\mathfrak{Q}_{12}^{(1)}$. From the fact that all constants $\mathfrak{Q}$ commute with group generators we then derive

$$\begin{aligned}
[\mathfrak{P}_3, \mathfrak{P}_1] &= -\frac{i}{\hbar}[([\mathfrak{J}_1, \mathfrak{P}_2] - i\mathfrak{Q}_{12}^{(1)}), \mathfrak{P}_1] = -\frac{i}{\hbar}[[\mathfrak{J}_1, \mathfrak{P}_2], \mathfrak{P}_1] \\
&= -\frac{i}{\hbar}[[\mathfrak{P}_1, \mathfrak{P}_2], \mathfrak{J}_1] - \frac{i}{\hbar}[[\mathfrak{J}_1, \mathfrak{P}_1], \mathfrak{P}_2] \\
&= \frac{1}{\hbar}[\mathfrak{Q}_{12}^{(4)}, \mathfrak{J}_1] + \frac{1}{\hbar}[\mathfrak{Q}_{11}^{(1)}, \mathfrak{P}_2] = 0.
\end{aligned}$$

Substituting this equality into (3.31), we obtain $\mathfrak{Q}_{31}^{(4)} = 0$. Similarly, we can show that $\mathfrak{Q}_{ij}^{(4)} = \mathfrak{Q}_i^{(5)} = \mathfrak{Q}_i^{(6)} = 0$ for all values of the indices $i, j = 1, 2, 3$.

We use the Jacobi identity, again, to write

$$\begin{aligned}
i\hbar[\mathfrak{J}_3, \mathfrak{P}_3] &= [[\mathfrak{J}_1, \mathfrak{J}_2], \mathfrak{P}_3] = [[\mathfrak{P}_3, \mathfrak{J}_2], \mathfrak{J}_1] + [[\mathfrak{J}_1, \mathfrak{P}_3], \mathfrak{J}_2] \\
&= i\hbar[\mathfrak{J}_1, \mathfrak{P}_1] + i\hbar[\mathfrak{J}_2, \mathfrak{P}_2]
\end{aligned}$$

and similarly

$$i\hbar[\mathfrak{J}_1, \mathfrak{P}_1] = i\hbar[\mathfrak{J}_2, \mathfrak{P}_2] + i\hbar[\mathfrak{J}_3, \mathfrak{P}_3].$$

Adding these two equalities, we see that

$$[\mathfrak{J}_2, \mathfrak{P}_2] = 0. \tag{3.37}$$

In the same way we get $[\mathfrak{J}_1, \mathfrak{P}_1] = [\mathfrak{J}_3, \mathfrak{P}_3] = 0$, which means

$$\mathfrak{Q}_{ii}^{(1)} = 0. \tag{3.38}$$

Applying the Jacobi identity once again, we have

$$i\hbar[\mathfrak{J}_2, \mathfrak{P}_3] = [[\mathfrak{J}_3, \mathfrak{J}_1], \mathfrak{P}_3] = [[\mathfrak{P}_3, \mathfrak{J}_1], \mathfrak{J}_3] + [[\mathfrak{J}_3, \mathfrak{P}_3], \mathfrak{J}_1] = -i\hbar[\mathfrak{J}_3, \mathfrak{P}_2].$$

This property of antisymmetry is valid for any pair of indices $i, j = 1, 2, 3$; $i \neq j$, so

$$[\mathfrak{J}_i, \mathfrak{P}_j] = -[\mathfrak{J}_j, \mathfrak{P}_i]. \tag{3.39}$$

Combining (3.37) and (3.39), we see that the tensor $[\mathfrak{J}_i, \mathfrak{P}_j]$ is antisymmetric. Hence we can define a 3-vector $\mathfrak{Q}_k^{(1)}$ such that (see Table D.1)

$$\mathfrak{Q}_{ij}^{(1)} = -\hbar \sum_{i=1}^{3} \epsilon_{ijk} \mathfrak{Q}_k^{(1)}$$

and

$$[\mathfrak{J}_i, \mathfrak{P}_j] = i\hbar \sum_{i=1}^{3} \epsilon_{ijk} (\mathfrak{P}_k - \mathfrak{Q}_k^{(1)}). \tag{3.40}$$

Similarly, we prove that $\mathfrak{Q}_{ii}^{(3)} = 0$ and

$$[\mathfrak{J}_i, \mathfrak{K}_j] = i\hbar \sum_{i=1}^{3} \epsilon_{ijk} (\mathfrak{K}_k - \mathfrak{Q}_k^{(3)}).$$

Taking into account the above results, the commutation relations (3.28)–(3.36) assume a simplified form. We have

$$[\mathfrak{J}_i, \mathfrak{P}_j] = i\hbar \sum_{k=1}^{3} \epsilon_{ijk} (\mathfrak{P}_k - \mathfrak{Q}_k^{(1)}), \tag{3.41}$$

$$[\mathfrak{J}_i, \mathfrak{J}_j] = i\hbar \sum_{k=1}^{3} \epsilon_{ijk} (\mathfrak{J}_k - \mathfrak{Q}_k^{(2)}), \tag{3.42}$$

$$[\mathfrak{J}_i, \mathfrak{K}_j] = i\hbar \sum_{k=1}^{3} \epsilon_{ijk} (\mathfrak{K}_k - \mathfrak{Q}_k^{(3)}), \tag{3.43}$$

$$[\mathfrak{P}_i, \mathfrak{P}_j] = [\mathfrak{J}_i, \mathfrak{H}] = [\mathfrak{P}_i, \mathfrak{H}] = 0, \tag{3.44}$$

$$[\mathfrak{K}_i, \mathfrak{K}_j] = -i\frac{\hbar}{c^2} \sum_{k=1}^{3} \epsilon_{ijk} (\mathfrak{J}_k - \mathfrak{Q}_k^{(7)}), \tag{3.45}$$

$$[\mathfrak{K}_i, \mathfrak{P}_j] = -i\frac{\hbar}{c^2} \mathfrak{H} \delta_{ij} + i\mathfrak{Q}_{ij}^{(8)}, \tag{3.46}$$

$$[\mathfrak{K}_i, \mathfrak{H}] = -i\hbar \mathfrak{P}_i + i\mathfrak{Q}_i^{(9)}, \tag{3.47}$$

where $\mathfrak{Q}$ on the right-hand sides are certain real constants.

3.2.4 Cancellation of central charges

As the next step we would like to cancel the central charges $\mathfrak{Q}$ on the right-hand sides of (3.41)–(3.47) by using the freedom to choose unimodular factors $\beta(\boldsymbol{\zeta})$ in front of the representation operators $\mathfrak{U}_\zeta$ (3.27). Accordingly, the choice of generators $\mathfrak{T}_a$ also has a certain degree of arbitrariness. Taking into account that $|\beta(\boldsymbol{\zeta})| = 1$ and assuming the smoothness of the function $\beta(\boldsymbol{\zeta})$, we write its Taylor expansion

$$\beta(\boldsymbol{\zeta}) = \exp(i\gamma(\boldsymbol{\zeta})) \approx 1 + i \sum_{a=1}^{10} G_a \zeta^a,$$

where $\gamma(\boldsymbol{\zeta})$ is a real function, which we are free to choose for our purposes. Hence, in the first order, the presence of the factors $\beta(\boldsymbol{\zeta})$ is equivalent to adding some real constants G_a to the generators $\mathfrak{T}_a$. Our goal is to show that by a suitable choice of these constants we can make all central charges to vanish.

Using the mentioned freedom, we now add (as yet undefined) real constants G_j to the generators $\mathfrak{P}_j$, $\mathfrak{J}_j$ and $\mathfrak{K}_j$ and denote the redefined generators as

$$\begin{aligned} P_j &\equiv \mathfrak{P}_j + G_j^{(1)}, \\ J_j &\equiv \mathfrak{J}_j + G_j^{(2)}, \\ K_j &\equiv \mathfrak{K}_j + G_j^{(3)}. \end{aligned}$$

Then the commutator (3.42) takes the form

$$[J_i, J_j] = [\mathfrak{J}_i + G_i^{(2)}, \mathfrak{J}_j + G_j^{(2)}] = [\mathfrak{J}_i, \mathfrak{J}_j] = i\hbar \sum_{k=1}^{3} \epsilon_{ijk}(\mathfrak{J}_k - \mathfrak{Q}_k^{(2)}).$$

Hence, if we choose $G_k^{(2)} = -\mathfrak{Q}_k^{(2)}$, then

$$[J_i, J_j] = i\hbar \sum_{k=1}^{3} \epsilon_{ijk} J_k$$

and the central charges have disappeared from this commutator.

Similarly, the central charges are removed from the two other commutators by choosing $G_k^{(1)} = -\mathfrak{Q}_k^{(1)}$ and $G_k^{(3)} = -\mathfrak{Q}_k^{(3)}$, so

$$\begin{aligned} [J_i, P_j] &= i\hbar \sum_{k=1}^{3} \epsilon_{ijk} P_k, \\ [J_i, K_j] &= i\hbar \sum_{k=1}^{3} \epsilon_{ijk} K_k. \end{aligned} \tag{3.48}$$

From equation (3.48) we then obtain

$$\begin{aligned} [K_1, K_2] &= -\frac{i}{\hbar}[[J_2, K_3], K_2] = -\frac{i}{\hbar}[[J_2, K_2], K_3] - \frac{i}{\hbar}[[K_2, K_3], J_2] \\ &= -\frac{i}{\hbar}\left[-\frac{i\hbar}{c^2}(J_1 - \mathfrak{Q}_1^{(7)}), J_2\right] = -\frac{i\hbar}{c^2} J_3. \end{aligned}$$

This means that our choice of the constants $G_k^{(1)}$, $G_k^{(2)}$ and $G_k^{(3)}$ also cancels the central charge $\mathfrak{Q}_i^{(7)}$.

Further, from equation (3.48) we have

$$[K_3, \mathfrak{H}] = -\frac{i}{\hbar}[[J_1, K_2], \mathfrak{H}] = -\frac{i}{\hbar}[[\mathfrak{H}, K_2], J_1] - \frac{i}{\hbar}[[J_1, \mathfrak{H}], K_2]$$
$$= -[J_1, P_2] = -i\hbar P_3,$$

which implies that the central charge $\mathfrak{Q}^{(9)}$ also cancels out. Finally

$$[K_1, P_2] = -\frac{i}{\hbar}[[J_2, K_3], P_2] = -\frac{i}{\hbar}[[J_2, P_2], K_3] + \frac{i}{\hbar}[[K_3, P_2], J_3] = 0,$$
$$[K_1, P_1] = -\frac{i}{\hbar}[[J_2, K_3], P_1] = -\frac{i}{\hbar}[[J_2, P_1], K_3] + \frac{i}{\hbar}[[K_3, P_1], J_3] = [K_3, P_3].$$

This means that $\mathfrak{Q}_{ij}^{(8)} = 0$ if $i \neq j$, and we can introduce a real scalar $\mathfrak{Q}^{(8)}$ such that

$$\mathfrak{Q}_{11}^{(8)} = \mathfrak{Q}_{22}^{(8)} = \mathfrak{Q}_{33}^{(8)} \equiv -\frac{\hbar}{c^2}\mathfrak{Q}^{(8)}$$

and equation (3.46) can be rewritten as

$$[K_i, P_i] = -\frac{i\hbar}{c^2}\delta_{ij}(\mathfrak{H} + \mathfrak{Q}^{(8)}).$$

At the conclusion of this procedure, we redefine the time translation generator $H \equiv \mathfrak{H} + \mathfrak{Q}^{(8)}$ and, thus, finally remove all central charges from commutation relations of the Lie algebra of the Poincaré group. These commutators now take the form

$$[J_i, P_j] = i\hbar \sum_{k=1}^{3} \epsilon_{ijk} P_k, \tag{3.49}$$

$$[J_i, J_j] = i\hbar \sum_{k=1}^{3} \epsilon_{ijk} J_k, \tag{3.50}$$

$$[J_i, K_j] = i\hbar \sum_{k=1}^{3} \epsilon_{ijk} K_k, \tag{3.51}$$

$$[P_i, P_j] = [J_i, H] = [P_i, H] = 0, \tag{3.52}$$

$$[K_i, K_j] = -\frac{i\hbar}{c^2} \sum_{k=1}^{3} \epsilon_{ijk} J_k, \tag{3.53}$$

$$[K_i, P_j] = -\frac{i\hbar}{c^2} H \delta_{ij}, \tag{3.54}$$

$$[K_i, H] = -i\hbar P_i. \tag{3.55}$$

Thus, the Hermitian operators H, $\boldsymbol{P}$, $\boldsymbol{J}$ and $\boldsymbol{K}$ provide a linear representation of the Poincaré Lie algebra, and the redefined unitary operators $U_g \equiv \beta(g)\mathfrak{U}_g$ form the desired *unique*[7] unitary representation of the Poincaré group, which is equivalent to the projective representation $\mathfrak{U}_g$ that was initially given to us near the group's identity element. Working with the unitary representation U_g is much easier than with the projective representation $\mathfrak{U}_g$.

The commutators (3.49)–(3.55) are, perhaps, the most important equalities of the relativistic quantum theory. Throughout this book, we will have many opportunities to appreciate the deep physical meaning of these formulas.

3.2.5 Single-valued and double-valued representations

In the previous subsection we removed phase factors $\alpha(g_2, g_1)$ from equation (3.16) by resorting to arguments related to the group's Lie algebra. However, strictly speaking, these arguments are valid only near the identity element of the group, where the Taylor series are applicable. It cannot be excluded that nontrivial phase factors can reappear in the product law (3.16), when the group manifold has a nontrivial topology, and the multiplied elements are "far" from the group's identity e.

In Appendix I.4, we find out that this possibility is realized in the case of the group of rotations whose manifold is doubly connected. This means that in quantum-mechanical applications one must consider both single-valued and double-valued[8] representations of this group. Since the group of rotations is a subgroup of the Poincaré group, the same conclusion applies to the latter: one must consider both single-valued and two-valued representations of the Poincaré group.[9] In Chapter 5 we will see that these two cases correspond to systems with integer and half-integer spin, respectively.

3.2.6 Fundamental statement of relativistic quantum theory

The most important result of this chapter is the establishment of a connection between the principle of relativity and quantum mechanics, which is expressed by the following statement (see, for example, [95]).

Statement 3.2 (Unitary representations of the Poincaré group). *In the relativistic quantum description, inertial transformations are expressed by operators U_g forming a unitary (single-valued or double-valued) representation of the Poincaré group in the Hilbert space of the physical system.*

---

7 The set of operators $e^{i\gamma} U_g$, where γ is an arbitrary real constant, also forms a unitary representation of the Poincaré group. Physically, it is absolutely equivalent to the representation U_g.

8 Where $\alpha(g_2, g_1) = \pm 1$.

9 Equivalently, one can consider only single-valued representations of the so-called *universal covering group*.

It is important to emphasize that this statement is absolutely general. The Hilbert space of *any* isolated physical system (no matter how complex) must carry a unitary representation of the Poincaré group. As we shall see later, the knowledge of this representation almost completely determines the entire physics of the isolated system. Construction of Hilbert spaces and unitary representations of the Poincaré group in them should be the main concern of theoretical physicists. The remainder of this book will be devoted to solving these difficult problems for specific systems.

Elementary inertial transformations from the Poincaré group are represented in the Hilbert space by unitary operators: $e^{-\frac{i}{\hbar}\boldsymbol{P}\cdot\boldsymbol{r}}$ for spatial translations, $e^{-\frac{i}{\hbar}\boldsymbol{J}\cdot\boldsymbol{\varphi}}$ for rotations, $e^{-\frac{ic}{\hbar}\boldsymbol{K}\cdot\boldsymbol{\theta}}$ for boosts and $e^{\frac{i}{\hbar}Ht}$ for time translations. A general inertial transformation $g = \{\boldsymbol{v}(\boldsymbol{\theta}); \boldsymbol{\varphi}; \boldsymbol{r}; t\}$ can be written as the following product[10]

$$U_g = e^{-\frac{ic}{\hbar}\boldsymbol{K}\cdot\boldsymbol{\theta}} e^{-\frac{i}{\hbar}\boldsymbol{J}\cdot\boldsymbol{\varphi}} e^{-\frac{i}{\hbar}\boldsymbol{P}\cdot\boldsymbol{r}} e^{\frac{i}{\hbar}Ht}. \tag{3.56}$$

We will also write this operator in alternative forms, e. g.,

$$U_g \equiv U(\boldsymbol{\theta}; \boldsymbol{\varphi}; \boldsymbol{r}; t) \equiv U(\Lambda; \boldsymbol{r}; t) \equiv U(\Lambda; \tilde{a}), \tag{3.57}$$

where $\Lambda \equiv \boldsymbol{\theta} \circ \boldsymbol{\varphi}$ is an inertial transformation from the Lorentz subgroup, which combines a boost $\boldsymbol{\theta}$ and a rotation $\boldsymbol{\varphi}$, and $\tilde{a} \equiv (ct, \boldsymbol{r})$ is a 4-vector of space–time translation. Then, in the Schrödinger picture the state vector is transformed between different frames of reference as

$$|\Psi'\rangle = U_g|\Psi\rangle. \tag{3.58}$$

In the Heisenberg picture, inertial transformations of observables have the form

$$F' = U_g F U_g^{-1}. \tag{3.59}$$

For example, the equation describing the time evolution of the observable F (see equation (E.14)),

$$F(t) = e^{\frac{i}{\hbar}Ht} F e^{-\frac{i}{\hbar}Ht} \tag{3.60}$$

$$= F + \frac{i}{\hbar}[H, F]t - \frac{1}{2\hbar^2}[H, [H, F]]t^2 + O(t^3), \tag{3.61}$$

can be also written in the differential form

$$\frac{dF(t)}{dt} = \frac{i}{\hbar}[H, F], \tag{3.62}$$

which is known as the *Heisenberg equation.*

---

10 We write the factors in the "canonical" order (boost) × (rotation) × (spatial translation) × (time shift). Compare with formula (2.48).

Note that similar "Heisenberg equations" can be written also for transformations of observables with respect to space translations, rotations and boosts:

$$\begin{aligned}\frac{dF(\boldsymbol{r})}{d\boldsymbol{r}} &= -\frac{i}{\hbar}[\boldsymbol{P}, F],\\ \frac{dF(\boldsymbol{\varphi})}{d\boldsymbol{\varphi}} &= -\frac{i}{\hbar}[\boldsymbol{J}, F],\\ \frac{dF(\boldsymbol{\theta})}{d\boldsymbol{\theta}} &= -\frac{ic}{\hbar}[\boldsymbol{K}, F].\end{aligned} \tag{3.63}$$

3.2.7 Time evolution in moving frame

Quite often we are interested in the time dependence of an observable F in the moving frame of reference. If in the rest frame O the time evolution (3.60) is determined by the Hamiltonian H, then in the moving frame O' we should (i) consider the transformed observable

$$F(\boldsymbol{\theta}) = e^{-\frac{ic}{\hbar}\boldsymbol{K}\cdot\boldsymbol{\theta}} F e^{\frac{ic}{\hbar}\boldsymbol{K}\cdot\boldsymbol{\theta}}$$

and (ii) use the transformed Hamiltonian

$$H(\boldsymbol{\theta}) = e^{-\frac{ic}{\hbar}\boldsymbol{K}\cdot\boldsymbol{\theta}} H e^{\frac{ic}{\hbar}\boldsymbol{K}\cdot\boldsymbol{\theta}} \tag{3.64}$$

as our generator of time translations. Therefore,[11]

$$\begin{aligned}F(\boldsymbol{\theta}, t') &= e^{\frac{i}{\hbar}H(\boldsymbol{\theta})t'} F(\boldsymbol{\theta}) e^{-\frac{i}{\hbar}H(\boldsymbol{\theta})t'}\\ &= e^{\frac{i}{\hbar}H(\boldsymbol{\theta})t'} e^{-\frac{ic}{\hbar}\boldsymbol{K}\cdot\boldsymbol{\theta}} F e^{\frac{ic}{\hbar}\boldsymbol{K}\cdot\boldsymbol{\theta}} e^{-\frac{i}{\hbar}H(\boldsymbol{\theta})t'}\\ &= (e^{-\frac{ic}{\hbar}\boldsymbol{K}\cdot\boldsymbol{\theta}} e^{\frac{i}{\hbar}Ht'} e^{\frac{ic}{\hbar}\boldsymbol{K}\cdot\boldsymbol{\theta}}) e^{-\frac{ic}{\hbar}\boldsymbol{K}\cdot\boldsymbol{\theta}} F e^{\frac{ic}{\hbar}\boldsymbol{K}\cdot\boldsymbol{\theta}} (e^{-\frac{ic}{\hbar}\boldsymbol{K}\cdot\boldsymbol{\theta}} e^{-\frac{i}{\hbar}Ht'} e^{\frac{ic}{\hbar}\boldsymbol{K}\cdot\boldsymbol{\theta}})\\ &= e^{-\frac{ic}{\hbar}\boldsymbol{K}\cdot\boldsymbol{\theta}} e^{\frac{i}{\hbar}Ht'} F e^{-\frac{i}{\hbar}Ht'} e^{\frac{ic}{\hbar}\boldsymbol{K}\cdot\boldsymbol{\theta}}.\end{aligned} \tag{3.65}$$

11 t' is time measured by the clock of the observer O'.

4 Observables

Throwing pebbles into the water, look at the ripples they form on the surface, otherwise, such occupation becomes an idle pastime.
Kozma Prutkov

So far, we have seen that in quantum theory, states of a physical system are described by vectors in the complex Hilbert space $\mathscr{H}$, and observables are represented by Hermitian operators in $\mathscr{H}$. We also learned that there exists a unitary representation U_g of the Poincaré group in $\mathscr{H}$, which determines how the state vectors and observables change under inertial transformations of the preparation device and/or measuring apparatus. Our next goal is to clarify the structure of the set of all observables. In particular, we want to know which operators correspond to known observables, such as velocity, momentum, mass, position, etc., what are their spectra and what are the relationships between these quantities.

It should be emphasized that physical systems considered in this chapter are absolutely arbitrary: they can be elementary particles, or composite systems of many particles, or even unstable systems in which the number of particles is not fixed. The only important requirement is that the system should be isolated, i. e., its interaction with the rest of the universe can be neglected.

4.1 Basic observables

4.1.1 Energy, momentum and angular momentum

Generators of any unitary representation of the Poincaré group in the Hilbert space of any system are Hermitian operators H, $\boldsymbol{P}$, $\boldsymbol{J}$ and $\boldsymbol{K}$, so it is reasonable to assume that they are related to some observables. What are these observables? Let us first postulate that the parameter $\hbar$, introduced in equation (3.20), is the famous *Planck constant*

$$\hbar = 1.055 \times 10^{-34} \frac{\text{kg}\cdot\text{m}^2}{\text{s}} = 6.582 \times 10^{-16}\,\text{eV}\cdot\text{s}, \tag{4.1}$$

whose physical dimensionality is $<\hbar> = <\text{mass}><\text{speed}><\text{distance}>$. Then the dimensionalities of the generators can be found from the condition that the arguments of exponents in (3.56) must be dimensionless, so

$$\begin{aligned}
<H> &= \frac{<\hbar>}{<\text{time}>} = <\text{mass}><\text{speed}>^2, \\
<\boldsymbol{P}> &= \frac{<\hbar>}{<\text{distance}>} = <\text{mass}><\text{speed}>, \\
<\boldsymbol{J}> &= <\hbar> = <\text{mass}><\text{speed}><\text{distance}>, \\
<\boldsymbol{K}> &= \frac{<\hbar>}{<\text{speed}>} = <\text{mass}><\text{distance}>.
\end{aligned}$$

https://doi.org/10.1515/9783110492132-004

From these dimensionalities, one can guess that we are dealing with *energy* (or *Hamiltonian*) H, *momentum* $\boldsymbol{P}$ and *angular momentum* $\boldsymbol{J}$ of the system.[1] We will call them *basic observables*. The operators H, $\boldsymbol{P}$ and $\boldsymbol{J}$ generate transformations of the system as a whole, so we can assume that these quantities refer to the entire physical system, i. e., these are the *total* energy, the *total* momentum and the *total* angular momentum. Of course, these considerations are not a proof. Our assignment of the physical meaning to $\{H, \boldsymbol{P}, \boldsymbol{J}, \boldsymbol{K}\}$ will be confirmed later, when we study in more detail properties of these operators and relationships between them.

From the Poincaré commutators (3.49)–(3.55), we immediately conclude which observables are compatible and can be measured simultaneously. For example, we see from (3.52) that energy is simultaneously measurable with the momentum and the angular momentum. From (3.50) it is also clear that different components of the angular momentum cannot be measured simultaneously. These facts are well known in nonrelativistic quantum mechanics. Now we have obtained them as a direct consequence of the principle of relativity and the structure of the Poincaré group.

From the commutators (3.49)–(3.55) we can also find transformations of H, $\boldsymbol{P}$, $\boldsymbol{J}$ and $\boldsymbol{K}$ from one frame to another. For example, every 3-vector observable $\boldsymbol{F} = \boldsymbol{P}, \boldsymbol{J}$ or $\boldsymbol{K}$ is transformed under rotations as (D.22). We have

$$\begin{aligned}\boldsymbol{F}(\boldsymbol{\varphi}) &= e^{-\frac{i}{\hbar}\boldsymbol{J}\cdot\boldsymbol{\varphi}}\boldsymbol{F}e^{\frac{i}{\hbar}\boldsymbol{J}\cdot\boldsymbol{\varphi}} \\ &= \boldsymbol{F}\cos\varphi + \frac{\boldsymbol{\varphi}}{\varphi}\left(\boldsymbol{F}\cdot\frac{\boldsymbol{\varphi}}{\varphi}\right)(1-\cos\varphi) + \left[\boldsymbol{F}\times\frac{\boldsymbol{\varphi}}{\varphi}\right]\sin\varphi. \end{aligned} \tag{4.2}$$

Boost transformations of the energy and momentum operators are (see equations (2.52) and (2.53))

$$\boldsymbol{P}(\boldsymbol{\theta}) = e^{-\frac{ic}{\hbar}\boldsymbol{K}\cdot\boldsymbol{\theta}}\boldsymbol{P}e^{\frac{ic}{\hbar}\boldsymbol{K}\cdot\boldsymbol{\theta}} = \boldsymbol{P} + \frac{\boldsymbol{\theta}}{\theta}\left(\boldsymbol{P}\cdot\frac{\boldsymbol{\theta}}{\theta}\right)(\cosh\theta - 1) - \frac{\boldsymbol{\theta}}{c\theta}H\sinh\theta, \tag{4.3}$$

$$H(\boldsymbol{\theta}) = e^{-\frac{ic}{\hbar}\boldsymbol{K}\cdot\boldsymbol{\theta}}He^{\frac{ic}{\hbar}\boldsymbol{K}\cdot\boldsymbol{\theta}} = H\cosh\theta - c\left(\boldsymbol{P}\cdot\frac{\boldsymbol{\theta}}{\theta}\right)\sinh\theta. \tag{4.4}$$

We will also need boost transformations of the boost generators. For example, the transformation of the component K_y due to a boost along the x-axis is obtained using equations (E.14), (3.51) and (3.53). Then we have

$$\begin{aligned}K_y(\theta) &= e^{-\frac{ic}{\hbar}K_x\theta}K_ye^{\frac{ic}{\hbar}K_x\theta} \\ &= K_y - \frac{ic\theta}{\hbar}[K_x, K_y] - \frac{c^2\theta^2}{2!\hbar^2}[K_x,[K_x,K_y]] + \frac{ic^3\theta^3}{3!\hbar^3}[K_x,[K_x,[K_x,K_y]]] + O(\theta^4) \\ &= K_y - \frac{\theta}{c}J_z + \frac{\theta^2}{2!}K_y - \frac{\theta^3}{3!c}J_z + O(\theta^4) = K_y\cosh\theta - \frac{J_z}{c}\sinh\theta. \end{aligned} \tag{4.5}$$

1 The observable corresponding to the boost generator $\boldsymbol{K}$ has no special name, but we will see later that $\boldsymbol{K}$ is closely related to the position and spin of the system.

Also, from (3.52) it follows that the total energy H, the total momentum $\boldsymbol{P}$ and the total angular momentum $\boldsymbol{J}$ are independent of time, i. e., these are *conserved observables.*

4.1.2 Operator of velocity

The operator of velocity is defined as [3, 41][2]

$$\boldsymbol{V} \equiv \frac{\boldsymbol{P}c^2}{H}. \tag{4.6}$$

Denoting by $\boldsymbol{V}(\theta)$ the velocity measured in the reference frame moving along the x-axis with the speed $v = c\tanh\theta$, we obtain

$$\begin{aligned} V_x(\theta) &= e^{-\frac{ic}{\hbar}K_x\theta}\frac{P_xc^2}{H}e^{\frac{ic}{\hbar}K_x\theta} = \frac{c^2P_x\cosh\theta - cH\sinh\theta}{H\cosh\theta - cP_x\sinh\theta} \\ &= \frac{c^2P_xH^{-1} - c\tanh\theta}{1 - cP_xH^{-1}\tanh\theta} = \frac{V_x - v}{1 - V_xv/c^2}, \end{aligned} \tag{4.7}$$

$$V_y(\theta) = \frac{V_y}{(1 - \frac{V_x}{c}\tanh\theta)\cosh\theta} = \frac{V_y\sqrt{1 - v^2/c^2}}{1 - V_xv/c^2}, \tag{4.8}$$

$$V_z(\theta) = \frac{V_z}{(1 - \frac{V_x}{c}\tanh\theta)\cosh\theta} = \frac{V_z\sqrt{1 - v^2/c^2}}{1 - V_xv/c^2}. \tag{4.9}$$

This is the usual relativistic *law of addition of velocities*. In the limit $c \to \infty$ it reduces to the familiar nonrelativistic form

$$\begin{aligned} V_x(v) &= V_x - v, \\ V_y(v) &= V_y, \\ V_z(v) &= V_z. \end{aligned}$$

Obviously, $[\boldsymbol{V}, H] = 0$, so velocity is a conserved quantity.

4.2 Casimir operators

Observables $H, \boldsymbol{P}, \boldsymbol{J}$ and $\boldsymbol{V}$ depend on the observer, and therefore they do not express intrinsic fundamental properties of the physical system that are independent of the observer and the state. But do such truly *invariant* properties of the system really exist? If they do exist, then their operators (called *Casimir operators*) must commute with all

2 The ratio $\boldsymbol{P}/H$ is well defined, because these two operators commute.

generators of the Poincaré group. It can be shown that the Poincaré group has only two independent Casimir operators [31], which are referred to as the *mass* and *spin*. Roughly speaking, these properties are related to the system's amount of matter and the speed of rotation about its own axis, respectively.

4.2.1 4-Vectors

Before we deal with Casimir operators, let us introduce several useful definitions. We will call the four operators $\tilde{\mathcal{A}} \equiv (\mathcal{A}_0, \mathcal{A}_x, \mathcal{A}_y, \mathcal{A}_z)$ a *4-vector*[3] if $(\mathcal{A}_x, \mathcal{A}_y, \mathcal{A}_z)$ is a 3-vector, $\mathcal{A}_0$ is a 3-scalar and their commutators with boost generators are the following:

$$[K_i, \mathcal{A}_j] = -\frac{i\hbar}{c}\mathcal{A}_0\delta_{ij} \quad (i,j = x,y,z), \tag{4.10}$$

$$[\boldsymbol{K}, \mathcal{A}_0] = -\frac{i\hbar}{c}\boldsymbol{\mathcal{A}}. \tag{4.11}$$

It is not difficult to show that the *4-square* $\tilde{\mathcal{A}}^2 \equiv \mathcal{A}_0^2 - \mathcal{A}_x^2 - \mathcal{A}_y^2 - \mathcal{A}_z^2 \equiv \mathcal{A}_0^2 - \mathcal{A}^2$ of a 4-vector is a *4-scalar*, i. e., it commutes with both rotations and boosts, and therefore it does not change under inertial transformations from the Lorentz subgroup. For example,

$$\begin{aligned}[K_x, \tilde{\mathcal{A}}^2] &= [K_x, \mathcal{A}_0^2 - \mathcal{A}_x^2 - \mathcal{A}_y^2 - \mathcal{A}_z^2] \\ &= -\frac{i\hbar}{c}(\mathcal{A}_x\mathcal{A}_0 + \mathcal{A}_0\mathcal{A}_x - \mathcal{A}_0\mathcal{A}_x - \mathcal{A}_x\mathcal{A}_0) = 0.\end{aligned}$$

Hence, in order to find the Casimir operators of the Poincaré algebra, it is sufficient to find two 4-vector functions of Poincaré generators that commute with H and $\boldsymbol{P}$. Then the 4-squares of these 4-vectors will be invariant with respect to all inertial transformations.

4.2.2 Mass operator

It follows immediately from (3.49), (3.52), (3.54), (3.55) that four operators $(H, c\boldsymbol{P})$ satisfy the conditions formulated for 4-vectors in Subsection 4.2.1. This quadruple is called the *4-vector of energy-momentum*. Then, from its 4-square we can construct the first Casimir invariant, called the *mass operator*, as follows[4]:

$$M \equiv +\frac{1}{c^2}\sqrt{H^2 - P^2c^2}.$$

3 We will mark 4-vectors by the tilde; see also Appendix J.

4 We choose the positive value of the square root, because masses of all known physical systems are nonnegative.

This formula can be also rewritten as an expression for the energy (the *Hamiltonian*) of the system. Then we have

$$H = +\sqrt{P^2c^2 + M^2c^4}. \tag{4.12}$$

In the nonrelativistic limit $c \to \infty$ we obtain (up to small terms of the order $(v/c)^2$)

$$H = Mc^2 + \frac{P^2}{2M} + \cdots,$$

which is the sum of the famous Einstein *rest energy* $E = Mc^2$ and the usual classical kinetic energy $P^2/(2M)$.

4.2.3 Pauli–Lubanski 4-vector

The second 4-vector $\tilde{W}$ commuting with H and $\boldsymbol{P}$ is the *Pauli–Lubanski operator*, whose components are defined as[5]

$$W_0 = (\boldsymbol{P} \cdot \boldsymbol{J}), \tag{4.13}$$

$$\boldsymbol{W} = \frac{H\boldsymbol{J}}{c} - c[\boldsymbol{P} \times \boldsymbol{K}]. \tag{4.14}$$

Let us now verify that all 4-vector properties are really satisfied for the four operators $(W^0, \boldsymbol{W})$. It is immediately evident that W_0 is a 3-scalar, i. e.,

$$[\boldsymbol{J}, W_0] = 0.$$

Moreover, W_0 changes its sign when the sign of $\boldsymbol{P}$ changes, so W_0 is a pseudo-scalar. The operator $\boldsymbol{W}$ is a pseudo-vector because its sign does not change under inversion (= simultaneous change of signs of of $\boldsymbol{K}$ and $\boldsymbol{P}$) and

$$[J_i, W_j] = i\hbar \sum_{k=1}^{3} \epsilon_{ijk} W_k.$$

Now we verify the commutators with boost generators,

$$\begin{aligned}
[K_x, W_0] &= [K_x, P_x J_x + P_y J_y + P_z J_z] \\
&= -i\hbar\left(\frac{HJ_x}{c^2} - P_y K_z + P_z K_y\right) = -\frac{i\hbar}{c} W_x,
\end{aligned} \tag{4.15}$$

5 The products of operators in these definitions are well defined, because the factors commute. This implies that operators W_0 and $\boldsymbol{W}$ are guaranteed to be Hermitian.

$$
\begin{aligned}
[K_x, W_x] &= \left[K_x, \frac{HJ_x}{c} - cP_yK_z + cP_zK_y\right] \\
&= \frac{i\hbar}{c}(-P_xJ_x - P_yJ_y - P_zJ_z) = -\frac{i\hbar}{c}W_0, &(4.16)\\
[K_x, W_y] &= \left[K_x, \frac{HJ_y}{c} - cP_zK_x + cP_xK_z\right] \\
&= \frac{i\hbar}{c}(HK_z - P_xJ_y - HK_z + P_xJ_y) = 0, &(4.17)\\
[K_x, W_z] &= 0. &(4.18)
\end{aligned}
$$

Taken together, equations (4.15)–(4.18) form the characteristic 4-vector relations (4.10)–(4.11). We have

$$
\begin{aligned}
[\boldsymbol{K}, W_0] &= -\frac{i\hbar}{c}\boldsymbol{W}, &(4.19)\\
[K_i, W_j] &= -\frac{i\hbar}{c}\delta_{ij}W_0. &(4.20)
\end{aligned}
$$

Next, we have to check the vanishing commutators with generators of translations,

$$
\begin{aligned}
[W_0, H] &= [\boldsymbol{P}\cdot\boldsymbol{J}, H] = 0,\\
[W_0, P_x] &= [J_xP_x + J_yP_y + J_zP_z, P_x] = P_y[J_y, P_x] + P_z[J_z, P_x]\\
&= -i\hbar P_yP_z + i\hbar P_zP_y = 0,\\
[\boldsymbol{W}, H] &= -c[[\boldsymbol{P}\times\boldsymbol{K}], H] = -c[[\boldsymbol{P}, H]\times\boldsymbol{K}] - c[\boldsymbol{P}\times[\boldsymbol{K}, H]] = 0,\\
[W_x, P_x] &= \frac{1}{c}[HJ_x, P_x] - c[[\boldsymbol{P}\times\boldsymbol{K}]_x, P_x] = -c[P_yK_z - P_zK_y, P_x] = 0,\\
[W_x, P_y] &= \frac{1}{c}[HJ_x, P_y] - c[[\boldsymbol{P}\times\boldsymbol{K}]_x, P_y] = \frac{i\hbar}{c}HP_z - c[P_yK_z - P_zK_y, P_y]\\
&= \frac{i\hbar}{c}HP_z - \frac{i\hbar}{c}HP_z = 0.
\end{aligned}
$$

This completes the proof that the 4-square of the Pauli–Lubanski 4-vector

$$\tilde{W}^2 = W_0^2 - W^2$$

is a Casimir operator. Although operators $(W_0, \boldsymbol{W})$ do not have a direct physical interpretation, in Section 4.3 we will see that they are very useful for deriving the operators of position $\boldsymbol{R}$ and spin $\boldsymbol{S}$. For these calculations, we will need commutators between components of $\tilde{W}$. For example,

$$
\begin{aligned}
[W_x, W_y] &= \left[W_x, \frac{HJ_y}{c} + cP_xK_z - cP_zK_x\right] = i\hbar\left(\frac{HW_z}{c} - W^0P_z\right),\\
[W_0, W_x] &= \left[W_0, \frac{HJ_x}{c} - cP_yK_z + cP_zK_y\right] = -i\hbar P_yW_z + i\hbar P_zW_y = -i\hbar[\boldsymbol{P}\times\boldsymbol{W}]_x.
\end{aligned}
$$

These equalities can be easily generalized, so we have

$$[W_i, W_j] = \frac{i\hbar}{c} \sum_{k=1}^{3} \epsilon_{ijk}(HW_k - cW_0P_k), \tag{4.21}$$

$$[W_0, W_j] = -i\hbar[\boldsymbol{P} \times \boldsymbol{W}]_j. \tag{4.22}$$

4.3 Operators of spin and position

Now we are ready to look for expressions for the operators of spin and position as functions of Poincaré generators [69, 62, 9, 16, 43].

4.3.1 Physical requirements

We will demand that operators of the total *spin* $\boldsymbol{S}$ and the center-of-energy[6] *position* $\boldsymbol{R}$ have the following natural properties:

(I) Because of the similarity between spin and angular momentum, we require that $\boldsymbol{S}$ is a pseudo-vector (like $\boldsymbol{J}$), so

$$[J_j, S_i] = i\hbar \sum_{k=1}^{3} \epsilon_{ijk} S_k.$$

(II) We also require that spin's components satisfy the same commutation relations (3.50) as the components of $\boldsymbol{J}$, so

$$[S_i, S_j] = i\hbar \sum_{k=1}^{3} \epsilon_{ijk} S_k. \tag{4.23}$$

(III) We want spin to be measurable simultaneously with momentum, so

$$[\boldsymbol{P}, \boldsymbol{S}] = 0.$$

(IV) We also want spin to be measurable simultaneously with position, so

$$[\boldsymbol{R}, \boldsymbol{S}] = 0. \tag{4.24}$$

(V) From the physical meaning of $\boldsymbol{R}$, it follows that spatial translations simply shift its components, as follows:

$$e^{-\frac{i}{\hbar}P_x a} R_x e^{\frac{i}{\hbar}P_x a} = R_x - a,$$

6 The name "center-of-energy" is appropriate, as explained in Subsection 6.3.3.

$$e^{-\frac{i}{\hbar}P_x a} R_y e^{\frac{i}{\hbar}P_x a} = R_y,$$
$$e^{-\frac{i}{\hbar}P_x a} R_z e^{\frac{i}{\hbar}P_x a} = R_z.$$

The "canonical" commutation relations follow immediately from this requirement. We have

$$[R_i, P_j] = i\hbar\delta_{ij}. \tag{4.25}$$

(VI) Finally, components of $\boldsymbol{R}$ obviously change their signs under space inversion, so we assume that they form a true 3-vector, i. e.,

$$[J_i, R_j] = i\hbar \sum_{k=1}^{3} \epsilon_{ijk} R_k. \tag{4.26}$$

Next, we are going to write explicit expressions for the operators $\boldsymbol{S}$ and $\boldsymbol{R}$ with the above properties. We will also prove that these expressions are unique.

4.3.2 Spin operator

To begin with, suppose that the spin operator has the form[7]

$$\boldsymbol{S} = \frac{\boldsymbol{W}}{Mc} - \frac{W_0 \boldsymbol{P}}{M(Mc^2 + H)}, \tag{4.27}$$

which is a pseudo-vector commuting with $\boldsymbol{P}$, as required by conditions (I) and (III).

The next step is to verify condition (II). To calculate the commutator (4.23), we introduce the notation

$$Y \equiv -\frac{1}{M(Mc^2 + H)}. \tag{4.28}$$

We also use commutators (4.21) and (4.22), the equality

$$(\boldsymbol{P} \cdot \boldsymbol{W}) = \frac{1}{c} H(\boldsymbol{P} \cdot \boldsymbol{J}) = \frac{1}{c} H W_0 \tag{4.29}$$

and formula (D.16). Then

$$[S_x, S_y] = \left[\left(YW_0P_x + \frac{W_x}{Mc}\right), \left(YW_0P_y + \frac{W_y}{Mc}\right)\right]$$

7 Note that the operator $\boldsymbol{S}$ contains the mass operator M in the denominator, so equation (4.27) has mathematical meaning only for systems with strictly positive mass spectrum. Thus, the concept of spin is not applicable to massless photons.

$$= i\hbar\left(-\frac{YP_x[\mathbf{P}\times\mathbf{W}]_y}{Mc} + \frac{YP_y[\mathbf{P}\times\mathbf{W}]_x}{Mc} + \frac{HW_z - cW_0P_z}{M^2c^3}\right)$$
$$= i\hbar\left(-\frac{Y[\mathbf{P}\times[\mathbf{P}\times\mathbf{W}]]_z}{Mc} + \frac{HW_z - cW_0P_z}{M^2c^3}\right)$$
$$= i\hbar\left(-\frac{Y(P_z(\mathbf{P}\cdot\mathbf{W}) - W_zP^2)}{Mc} + \frac{HW_z - cW_0P_z}{M^2c^3}\right)$$
$$= i\hbar\left(-\frac{Y(P_zHW_0c^{-1} - W_zP^2)}{Mc} + \frac{HW_z - cW_0P_z}{M^2c^3}\right)$$
$$= i\hbar W_z\left(\frac{P^2Y}{Mc} + \frac{H}{M^2c^3}\right) + i\hbar P_zW_0\left(-\frac{HY}{Mc^2} - \frac{1}{M^2c^2}\right).$$

For the expressions in parentheses we have

$$\frac{P^2Y}{Mc} + \frac{H}{M^2c^3} = -\frac{P^2}{M^2c(Mc^2+H)} + \frac{H}{M^2c^3} = \frac{H(Mc^2+H) - P^2c^2}{M^2c^3(Mc^2+H)}$$
$$= \frac{H(Mc^2+H) - (Mc^2+H)(H-Mc^2)}{M^2c^3(Mc^2+H)} = \frac{1}{Mc},$$
$$-\frac{HY}{Mc^2} - \frac{1}{M^2c^2} = \frac{H}{M^2c^2(Mc^2+H)} - \frac{1}{M^2c^2}$$
$$= \frac{H - (Mc^2+H)}{M^2c^2(Mc^2+H)} = -\frac{1}{M(Mc^2+H)} = Y,$$

whence follows the commutator

$$[S_x, S_y] = i\hbar\left(\frac{W_z}{Mc} + YW_0P_z\right) = i\hbar S_z,$$

which is (4.23). So, we have confirmed that the operator (4.27) satisfies our requirements (I)–(III). In Subsection 4.3.8 we will demonstrate the uniqueness of this spin operator.

4.3.3 Position operator

Now we turn our attention to the position operator $\boldsymbol{R}$. We follow a path similar to the derivation of spin: first we "guess" the form of $\boldsymbol{R}$, which satisfies all requirements from Subsection 4.3.1, and then in Subsection 4.3.9 we will prove that this expression is unique. As the initial guess for $\boldsymbol{R}$, we choose the well-known *Newton–Wigner operator*[8] [69, 62, 9, 16, 43], which we write in three equivalent forms:

$$\boldsymbol{R} = -\frac{c^2}{2}(H^{-1}\boldsymbol{K} + \boldsymbol{K}H^{-1}) - \frac{c^2[\mathbf{P}\times\mathbf{S}]}{H(Mc^2+H)} \tag{4.30}$$

8 Similar to the spin operator, $\boldsymbol{R}$ is defined only for systems with strictly positive spectrum of the mass operator. We had to use the symmetrized product $H^{-1}\boldsymbol{K} + \boldsymbol{K}H^{-1}$ of noncommuting operators to keep our $\boldsymbol{R}$ Hermitian. We also used $[\boldsymbol{K}, H^{-1}] = -H^{-2}[\boldsymbol{K}, H] = i\hbar\mathbf{P}/H^2$.

$$= -\frac{c^2}{2}(H^{-1}\boldsymbol{K} + \boldsymbol{K}H^{-1}) - \frac{c[\boldsymbol{P} \times \boldsymbol{W}]}{MH(Mc^2 + H)} \tag{4.31}$$

$$= -c^2H^{-1}\boldsymbol{K} - \frac{i\hbar c^2\boldsymbol{P}}{2H^2} + \frac{cY[\boldsymbol{P} \times \boldsymbol{W}]}{H}. \tag{4.32}$$

Here $\boldsymbol{R}$ is a true vector with properties (V) and (VI). For example,

$$[R_x, P_x] = -\frac{c^2}{2}[(H^{-1}K_x + K_xH^{-1}), P_x] = \frac{i\hbar}{2}(H^{-1}H + HH^{-1}) = i\hbar,$$

$$[R_x, P_y] = -\frac{c^2}{2}[(H^{-1}K_x + K_xH^{-1}), P_y] = 0.$$

Next, we calculate[9]

$$\begin{aligned}
\boldsymbol{J} - [\boldsymbol{R} \times \boldsymbol{P}] &= \boldsymbol{J} + \frac{c^2}{H}[\boldsymbol{K} \times \boldsymbol{P}] + \frac{c^2[[\boldsymbol{P} \times \boldsymbol{S}] \times \boldsymbol{P}]}{H(Mc^2 + H)} \\
&= \boldsymbol{J} + \frac{c^2}{H}[\boldsymbol{K} \times \boldsymbol{P}] - \frac{c^2(\boldsymbol{P}(\boldsymbol{P} \cdot \boldsymbol{S}) - \boldsymbol{S}P^2)}{H(Mc^2 + H)} \\
&= \boldsymbol{J} + \frac{c^2}{H}[\boldsymbol{K} \times \boldsymbol{P}] - \frac{c^2\boldsymbol{P}(\boldsymbol{P} \cdot \boldsymbol{S}) - \boldsymbol{S}(H - Mc^2)(H + Mc^2)}{H(Mc^2 + H)} \\
&= \boldsymbol{J} + \frac{c^2}{H}[\boldsymbol{K} \times \boldsymbol{P}] + \boldsymbol{S} - \frac{c^2\boldsymbol{P}(\boldsymbol{P} \cdot \boldsymbol{S})}{H(Mc^2 + H)} - \frac{Mc^2}{H}\boldsymbol{S} \\
&= \boldsymbol{J} + \frac{c^2}{H}[\boldsymbol{K} \times \boldsymbol{P}] + \boldsymbol{S} - \frac{c^2\boldsymbol{P}(\boldsymbol{P} \cdot \boldsymbol{S})}{H(Mc^2 + H)} - \boldsymbol{J} + \frac{c^2\boldsymbol{P}(\boldsymbol{P} \cdot \boldsymbol{J})}{H(Mc^2 + H)} + \frac{c^2}{H}[\boldsymbol{P} \times \boldsymbol{K}] \\
&= \boldsymbol{S}.
\end{aligned}$$

From this result it follows that, like in classical nonrelativistic physics, the total angular momentum is the sum of two parts

$$\boldsymbol{J} = [\boldsymbol{R} \times \boldsymbol{P}] + \boldsymbol{S}:$$

the *orbital angular momentum* $[\boldsymbol{R} \times \boldsymbol{P}]$ and the *internal angular momentum* or spin $\boldsymbol{S}$.[10] It is easy to verify that condition (IV) is also satisfied. For example,

$$[S_x, R_y] = [J_x - [\boldsymbol{R} \times \boldsymbol{P}]_x, R_y] = i\hbar R_z - [P_yR_z - P_zR_y, R_y] = i\hbar R_z - i\hbar R_z = 0.$$

---

9 Note that the expression $[\boldsymbol{R} \times \boldsymbol{P}]$ contains only products of commuting components R_i and P_j. Therefore $[\boldsymbol{R} \times \boldsymbol{P}]$ is an Hermitian operator.

10 It is often said that spin is a purely quantum-mechanical observable, which does not have a classical analog. We do not share this view. The most significant difference between the classical and quantum internal angular momentum (spin) is that the latter has a discrete spectrum, whereas the former is continuous.

4.3.4 Commutators of position

From the commutator (4.25) it follows that

$$[R_x, P_x^n] = i\hbar n P_x^{n-1} = i\hbar \frac{\partial P_x^n}{\partial P_x}. \tag{4.33}$$

Then for any momentum function $f(P_x, P_y, P_z)$ we have

$$[\boldsymbol{R}, f(P_x, P_y, P_z)] = i\hbar \frac{\partial f(P_x, P_y, P_z)}{\partial \boldsymbol{P}}. \tag{4.34}$$

For example,

$$\begin{aligned}[\boldsymbol{R}, H] &= [\boldsymbol{R}, \sqrt{P^2c^2 + M^2c^4}] = i\hbar \frac{\partial \sqrt{P^2c^2 + M^2c^4}}{\partial \boldsymbol{P}} \\ &= \frac{i\hbar \boldsymbol{P}c^2}{\sqrt{P^2c^2 + M^2c^4}} = i\hbar \frac{\boldsymbol{P}c^2}{H} = i\hbar \boldsymbol{V},\end{aligned} \tag{4.35}$$

where $\boldsymbol{V}$ is the relativistic operator of velocity (4.6). Therefore, an observer shifted in time by t sees the position of the physical system as shifted by $\boldsymbol{V}t$:

$$\boldsymbol{R}(t) = e^{\frac{i}{\hbar}Ht} \boldsymbol{R} e^{-\frac{i}{\hbar}Ht} = \boldsymbol{R} + \frac{i}{\hbar}[H, \boldsymbol{R}]t = \boldsymbol{R} + \boldsymbol{V}t. \tag{4.36}$$

In other words, the center-of-energy $\boldsymbol{R}$ of any isolated system moves with a constant velocity, as expected. This result does not depend on the internal structure of the system or on interactions between its parts.

Theorem 4.1. *Components of the Newton–Wigner position operator commute with each other:* $[R_i, R_j] = 0$.

Proof. To begin with, we calculate the commutator $[HR_x, HR_y]$, which is related to the desired expression $[R_x, R_y]$ by the formula (here we used (E.12) and (4.35))

$$\begin{aligned}[HR_x, HR_y] &= [HR_x, H]R_y + H[HR_x, R_y] \\ &= H[R_x, H]R_y + H[H, R_y]R_x + H^2[R_x, R_y] \\ &= i\hbar c^2(P_xR_y - R_yP_x) + H^2[R_x, R_y] = i\hbar c^2[\boldsymbol{P} \times \boldsymbol{R}]_z + H^2[R_x, R_y] \\ &= -i\hbar c^2 J_z + i\hbar c^2 S_z + H^2[R_x, R_y].\end{aligned} \tag{4.37}$$

From equation (4.32) we get

$$[HR_x, HR_y] = \left[\left(-c^2K_x - \frac{i\hbar c^2 P_x}{2H} + cY[\boldsymbol{P} \times \boldsymbol{W}]_x\right), \left(-c^2K_y - \frac{i\hbar c^2 P_y}{2H} + cY[\boldsymbol{P} \times \boldsymbol{W}]_y\right)\right].$$

Nonzero contributions to this commutator are

$$[-c^2K_x, -c^2K_y] = -i\hbar c^2 J_z, \tag{4.38}$$

$$\left[-\frac{i\hbar c^2 P_x}{2H}, -c^2 K_y\right] = -\frac{i\hbar c^4}{2}\left[K_y, \frac{P_x}{H}\right] = \frac{\hbar^2 c^4 P_y P_x}{2H^2}, \tag{4.39}$$

$$\left[-c^2 K_x, -\frac{i\hbar c^2 P_y}{2H}\right] = -\frac{\hbar^2 c^4 P_y P_x}{2H^2}, \tag{4.40}$$

$$\begin{aligned}
[-c^2 K_x, cY[\boldsymbol{P} \times \boldsymbol{W}]_y] &= \frac{c^3}{M}\left[K_x, \frac{P_z W_x - P_x W_z}{H + Mc^2}\right] \\
&= \frac{c^3}{M}\left(-\frac{P_z W_x - P_x W_z}{(H + Mc^2)^2}[K_x, H] + \frac{P_z[K_x, W_x]}{H + Mc^2} - \frac{[K_x, P_x]W_z}{H + Mc^2}\right) \\
&= i\hbar c^3 (MY^2(P_z W_x - P_x W_z)P_x + YP_z W_0 c^{-1} - YHW_z c^{-2}),
\end{aligned}$$

$$[cY[\boldsymbol{P} \times \boldsymbol{W}]_x, -c^2 K_y] = i\hbar c^3 (MY^2(P_y W_z - P_z W_y)P_y + YP_z W_0 c^{-1} - YHW_z c^{-2}).$$

Combining the last two results and using (4.29), we obtain

$$\begin{aligned}
&[-c^2 K_x, cY[\boldsymbol{P} \times \boldsymbol{W}]_y] + [cY[\boldsymbol{P} \times \boldsymbol{W}]_x, -c^2 K_y] \\
&\quad = i\hbar c^3 (MY^2[\boldsymbol{P} \times [\boldsymbol{P} \times \boldsymbol{W}]]_z + 2YP_z W_0 c^{-1} - 2YHW_z c^{-2}) \\
&\quad = i\hbar c^3 (MY^2(P_z(\boldsymbol{P} \cdot \boldsymbol{W}) - W_z P^2) + 2YP_z W_0 c^{-1} - 2YHW_z c^{-2}) \\
&\quad = i\hbar c^3 (MY^2(P_z H W_0 c^{-1} - W_z P^2) + 2YP_z W_0 c^{-1} - 2YHW_z c^{-2}) \\
&\quad = i\hbar c^2 MY^2 P_z W_0 (H - 2(H + Mc^2)) \\
&\qquad + i\hbar c MY^2 W_z (-(H - Mc^2)(H + Mc^2) + 2H(H + Mc^2)) \\
&\quad = i\hbar c^3 P_z W_0 (Yc^{-1} - M^2 Y^2 c) + \frac{i\hbar c W_z}{M} = i\hbar c^2 S_z - i\hbar c^4 P_z W_0 M^2 Y^2.
\end{aligned} \tag{4.41}$$

Using the formula

$$(\boldsymbol{P} \cdot \boldsymbol{S}) = \frac{(\boldsymbol{P} \cdot \boldsymbol{J})H}{Mc^2} - \frac{P^2(\boldsymbol{P} \cdot \boldsymbol{J})(H - Mc^2)}{P^2 Mc^2} = (\boldsymbol{P} \cdot \boldsymbol{J}) = W_0,$$

we obtain one more commutator,

$$\begin{aligned}
&[cY[\boldsymbol{P} \times \boldsymbol{W}]_x, cY[\boldsymbol{P} \times \boldsymbol{W}]_y] \\
&\quad = M^2 Y^2 c^4 [[\boldsymbol{P} \times \boldsymbol{S}]_x, [\boldsymbol{P} \times \boldsymbol{S}]_y] = M^2 Y^2 c^4 [P_y S_z - P_z S_y, P_z S_x - P_x S_z] \\
&\quad = M^2 Y^2 c^4 (P_z P_y [S_z, S_x] - P_z^2 [S_y, S_x] + P_x P_z [S_y, S_z]) \\
&\quad = i\hbar M^2 Y^2 c^4 P_z (\boldsymbol{P} \cdot \boldsymbol{S}) = i\hbar c^4 P_z W_0 M^2 Y^2.
\end{aligned} \tag{4.42}$$

Now we can collect all contributions (4.38)–(4.42) and obtain

$$[HR_x, HR_y] = -i\hbar c^2 J_z + i\hbar c^2 S_z.$$

Comparison with equation (4.37) gives

$$H^2[R_x, R_y] = 0.$$

Since we assumed that M is strictly positive, the operator $H^2 = M^2c^4 + P^2c^2$ has no zero eigenvalues. This leads us to the desired result,

$$[R_x, R_y] = 0. \qquad \square$$

4.3.5 Alternative set of basic operators

Until now, our plan was to construct operators of observables as functions of 10 Poincaré generators. However, calculations with these generators sometimes become unnecessarily complicated because of the complex commutation relations in the Lie algebra $\{\boldsymbol{P}, \boldsymbol{J}, \boldsymbol{K}, H\}$. For systems with strictly positive spectrum of the mass operator, another set of basic operators $\{\boldsymbol{P}, \boldsymbol{R}, \boldsymbol{S}, M\}$ with simpler commutators,

$$\begin{aligned}
[\boldsymbol{P}, M] &= [\boldsymbol{R}, M] = [\boldsymbol{S}, M] = [R_i, R_j] = [P_i, P_j] = 0,\\
[R_i, P_j] &= i\hbar\delta_{ij},\\
[\boldsymbol{P}, \boldsymbol{S}] &= [\boldsymbol{R}, \boldsymbol{S}] = 0,\\
[S_i, S_j] &= i\hbar \sum_{k=1}^{3} \epsilon_{ijk} S_k
\end{aligned}$$

may appear more convenient. Using our previous results, we can express operators in this set through original Poincaré generators, in particular[11]

$$\boldsymbol{R} = -\frac{c^2}{2}(H^{-1}\boldsymbol{K} + \boldsymbol{K}H^{-1}) - \frac{c[\boldsymbol{P} \times \boldsymbol{W}]}{MH(Mc^2 + H)}, \tag{4.43}$$

$$\boldsymbol{S} = \boldsymbol{J} - [\boldsymbol{R} \times \boldsymbol{P}], \tag{4.44}$$

$$M = +\frac{1}{c^2}\sqrt{H^2 - P^2c^2}. \tag{4.45}$$

Conversely, generators $\{\boldsymbol{P}, \boldsymbol{K}, \boldsymbol{J}, H\}$ can be expressed in terms of alternative operators $\{\boldsymbol{P}, \boldsymbol{R}, \boldsymbol{S}, M\}$. For the energy and angular momentum we obtain

$$H = +\sqrt{M^2c^4 + P^2c^2}, \tag{4.46}$$

$$\boldsymbol{J} = [\boldsymbol{R} \times \boldsymbol{P}] + \boldsymbol{S} \tag{4.47}$$

and the boost operator is

$$-\frac{1}{2c^2}(\boldsymbol{R}H + H\boldsymbol{R}) - \frac{[\boldsymbol{P} \times \boldsymbol{S}]}{Mc^2 + H}$$

11 The operator $\boldsymbol{P}$ is the same in both sets.

$$
\begin{aligned}
&= -\frac{1}{2}\left(-\frac{1}{2}(H^{-1}\boldsymbol{K}H + \boldsymbol{K}) - \frac{[\boldsymbol{P} \times \boldsymbol{S}]}{Mc^2 + H}\right) \\
&\quad - \frac{1}{2}\left(-\frac{1}{2}(\boldsymbol{K} + H\boldsymbol{K}H^{-1}) - \frac{[\boldsymbol{P} \times \boldsymbol{S}]}{Mc^2 + H}\right) - \frac{[\boldsymbol{P} \times \boldsymbol{S}]}{Mc^2 + H} \\
&= \frac{1}{4}(H^{-1}\boldsymbol{K}H + \boldsymbol{K} + \boldsymbol{K} + H\boldsymbol{K}H^{-1}) = \boldsymbol{K} - \frac{i\hbar}{4}(H^{-1}\boldsymbol{P} - \boldsymbol{P}H^{-1}) \\
&= \boldsymbol{K}.
\end{aligned} \tag{4.48}
$$

Each function of operators from the set $\{\boldsymbol{P}, \boldsymbol{J}, \boldsymbol{K}, H\}$ can be rewritten as a function of operators from the alternative set $\{\boldsymbol{P}, \boldsymbol{R}, \boldsymbol{S}, M\}$, and *vice versa*. Consequently, these two groups of operators provide equivalent descriptions of Poincaré invariant theories. These observations will be useful to us not only in this chapter, but also in Chapters 6 and 7.

4.3.6 Canonical order of operators

When we perform calculations with functions of Poincaré generators, we encounter a certain difficulty when the same operator can be written in several equivalent forms. For example, according to (3.55), $K_x H$ and $HK_x - i\hbar P_x$ are two forms of the same operator. To solve this nonuniqueness problem, we will always agree to write products of generators in the *canonical order*, such that their factors are positioned in the following order from left to right[12]:

$$
\mathfrak{C}(P_x, P_y, P_z, H), J_x, J_y, J_z, K_x, K_y, K_z. \tag{4.49}
$$

Consider, for example, the noncanonical product $K_y P_y J_x$. To bring it to the canonical form, we first move the factor P_y to the extreme left position, using commutator (3.54), so we have

$$
K_y P_y J_x = P_y K_y J_x + [K_y, P_y] J_x = P_y K_y J_x - \frac{i\hbar}{c^2} H J_x.
$$

The second term on the right-hand side is already in the canonical order, but the first term is not. There we need to swap the factors J_x and K_y, so we have

$$
K_y P_y J_x = P_y J_x K_y + P_y [K_y, J_x] - \frac{i\hbar}{c^2} H J_x = P_y J_x K_y - i\hbar P_y K_z - \frac{i\hbar}{c^2} H J_x. \tag{4.50}
$$

Now all the terms in (4.50) are written canonically.

12 Since H, P_x, P_y and P_z commute with each other, the part of the operator that depends on these factors can be written as an ordinary function of commuting arguments $\mathfrak{C}(P_x, P_y, P_z, H)$, whose order is not important.

The procedure for converting any operator to the canonical order is not much more complicated than the example described above. If we call the original operator *primary term*, then this procedure can be formalized as the following sequence of steps. First, we transform the primary term itself to the canonical order. We do this by changing the order of neighboring factors, if they happen to be in the "wrong order". Let us call these neighbors the "left factor" L and the "right factor" R. Now, if R commutes with L, then such a permutation has no side effects. If R does not commute with L, then the permutation results in $LR \to RL + [L, R]$. This means that, in addition to swapping $L \leftrightarrow R$, we have to add a *secondary term* to the original expression. This secondary term is obtained from the primary one by replacing the product LR with the commutator $[L, R]$.[13] So, we will proceed with all permutations $L \leftrightarrow R$ described above in the primary term until all its factors are aligned in the canonical order. If in the course of this process all the commutators of $[L, R]$ are zero, then our work is done. If nonzero commutators are encountered, then, along with the ordered primary term, we get a certain number of (still unordered) secondary terms. Now the procedure described above should be applied to all these secondary terms, which will lead to the appearance of *tertiary* terms, etc., until we have all terms in the canonical order.

As a result of this procedure, for each operator F we get a unique representation as a sum of canonically ordered terms,

$$\begin{aligned} F &= \mathfrak{C}^{00}(\boldsymbol{P}, H) + \sum_{i=1}^{3} \mathfrak{C}_i^{10}(\boldsymbol{P}, H) J_i + \sum_{i=1}^{3} \mathfrak{C}_i^{01}(\boldsymbol{P}, H) K_i \\ &+ \sum_{i,j=1}^{3} \mathfrak{C}_{ij}^{11}(\boldsymbol{P}, H) J_i K_j + \sum_{i,j=1; i \le j}^{3} \mathfrak{C}_{ij}^{02}(\boldsymbol{P}, H) K_i K_j + \cdots, \end{aligned} \tag{4.51}$$

where $\mathfrak{C}_{ijk\ldots}^{\alpha\beta}$ are functions of mutually commuting translation generators.

4.3.7 Power of operator

For our work in the rest of this chapter we will also need the notion of the *power* of operators. For monomial operators we denote by pow(A) the number of factors J and/or K in the term A. For example, the first term on the right-hand side of (4.51) has power 0. The second and third terms have power 1, and so on. The power of the general polynomial (4.51) is defined as the maximum power of its terms (= monomials). For operators considered earlier in this chapter, we have the following powers:

$$\begin{aligned} \mathrm{pow}(H) = \mathrm{pow}(M) = \mathrm{pow}(\boldsymbol{P}) = \mathrm{pow}(\boldsymbol{V}) = 0, \\ \mathrm{pow}(W^0) = \mathrm{pow}(\boldsymbol{W}) = \mathrm{pow}(\boldsymbol{S}) = \mathrm{pow}(\boldsymbol{R}) = 1. \end{aligned}$$

13 The second and third terms on the right-hand side of (4.50) are secondary.

Lemma 4.2. *If L and R are operators from the list (4.49) and $[L, R] \neq 0$, then*

$$\mathrm{pow}([L,R]) = \mathrm{pow}(L) + \mathrm{pow}(R) - 1. \tag{4.52}$$

Proof. This result follows from the fact that equation (4.52) is valid for all nonzero commutators in the Poincaré Lie algebra (3.49)–(3.55). □

It is easy to see that (4.52) is satisfied in more complex cases as well. For example, if $\mathfrak{C}$ and $\mathfrak{D}$ are two functions of P_x, P_y, P_z, H, then, using (E.12) and $[\mathfrak{C}, \mathfrak{D}] = 0$, we get

$$\begin{aligned}[\mathfrak{C}J_x, \mathfrak{D}J_y] &= [\mathfrak{C}J_x, \mathfrak{D}]J_y + \mathfrak{D}[\mathfrak{C}J_x, J_y] \\ &= \mathfrak{C}[J_x, \mathfrak{D}]J_y + \mathfrak{D}\mathfrak{C}[J_x, J_y] + \mathfrak{D}J_x[\mathfrak{C}, J_y].\end{aligned}$$

The power of the right-hand side is 1, according to Lemma 4.2. Generalizing this result, one can show that either $[\mathfrak{C}L, \mathfrak{D}R] = 0$ or $\mathrm{pow}([\mathfrak{C}L, \mathfrak{D}R]) = 1$, where L and R are any components of $\boldsymbol{J}$ or $\boldsymbol{K}$. This proves formula (4.52) for all operators L and R having power 0 or 1. Now we try to extend this result to general operators.[14]

Theorem 4.3. *For any two noncommuting monomials A and B,*

$$\mathrm{pow}([A,B]) = \mathrm{pow}(A) + \mathrm{pow}(B) - 1.$$

Proof. It is obvious that the primary term in the ordered product of two operators AB has the same number of Lorentz generators as the original product, i. e., $\mathrm{pow}(A) + \mathrm{pow}(B)$. Each secondary term is obtained by replacing products of two generators LR in the primary term by their commutator $[L, R]$. In accordance with Lemma 4.2, if $[L, R] \neq 0$, then such a replacement reduces the power of the term by one. Consequently, the power of the secondary term is $\mathrm{pow}(A)+\mathrm{pow}(B)-1$. Powers of tertiary and higher terms are lower than the power of the secondary term. Therefore, the power of any product AB is determined only by its primary term, so

$$\mathrm{pow}(AB) = \mathrm{pow}(BA) = \mathrm{pow}(A) + \mathrm{pow}(B).$$

In the commutator $AB - BA$, the primary term of AB cancels with the primary term of BA. If $[A, B] \neq 0$, then the secondary terms do not cancel out. Hence, there exists at least one nonzero secondary term whose power is $\mathrm{pow}(A) + \mathrm{pow}(B) - 1$. This is also the power of the whole commutator. □

Having at our disposal the alternative basic operators $\{\boldsymbol{P}, \boldsymbol{R}, \boldsymbol{S}, M\}$, we can form several Hermitian 3-scalars, 3-vectors and 3-tensors, which are classified in Table 4.1 in accordance with their true/pseudo character and powers. Note that $\mathrm{pow}(\boldsymbol{R}) = \mathrm{pow}(\boldsymbol{S}) = 1$ and $\mathrm{pow}(\boldsymbol{P}) = \mathrm{pow}(M) = 0$. Therefore, the power of operators in the alternative notation can be determined by the number of factors R_i and/or S_i.

14 This result was first used by Berg in [9].

Table 4.1: Scalar, vector and tensor functions of operators from the alternative set $\{\boldsymbol{P}, \boldsymbol{R}, \boldsymbol{S}, M\}$.

	power 0	power 1	power 2
True scalar	P^2; M	$\boldsymbol{P}\cdot\boldsymbol{R}+\boldsymbol{R}\cdot\boldsymbol{P}$	R^2; S^2
Pseudo-scalar		$\boldsymbol{P}\cdot\boldsymbol{S}$	$\boldsymbol{R}\cdot\boldsymbol{S}$
True vector	$\boldsymbol{P}$	$\boldsymbol{R}$; $[\boldsymbol{P}\times\boldsymbol{S}]$	$[\boldsymbol{R}\times\boldsymbol{S}]$
Pseudo-vector		$\boldsymbol{S}$; $[\boldsymbol{P}\times\boldsymbol{R}]$	
True tensor	P_iP_j	$\sum_{k=1}^3 \epsilon_{ijk}S_k$; $P_iR_j+R_jP_i$	$S_iS_j+S_jS_i$; R_iR_j
Pseudo-tensor	$\sum_{k=1}^3 \epsilon_{ijk}P_k$	$\sum_{k=1}^3 \epsilon_{ijk}R_k$; P_iS_j	R_iS_j

4.3.8 Uniqueness of spin operator

Let us now prove that (4.27) is the unique spin operator satisfying requirements (I)–(IV) from Subsection 4.3.1. Suppose that there is another operator $\boldsymbol{S}'$ satisfying the same conditions. Denoting the power of the spin components by $p \equiv \mathrm{pow}(S'_x) = \mathrm{pow}(S'_y) = \mathrm{pow}(S'_z)$, we get from (4.23) and Theorem 4.3 the equation

$$\begin{aligned}\mathrm{pow}([S'_x, S'_y]) &= \mathrm{pow}(S'_z),\\ 2p-1 &= p,\end{aligned}$$

with only solution $p = 1$. The most general form of a pseudo-vector operator with power 1 can be extracted from Table 4.1 and reads

$$\boldsymbol{S}' = b(M, P^2)\boldsymbol{S} + f(M, P^2)[\boldsymbol{P}\times\boldsymbol{R}] + e(M, P^2)(\boldsymbol{S}\cdot\boldsymbol{P})\boldsymbol{P},$$

where b, f and e are arbitrary real functions.[15] From condition (III) we obtain $f(M, P^2) = 0$. Comparing commutator (here we used (D.16))

$$\begin{aligned}[S'_x, S'_y] &= [(bS_x + e(\boldsymbol{S}\cdot\boldsymbol{P})P_x), (bS_y + e(\boldsymbol{S}\cdot\boldsymbol{P})P_y)]\\ &= b^2[S_x, S_y] + i\hbar ebP_x[\boldsymbol{S}\times\boldsymbol{P}]_y - i\hbar ebP_y[\boldsymbol{S}\times\boldsymbol{P}]_x\\ &= i\hbar b^2 S_z + i\hbar eb(\boldsymbol{P}\times[\boldsymbol{S}\times\boldsymbol{P}])_z\\ &= i\hbar(b^2S_z + ebP^2S_z - eb(\boldsymbol{S}\cdot\boldsymbol{P})P_z)\end{aligned}$$

with condition (II),

$$[S'_x, S'_y] = i\hbar S'_z = i\hbar(bS_z + e(\boldsymbol{S}\cdot\boldsymbol{P})P_z),$$

we get the system of equations

$$\begin{aligned}b^2 + ebP^2 &= b,\\ -eb &= e,\end{aligned}$$

15 To satisfy condition (I), these functions must depend only on the scalars P^2 and M.

with a single nontrivial solution $b = 1$ and $e = 0$. Therefore, our postulated spin operator is unique: $\boldsymbol{S}' = \boldsymbol{S}$. Its square is an invariant Casimir operator. Indeed

$$\begin{aligned}
\boldsymbol{S}^2 &= \left(\frac{\boldsymbol{W}}{Mc} + W_0\boldsymbol{P}Y\right)^2 = \frac{W^2}{M^2c^2} + \frac{2W_0Y(\boldsymbol{P}\cdot\boldsymbol{W})}{Mc} + W_0^2P^2Y^2 \\
&= \frac{W^2}{M^2c^2} + W_0^2Y\left(\frac{2H}{Mc^2} + P^2Y\right) = \frac{W^2}{M^2c^2} + W_0^2Y\frac{2H(Mc^2+H) - P^2c^2}{Mc^2(Mc^2+H)} \\
&= \frac{W^2}{M^2c^2} - W_0^2\frac{H^2 + 2HMc^2 + M^2c^4}{M^2c^2(Mc^2+H)^2} = \frac{W^2 - W_0^2}{M^2c^2} = -\frac{\tilde{W}^2}{M^2c^2}.
\end{aligned}$$

4.3.9 Uniqueness of position operator

Suppose that in addition to the Newton–Wigner position $\boldsymbol{R}$ (4.30) there exists another operator $\boldsymbol{R}'$ satisfying the requirements (IV)–(VI). Then, from (4.25) and Theorem 4.3 it follows that $\boldsymbol{R}'$ has power 1. The most general true vector with this property is

$$\boldsymbol{R}' = a(P^2,M)\boldsymbol{R} + d(P^2,M)[\boldsymbol{S}\times\boldsymbol{P}] + g(P^2,M)(\mathfrak{Z}\boldsymbol{P} + \boldsymbol{P}\mathfrak{Z}),$$

where a, d and g are arbitrary real functions and the operator

$$\mathfrak{Z} \equiv \frac{1}{2\hbar}(\boldsymbol{P}\cdot\boldsymbol{R} + \boldsymbol{R}\cdot\boldsymbol{P}) \tag{4.53}$$

has the following commutators:

$$\begin{aligned}
[\mathfrak{Z}, \boldsymbol{S}] &= 0, \\
[\mathfrak{Z}, \boldsymbol{P}] &= i\boldsymbol{P}.
\end{aligned}$$

Condition (IV) then implies

$$0 = [R'_x, S_y] = d(P^2,M)[(S_yP_z - S_zP_y), S_y] = i\hbar d(P^2,M)P_yS_x,$$

which means that $d(P^2,M) = 0$. From condition (V) we obtain

$$\begin{aligned}
i\hbar = [R'_x, P_x] &= a(P^2,M)[R_x,P_x] + g(P^2,M)[(\mathfrak{Z}P_x + P_x\mathfrak{Z}), P_x] \\
&= i\hbar a(P^2,M) + 2ig(P^2,M)P_x^2,
\end{aligned}$$

which implies $a(P^2,M) = 1$, $g(P^2,M) = 0$ and proves the desired uniqueness of the position operator $\boldsymbol{R}$.

So, from now on we will always use the Newton–Wigner operator (4.30)–(4.32) as the representative of the position observable.

4.3.10 Boost of position

Let us now find out how the position transforms with respect to boosts. For simplicity, we consider a massive system without spin, for which the Newton–Wigner operator in the rest frame O is

$$\boldsymbol{R} = -\frac{c^2}{2}(\boldsymbol{K}H^{-1} + H^{-1}\boldsymbol{K}). \tag{4.54}$$

Using formulas (4.5) and (4.4) we find that, in the frame O' moving along the x-axis, the components of $\boldsymbol{R}$ are [56]

$$R_x(\theta) = -\frac{c^2}{2}K_x(H\cosh\theta - cP_x\sinh\theta)^{-1} - \frac{c^2}{2}(H\cosh\theta - cP_x\sinh\theta)^{-1}K_x, \tag{4.55}$$

$$\begin{aligned} R_y(\theta) &= e^{-\frac{ic}{\hbar}K_x\theta}R_y e^{\frac{ic}{\hbar}K_x\theta} \\ &= -\frac{c^2}{2}e^{-\frac{ic}{\hbar}K_x\theta}(K_yH^{-1} + H^{-1}K_y)e^{\frac{ic}{\hbar}K_x\theta} \\ &= -\frac{c^2}{2}\left(K_y\cosh\theta - \frac{J_z}{c}\sinh\theta\right)(H\cosh\theta - cP_x\sinh\theta)^{-1} \\ &\quad - \frac{c^2}{2}(H\cosh\theta - cP_x\sinh\theta)^{-1}\left(K_y\cosh\theta - \frac{J_z}{c}\sinh\theta\right), \end{aligned} \tag{4.56}$$

$$\begin{aligned} R_z(\theta) &= -\frac{c^2}{2}\left(K_z\cosh\theta + \frac{J_y}{c}\sinh\theta\right)(H\cosh\theta - cP_x\sinh\theta)^{-1} \\ &\quad - \frac{c^2}{2}(H\cosh\theta - cP_x\sinh\theta)^{-1}\left(K_z\cosh\theta + \frac{J_y}{c}\sinh\theta\right). \end{aligned} \tag{4.57}$$

These formulas are very different from the usual Lorentz transformations in special relativity.[16] This is not surprising, since the Newton–Wigner position does not form the 3-vector part of any 4-vector quantity.[17]

In addition, we can find the time dependence of the position operator in the moving frame of reference. Applying formula (3.65), we obtain

$$\begin{aligned} \boldsymbol{R}(\theta, t') &= e^{-\frac{ic}{\hbar}K_x\theta}e^{\frac{i}{\hbar}Ht'}\boldsymbol{R}e^{-\frac{i}{\hbar}Ht'}e^{\frac{ic}{\hbar}K_x\theta} = e^{-\frac{ic}{\hbar}K_x\theta}(\boldsymbol{R} + \boldsymbol{V}t')e^{\frac{ic}{\hbar}K_x\theta} \\ &= \boldsymbol{R}(\theta) + \boldsymbol{V}(\theta)t', \end{aligned} \tag{4.58}$$

where the velocity $\boldsymbol{V}(\theta)$ in the frame O' is given by equations (4.7)–(4.9), as expected.

---

16 See Appendix A.1 in Volume 3.

17 In our formalism there is no "time operator" that could serve as the fourth component of such a 4-vector. In the third volume of this book, we will discuss in more detail the difference between the 4-tensor formalism of special relativity and our approach.

5 Elementary particles

The electron is as inexhaustible as the atom...
Vladimir I. Lenin

Our results from the previous chapter were universal and applicable to any isolated physical system, whether it is an electron or the solar system. We have not specified how this system was put together, and we considered only observables relating to the system as a whole. Our findings turned out to be unsurprising. In accordance with observations, we concluded that the total energy, the total momentum and the total angular momentum of any system are conserved, while the center of energy moves along a straight line at a constant speed.

We know that the internal structure of composite systems can undergo very dramatic changes due to collisions, reactions, decay, etc. The description of such events and processes is the most interesting and difficult part of theoretical physics. In order to understand this behavior, we must first understand how such compound physical systems are put together. The central idea of this book is that material objects consist of *elementary particles*, i. e., localizable, countable, indivisible, simplest systems that lack any internal structure.[1] In this chapter we will tackle these fundamental ingredients of nature.

In Section 3.2 we have established that the Hilbert space of any physical system carries a unitary representation of the Poincaré group. Each such representation can be decomposed into a direct sum of the simplest, so-called *irreducible*, representations (see Appendix I.1). Elementary particles are defined as systems for which this sum consists of only one term, i. e., the Hilbert space of a stable elementary particle carries an irreducible unitary representation of the Poincaré group. Hence, in a certain sense, elementary particles have simplest, indecomposable state spaces.

This simple consideration elegantly reduces the physical problem of classifying particles to the purely mathematical task of listing all irreducible unitary representations U_g of the Poincaré group and their Hilbert spaces $\mathscr{H}$. This task was fulfilled by Wigner [98] in 1939.[2]

From Schur's first lemma (see Appendix I.1) we can conclude that in any irreducible unitary representation of the Poincaré group, the actions of two Casimir operators M and $\boldsymbol{S}^2$ reduce to multiplication by constants. Hence, all different irreducible representations and, consequently, all allowed types of elementary particles are classified in accordance with the values of these two constants – the mass and the spin squared. Of course, there are many other parameters that describe elementary par-

1 In Volume 3 we will discuss how this idea corresponds with the widely held view that the fundamental components of nature are continuous fields.
2 A more modern exposition can be found in Section 2.5 of [95].

https://doi.org/10.1515/9783110492132-005

Table 5.1: Properties of stable elementary particles.

Particle	Symbol	Mass	Spin or helicity
Electron	e^-	$m_e = 0.511\ \text{MeV}/c^2$	$\hbar/2$
Proton	p^+	$m_p = 938.3\ \text{MeV}/c^2$	$\hbar/2$
Electron neutrino	ν_e	$< 1\ \text{eV}/c^2$	$\hbar/2$
Muon neutrino	ν_μ	$< 1\ \text{eV}/c^2$	$\hbar/2$
Tau neutrino	ν_τ	$< 1\ \text{eV}/c^2$	$\hbar/2$
Photon	γ	0	$\pm\hbar$

ticles, such as charge, magnetic moment and strangeness, but they all relate to the manner in which the particle participates in interactions. In a world without interactions, particles have only two intrinsic parameters: the mass and the spin.

Only six stable elementary particles are known, to which the above classification by mass and spin is applicable (see Table 5.1). However, this statement is not entirely accurate, and few explanations are required. First, for each particle in Table 5.1 (except the photon) there is a corresponding antiparticle having the same mass and spin, but opposite values of electrical, baryon and lepton charges.[3] Second, there are many particles, such as *muons*, *pions*, *neutrons*, etc., which are usually called elementary, but which are unstable and eventually break down into constituents listed in Table 5.1. This does not mean that unstable particles are "made of" the stable ones or that they are less elementary. Just the stable particles listed in the table have the lowest masses in their classes, and there are no lighter particles to which they could disintegrate without violating some conservation laws. Third, in Table 5.1 we did not include *quarks*, *gluons*, *gravitons* and other particles predicted theoretically, but never seen in experiments. Fourth, strictly speaking, the photon is not a truly elementary particle, because it is not described by an irreducible representation. In Subsection 5.4.4, we will see that the photon is described by a *reducible* representation of the Poincaré group, which is the direct sum of two massless irreducible representations with helicities $\pm\hbar$. Fifth, neutrinos are not purely elementary particles either. It has been experimentally established that the three neutrino flavors periodically transform into each other (oscillate) with time. Finally, it is entirely possible that protons are also not elementary. It is assumed that they consist of quarks and gluons.[4] This leaves us with only two truly stable, elementary and directly observable particles – the electron and the positron.

3 Definitions of charges will be given in Subsection 1.2.1 of Volume 2.

4 In the author's works [80, 81], it was proposed to revive Sakata's idea [76] that hadrons consist not of quarks and gluons, but of elementary hadrons, such as the proton, neutron and Λ-hyperon.

In what follows, we shall denote by m eigenvalues of the mass operator M and consider separately two cases: massive particles ($m > 0$) and massless particles ($m = 0$).[5]

5.1 Massive particles

In this section, we are going to construct unitary irreducible representations of the Poincaré group for massive particles. To do this, we first define a convenient basis in the Hilbert space $\mathscr{H}$ of states of the particle and then write explicitly representation operators U_g in this basis.

5.1.1 One-particle Hilbert space

Suppose that the Hilbert space $\mathscr{H}$ carries a unitary irreducible representation U_g of the Poincaré group, characterized by a positive eigenvalue m of the mass operator M. In this space, Hermitian generators $\{\boldsymbol{P}, \boldsymbol{J}, \boldsymbol{K}, H\}$ of the representation are defined in a natural way. As we saw in Section 4.3, the position operator $\boldsymbol{R}$ is also well defined. The components of position and momentum satisfy commutation relations of the *Heisenberg algebra*[6] $\mathfrak{h}_3$. We have

$$[P_i, P_j] = [R_i, R_j] = 0,$$
$$[R_i, P_j] = i\hbar\delta_{ij},$$

where $i, j = 1, 2, 3$. Obviously, $\mathscr{H}$ must be a representation space of the algebra $\mathfrak{h}_3$. According to Corollary I.3, $\mathscr{H}$ is a direct sum, $\mathscr{H} = \bigoplus_k \mathscr{H}_k$, where every irreducible component $\mathscr{H}_k$ is isomorphic to the space of the *Schrödinger* representation, i. e., the one in which vectors in $\mathscr{H}_k$ are expressed by normalizable functions $\psi(\boldsymbol{p}, k)$ on $\mathbb{R}^3$. In this representation, the momentum operator $\boldsymbol{P}$ multiplies wave functions by $\boldsymbol{p}$ and the position operator $\boldsymbol{R}$ differentiates them with respect to $\boldsymbol{p}$ as follows:[7]

$$\hat{\boldsymbol{P}}\psi(\boldsymbol{p}, k) = \boldsymbol{p}\psi(\boldsymbol{p}, k),$$
$$\hat{\boldsymbol{R}}\psi(\boldsymbol{p}, k) = i\hbar\frac{\partial}{\partial \boldsymbol{p}}\psi(\boldsymbol{p}, k).$$

Therefore, in order to construct the full representation space $\mathscr{H}$, we can define $n \geq 1$ copies of the momentum space $\mathbb{R}^3$, as shown in Figure 5.1. Then vectors in $\mathscr{H}$ are

5 The Wigner classification also covers irreducible representations with negative energies and imaginary masses. But there is no indication that such particles exist in nature. Therefore, we will not discuss them in this book.

6 See equations (3.52) and (4.25) and Theorem 4.1.

7 We will put a cap over the operator symbol to distinguish operators that act on wave functions (rather than on state vectors).

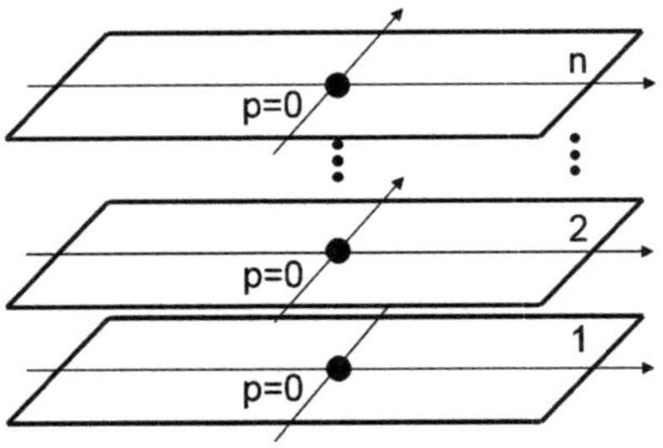

Figure 5.1: n copies of the momentum space $\mathbb{R}^3$, where wave functions $\psi(\boldsymbol{p}, k)$ $(k = 1, 2, \ldots, n)$ are defined. These functions are representatives of vectors in $\mathcal{H}$.

represented by normalizable functions $\psi(\boldsymbol{p}, k)$ on these copies, where $k = 1, 2, \ldots, n$. The normalization condition has the form

$$\sum_{k=1}^{n} \int d\boldsymbol{p} |\psi(\boldsymbol{p}, k)|^2 = 1.$$

The thick dots in Figure 5.1 denote n mutually orthogonal eigenvectors of the momentum with zero eigenvalue $\boldsymbol{p} = \mathbf{0}$.[8] The linear span of these vectors is the subspace $\mathcal{H}_{\mathbf{0}}$ of zero momentum. Similarly, n-dimensional eigensubspaces of the operator $\boldsymbol{P}$ with eigenvalues $\boldsymbol{p}$ will be denoted $\mathcal{H}_{\boldsymbol{p}}$. The complete one-particle Hilbert space $\mathcal{H}$ is a direct sum of such subspaces,

$$\mathcal{H} = \bigoplus_{\boldsymbol{p} \in \mathbb{R}^3} \mathcal{H}_{\boldsymbol{p}}. \tag{5.1}$$

5.1.2 Action of rotation subgroup in $\mathcal{H}_{\mathbf{0}}$

First, we focus on the n-dimensional zero-momentum subspace $\mathcal{H}_{\mathbf{0}}$. This subspace is invariant with respect to rotations, because for any vector $|\mathbf{0}\rangle \in \mathcal{H}_{\mathbf{0}}$ the result of arbitrary rotation $e^{-\frac{i}{\hbar}\boldsymbol{J}\cdot\boldsymbol{\varphi}}|\mathbf{0}\rangle$ lies in the same subspace $\mathcal{H}_{\mathbf{0}}$. Indeed, using equation (4.2)), we have

$$\begin{aligned}
\boldsymbol{P} e^{-\frac{i}{\hbar}\boldsymbol{J}\cdot\boldsymbol{\varphi}}|\mathbf{0}\rangle &= e^{-\frac{i}{\hbar}\boldsymbol{J}\cdot\boldsymbol{\varphi}} e^{\frac{i}{\hbar}\boldsymbol{J}\cdot\boldsymbol{\varphi}} \boldsymbol{P} e^{-\frac{i}{\hbar}\boldsymbol{J}\cdot\boldsymbol{\varphi}}|\mathbf{0}\rangle \\
&= e^{-\frac{i}{\hbar}\boldsymbol{J}\cdot\boldsymbol{\varphi}}\left(\boldsymbol{P}\cos\varphi + \left(\boldsymbol{P}\cdot\frac{\boldsymbol{\varphi}}{\varphi}\right)\frac{\boldsymbol{\varphi}}{\varphi}(1-\cos\varphi) - \left[\boldsymbol{P}\times\frac{\boldsymbol{\varphi}}{\varphi}\right]\sin\varphi\right)|\mathbf{0}\rangle \\
&= \mathbf{0}.
\end{aligned} \tag{5.2}$$

This means that representation U_g of the Poincaré group in $\mathcal{H}$ induces a unitary representation V_g of the rotation subgroup in $\mathcal{H}_{\mathbf{0}}$.

8 The corresponding wave functions $\psi(\boldsymbol{p}, k) = \delta(\boldsymbol{p})\delta_{ik}$ are momentum eigenfunctions.

In the entire Hilbert space $\mathscr{H}$, rotations are generated by the angular momentum operator $\boldsymbol{J}$. However, in the subspace $\mathscr{H}_{\mathbf{0}}$, the generator $\boldsymbol{J}$ can be replaced by the spin operator $\boldsymbol{S}$, because

$$S_z|\mathbf{0}\rangle = J_z|\mathbf{0}\rangle - [\boldsymbol{R}\times\boldsymbol{P}]_z|\mathbf{0}\rangle = J_z|\mathbf{0}\rangle - (R_xP_y - R_yP_x)|\mathbf{0} = J_z|\mathbf{0}\rangle.$$

We will show later that representation U_g of the full Poincaré group is irreducible if and only if the representation V_g of the rotation group in $\mathscr{H}_{\mathbf{0}}$ is irreducible. So, now we will be interested in exactly such irreducible unitary (both single-valued and double-valued) representations V_g of the group of rotations. Their classification is described in Appendix I.5. They are labeled by one integer or half-integer parameter s, which is called the *spin* of the particle. The one-dimensional representation is characterized by zero spin ($s = 0$) and corresponds to *spinless* particles. The two-dimensional (double-valued) representation corresponds to particles with spin ($s = 1/2$). The three-dimensional representation corresponds to particles with unit spin ($s = 1$), etc. In the general case, $n = \dim(\mathscr{H}_{\mathbf{0}}) = 2s + 1$.

It is convenient to choose in $\mathscr{H}_{\mathbf{0}}$ a basis consisting of $2s+1$ eigenvectors of the spin component S_z and denote these vectors $|\mathbf{0}, s_z\rangle$, i. e.,

$$\begin{aligned}\boldsymbol{P}|\mathbf{0}, s_z\rangle &= 0,\\ H|\mathbf{0}, s_z\rangle &= mc^2|\mathbf{0}, s_z\rangle,\\ S_z|\mathbf{0}, s_z\rangle &= \hbar s_z|\mathbf{0}, s_z\rangle,\end{aligned}$$

where $s_z = -s, -s+1, \ldots, s-1, s$.[9] As we already established, any rotation keeps these vectors inside $\mathscr{H}_{\mathbf{0}}$. We have

$$e^{-\frac{i}{\hbar}\boldsymbol{J}\cdot\boldsymbol{\varphi}}|\mathbf{0}, s_z\rangle = e^{-\frac{i}{\hbar}\boldsymbol{S}\cdot\boldsymbol{\varphi}}|\mathbf{0}, s_z\rangle = \sum_{s_z'=-s}^{s} \mathscr{D}^s_{s_z' s_z}(\boldsymbol{\varphi})|\mathbf{0}, s_z'\rangle, \tag{5.3}$$

where $\mathscr{D}^s$ are $(2s+1)\times(2s+1)$ matrices of the irreducible representation V_g described in Appendix I.5.

Note that in (5.3) the summation indices stand in an unusual order, $s_z' s_z \ldots s_z'$.[10] Such a notation is necessary for the matrices $\mathscr{D}^s$ to fulfill a representation of the rotation group, i. e., $\mathscr{D}^s(\boldsymbol{\varphi}_1 \circ \boldsymbol{\varphi}_2) = \mathscr{D}^s(\boldsymbol{\varphi}_1)\mathscr{D}^s(\boldsymbol{\varphi}_2)$. Indeed, it is not difficult to verify[11]

$$\begin{aligned}\mathscr{D}^s_{\sigma'\sigma}(\boldsymbol{\varphi}_1 \circ \boldsymbol{\varphi}_2)|\mathbf{0}, \sigma'\rangle &= e^{-\frac{i}{\hbar}\boldsymbol{J}\cdot(\boldsymbol{\varphi}_1\circ\boldsymbol{\varphi}_2)}|\mathbf{0}, \sigma\rangle = e^{-\frac{i}{\hbar}\boldsymbol{J}\cdot\boldsymbol{\varphi}_1}e^{-\frac{i}{\hbar}\boldsymbol{J}\cdot\boldsymbol{\varphi}_2}|\mathbf{0}, \sigma\rangle\\ &= e^{-\frac{i}{\hbar}\boldsymbol{J}\cdot\boldsymbol{\varphi}_1}\mathscr{D}^s_{\sigma''\sigma}(\boldsymbol{\varphi}_2)|\mathbf{0}, \sigma''\rangle = \mathscr{D}^s_{\sigma''\sigma}(\boldsymbol{\varphi}_2)e^{-\frac{i}{\hbar}\boldsymbol{J}\cdot\boldsymbol{\varphi}_1}|\mathbf{0}, \sigma''\rangle\\ &= \mathscr{D}^s_{\sigma''\sigma}(\boldsymbol{\varphi}_2)\mathscr{D}^s_{\sigma'\sigma''}(\boldsymbol{\varphi}_1)|\mathbf{0}, \sigma'\rangle = \mathscr{D}^s_{\sigma'\sigma''}(\boldsymbol{\varphi}_1)\mathscr{D}^s_{\sigma''\sigma}(\boldsymbol{\varphi}_2)|\mathbf{0}, \sigma'\rangle.\end{aligned}$$

---

9 Obviously, parameters s_z and $2s+1$ replace the previously introduced parameters k and n.

10 This indicates that we use *transposed* matrices $(\mathscr{D}^s)^T$.

11 For brevity, we omit summation signs and assume summation over repeated indices; see also the derivation of formula (2.5.9) in [95].

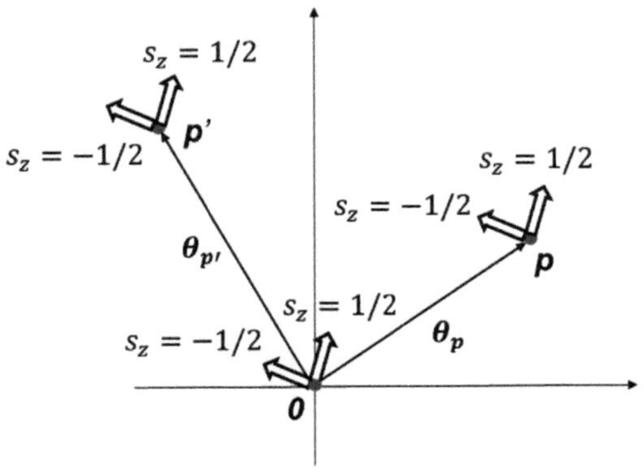

Figure 5.2: Construction of a momentum–spin basis for a particle with spin 1/2. Spin eigenvectors (with eigenvalues $s_z = -1/2, 1/2$) at zero momentum are translated to subspaces with nonzero momenta $\mathscr{H}_{\boldsymbol{p}}$ and $\mathscr{H}_{\boldsymbol{p}'}$ by pure boosts $\boldsymbol{\theta}_{\boldsymbol{p}}$ and $\boldsymbol{\theta}_{\boldsymbol{p}'}$, respectively.

5.1.3 Momentum–spin basis

In the previous subsection we constructed a basis $|\mathbf{0}, s_z\rangle$ in the subspace $\mathscr{H}_{\mathbf{0}}$. But we also need basis vectors $|\boldsymbol{p}, s_z\rangle$ in subspaces with nonzero momentum $\mathscr{H}_{\boldsymbol{p}}$ $(\boldsymbol{p} \neq \mathbf{0})$. We will construct them by transferring the vectors $|\mathbf{0}, s_z\rangle$ to other points in the three-dimensional momentum space using pure boost transformations. The only pure boost that takes the momentum $\mathbf{0}$ to $\boldsymbol{p}$ will be denoted by $\{\boldsymbol{v}(\boldsymbol{\theta}_{\boldsymbol{p}}); \mathbf{0}; \mathbf{0}; 0\}$, where $\boldsymbol{v}(\boldsymbol{\theta}_{\boldsymbol{p}})$ is the boost's velocity (see Figure 5.2) and

$$\sinh \theta_{\boldsymbol{p}} = \frac{p}{mc},$$
$$\boldsymbol{\theta}_{\boldsymbol{p}} = \frac{\boldsymbol{p}}{p}\theta_{\boldsymbol{p}},$$
$$\boldsymbol{v}(\boldsymbol{\theta}_{\boldsymbol{p}}) = \frac{\boldsymbol{p}c}{p}\tanh\theta.$$

In the notation (3.56)–(3.57), the corresponding unitary operator in the Hilbert space can be written as

$$U(\boldsymbol{\theta}_{\boldsymbol{p}}; \mathbf{0}; \mathbf{0}; 0) \equiv e^{-\frac{ic}{\hbar}\boldsymbol{K}\cdot\boldsymbol{\theta}_{p}}. \tag{5.4}$$

Therefore

$$|\boldsymbol{p}, s_z\rangle = N(\boldsymbol{p})U(\boldsymbol{\theta}_{\boldsymbol{p}}; \mathbf{0}; \mathbf{0}; 0)|\mathbf{0}, s_z\rangle = N(\boldsymbol{p})e^{-\frac{ic}{\hbar}\boldsymbol{K}\cdot\boldsymbol{\theta}_{p}}|\mathbf{0}, s_z\rangle, \tag{5.5}$$

where $N(\boldsymbol{p})$ is a normalization factor. The explicit expression for $N(\boldsymbol{p})$ will be derived in (5.29).

One can verify that vector (5.5) is, indeed, an eigenvector of the momentum and energy operators with eigenvalues $\boldsymbol{p}$ and

$$\omega_{\boldsymbol{p}} \equiv \sqrt{m^2c^4 + p^2c^2}, \tag{5.6}$$

respectively. For this proof we use equation (4.3) and obtain

$$\begin{aligned}
P|\boldsymbol{p}, s_z\rangle &= N(\boldsymbol{p})\boldsymbol{P}e^{-\frac{ic}{\hbar}\boldsymbol{K}\cdot\boldsymbol{\theta}_p}|\boldsymbol{0}, s_z\rangle = N(\boldsymbol{p})e^{-\frac{ic}{\hbar}\boldsymbol{K}\cdot\boldsymbol{\theta}_p}e^{\frac{ic}{\hbar}\boldsymbol{K}\cdot\boldsymbol{\theta}_p}\boldsymbol{P}e^{-\frac{ic}{\hbar}\boldsymbol{K}\cdot\boldsymbol{\theta}_p}|\boldsymbol{0}, s_z\rangle \\
&= N(\boldsymbol{p})\frac{\boldsymbol{\theta}_p \sinh\theta_p}{c\theta_p}e^{-\frac{ic}{\hbar}\boldsymbol{K}\cdot\boldsymbol{\theta}_p}H|\boldsymbol{0}, s_z\rangle = N(\boldsymbol{p})\frac{mc\boldsymbol{\theta}_p \sinh\theta_p}{\theta_p}e^{-\frac{ic}{\hbar}\boldsymbol{K}\cdot\boldsymbol{\theta}_p}|\boldsymbol{0}, s_z\rangle \\
&= N(\boldsymbol{p})\boldsymbol{p}e^{-\frac{ic}{\hbar}\boldsymbol{K}\cdot\boldsymbol{\theta}_p}|\boldsymbol{0}, s_z\rangle = \boldsymbol{p}|\boldsymbol{p}, s_z\rangle, \qquad (5.7)\\
H|\boldsymbol{p}, s_z\rangle &= \sqrt{M^2c^4 + P^2c^2}|\boldsymbol{p}, s_z\rangle = \sqrt{m^2c^4 + p^2c^2}|\boldsymbol{p}, s_z\rangle.
\end{aligned}$$

Let us now find out how the z-component of spin acts on our basis vectors $|\boldsymbol{p}, s_z\rangle$. We have[12]

$$\begin{aligned}
S_z|\boldsymbol{p}, s_z\rangle &= N(\boldsymbol{p})S_z e^{-\frac{ic}{\hbar}\boldsymbol{K}\boldsymbol{\theta}_p}|\boldsymbol{0}, s_z\rangle \\
&= N(\boldsymbol{p})e^{-\frac{ic}{\hbar}\boldsymbol{K}\cdot\boldsymbol{\theta}_p}e^{\frac{ic}{\hbar}\boldsymbol{K}\cdot\boldsymbol{\theta}_p}\left(\frac{W_z}{Mc} - \frac{W_0 P_z}{M(Mc^2 + H)}\right)e^{-\frac{ic}{\hbar}\boldsymbol{K}\cdot\boldsymbol{\theta}_p}|\boldsymbol{0}, s_z\rangle \\
&= N(\boldsymbol{p})e^{-\frac{ic}{\hbar}\boldsymbol{K}\cdot\boldsymbol{\theta}_p}\left(\frac{W_z + \frac{\theta_z}{\theta}[(\boldsymbol{W}\cdot\frac{\boldsymbol{\theta}}{\theta})(\cosh\theta - 1) + W_0\sinh\theta]}{Mc}\right. \\
&\quad - \left[W_0\cosh\theta + \left(\boldsymbol{W}\cdot\frac{\boldsymbol{\theta}}{\theta}\right)\sinh\theta\right] \\
&\quad \left.\times \frac{P_z + \frac{\theta_z}{\theta}[(\boldsymbol{P}\cdot\frac{\boldsymbol{\theta}}{\theta})(\cosh\theta - 1) + \frac{1}{c}H\sinh\theta]}{M(Mc^2 + H\cosh\theta + c(\boldsymbol{P}\cdot\frac{\boldsymbol{\theta}}{\theta})\sinh\theta)}\right)|\boldsymbol{0}, s_z\rangle \\
&= N(\boldsymbol{p})e^{-\frac{ic}{\hbar}\boldsymbol{K}\cdot\boldsymbol{\theta}_p} \\
&\quad \times\left(\frac{W_z + \frac{\theta_z}{\theta}(\boldsymbol{W}\cdot\frac{\boldsymbol{\theta}}{\theta})(\cosh\theta - 1)}{Mc} - \frac{(\boldsymbol{W}\cdot\frac{\boldsymbol{\theta}}{\theta})\sinh\theta(\frac{\theta_z}{\theta}Mc\sinh\theta)}{M(Mc^2 + Mc^2\cosh\theta)}\right)|\boldsymbol{0}, s_z\rangle \\
&= N(\boldsymbol{p})e^{-\frac{ic}{\hbar}\boldsymbol{K}\cdot\boldsymbol{\theta}_p} \\
&\quad \times\left(\frac{W_z}{Mc} + \frac{\theta_z}{\theta}\left(\boldsymbol{W}\cdot\frac{\boldsymbol{\theta}}{\theta}\right)\left(\frac{\cosh\theta - 1}{Mc} - \frac{\sinh^2\theta}{Mc(1 + \cosh\theta)}\right)\right)|\boldsymbol{0}, s_z\rangle \\
&= N(\boldsymbol{p})e^{-\frac{ic}{\hbar}\boldsymbol{K}\cdot\boldsymbol{\theta}_p}\frac{W_z}{Mc}|\boldsymbol{0}, s_z\rangle = N(\boldsymbol{p})e^{-\frac{ic}{\hbar}\boldsymbol{K}\cdot\boldsymbol{\theta}_p}S_z|\boldsymbol{0}, s_z\rangle \\
&= N(\boldsymbol{p})e^{-\frac{ic}{\hbar}\boldsymbol{K}\cdot\boldsymbol{\theta}_p}\hbar s_z|\boldsymbol{0}, s_z\rangle = \hbar s_z|\boldsymbol{p}, s_z\rangle.
\end{aligned}$$

Hence, as expected, $|\boldsymbol{p}, s_z\rangle$ are common eigenvectors of the complete set of mutually commuting operators $\{\boldsymbol{P}, H, S_z\}$.

The common spectrum of energy momentum eigenvalues $(\omega_{\boldsymbol{p}}, \boldsymbol{p}c)$ can be represented by points on the *mass hyperboloid* in a four-dimensional space (see Figure 5.3).

---

12 Here we used (4.27) and took into account that $W_0|\boldsymbol{0}, s_z\rangle = 0$, $\boldsymbol{P}|\boldsymbol{0}, s_z\rangle = \boldsymbol{0}$ and $H|\boldsymbol{0}, s_z\rangle = Mc^2|\boldsymbol{0}, s_z\rangle$. We also used formulas (4.3)–(4.4) for boost transformations of the energy momentum 4-vector $(H, c\boldsymbol{P})$ and similar formulas for the Pauli–Lubanski 4-vector $(W_0, \boldsymbol{W})$. For brevity, we denoted by θ_z the z-component of the vector $\boldsymbol{\theta}_p$ and by $\theta \equiv |\boldsymbol{\theta}_p|$ its length.

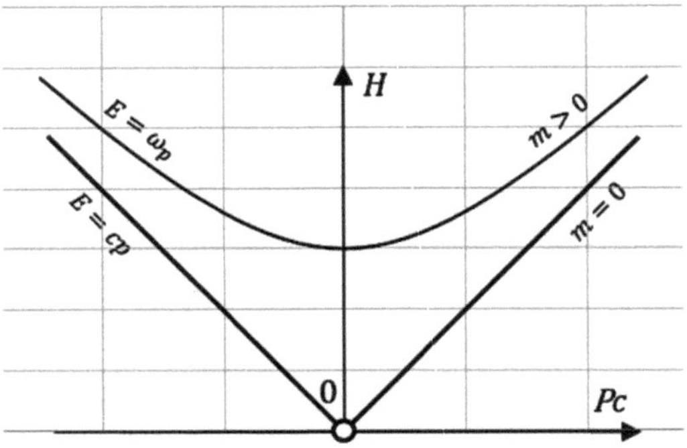

Figure 5.3: Mass hyperboloid in the energy momentum space for massive particles and the zero-mass cone for $m = 0$.

For massive particles, the spectrum of the velocity operator $\boldsymbol{V} = \boldsymbol{P}c^2/H$ is the interior of a sphere $|\boldsymbol{v}| < c$ in the three-dimensional velocity space. This spectrum does not include the surface of the sphere, so massive particles cannot reach the speed of light.[13]

5.1.4 Nonuniqueness of momentum–spin basis

Note that in the preceding subsection, the eigenvectors of spin $\boldsymbol{S}$ were obtained by applying pure boosts to basis vectors at the point $\boldsymbol{p} = \boldsymbol{0}$. Of course, this choice was purely arbitrary. We could also choose to construct another, but equivalent, basis, for example, by moving vectors from $\boldsymbol{0}$ to $\boldsymbol{p}$ by means of boosts coupled with rotations. However, once the basis is fixed, all formulas must be written with respect to it.

Obviously, different choices of the basis $|\boldsymbol{p}, s_z\rangle$ translate into different choices for the spin operator $\boldsymbol{S}$ [46]. Does this contradict our statement from Subsection 4.3.8 about the uniqueness of the spin operator? No. The bottom line is that the alternative spin operator $\boldsymbol{S}'$ (and the corresponding alternative position operator $\boldsymbol{R}'$) will not be expressed as a function of the basic generators $\{H, \boldsymbol{P}, \boldsymbol{J}, \boldsymbol{K}\}$ of the Poincaré group. Being functions of the basic generators was an important condition in our proof of the uniqueness of the operators $\boldsymbol{S}$ and $\boldsymbol{R}$ in Section 4.3.

5.1.5 Action of translations and rotations on basis vectors

Now we can determine the action of Poincaré group elements on the basis vectors $|\boldsymbol{p}, s_z\rangle$ constructed above.[14] Obviously, translations are represented simply by multi-

---

13 Note that our description of a particle's velocity is more natural than the "zitterbewegung" in Dirac's theory of the electron [11].

14 We are working in the Schrödinger picture.

plications, so we have

$$e^{-\frac{i}{\hbar}\boldsymbol{P}\cdot\boldsymbol{a}}|\boldsymbol{p}, s_z\rangle = e^{-\frac{i}{\hbar}\boldsymbol{p}\cdot\boldsymbol{a}}|\boldsymbol{p}, s_z\rangle, \tag{5.8}$$

$$e^{\frac{i}{\hbar}Ht}|\boldsymbol{p}, s_z\rangle = e^{\frac{i}{\hbar}\omega_{\boldsymbol{p}}t}|\boldsymbol{p}, s_z\rangle. \tag{5.9}$$

Let us now apply a rotation $e^{-\frac{i}{\hbar}\boldsymbol{J}\cdot\boldsymbol{\varphi}}$ to the vector $|\boldsymbol{p}, s_z\rangle$. We use equations (5.3), (D.7) and assume that the normalization factor is rotationally invariant, i. e., $N(\boldsymbol{\varphi}\boldsymbol{p}) = N(\boldsymbol{p})$,[15] to obtain

$$\begin{aligned}
e^{-\frac{i}{\hbar}\boldsymbol{J}\cdot\boldsymbol{\varphi}}|\boldsymbol{p}, s_z\rangle &= N(\boldsymbol{p})e^{-\frac{i}{\hbar}\boldsymbol{J}\cdot\boldsymbol{\varphi}}e^{-\frac{ic}{\hbar}\boldsymbol{K}\cdot\boldsymbol{\theta}_p}|\boldsymbol{0}, s_z\rangle \\
&= N(\boldsymbol{p})e^{-\frac{i}{\hbar}\boldsymbol{J}\cdot\boldsymbol{\varphi}}e^{-\frac{ic}{\hbar}\boldsymbol{K}\cdot\boldsymbol{\theta}_p}e^{\frac{i}{\hbar}\boldsymbol{J}\cdot\boldsymbol{\varphi}}e^{-\frac{i}{\hbar}\boldsymbol{J}\cdot\boldsymbol{\varphi}}|\boldsymbol{0}, s_z\rangle \\
&= N(\boldsymbol{p})e^{-\frac{ic}{\hbar}(\boldsymbol{\varphi}^{-1}\boldsymbol{K})\cdot\boldsymbol{\theta}_p}\sum_{s_z'=-s}^{s}\mathscr{D}^s_{s_z's_z}(\boldsymbol{\varphi})|\boldsymbol{0}, s_z'\rangle \\
&= N(\boldsymbol{p})e^{-\frac{ic}{\hbar}\boldsymbol{K}\cdot\boldsymbol{\varphi}\boldsymbol{\theta}_p}\sum_{s_z'=-s}^{s}\mathscr{D}^s_{s_z's_z}(\boldsymbol{\varphi})|\boldsymbol{0}, s_z'\rangle \\
&= N(\boldsymbol{p})\sum_{s_z'=-s}^{s}\mathscr{D}^s_{s_z's_z}(\boldsymbol{\varphi})[e^{-\frac{ic}{\hbar}\boldsymbol{K}\cdot\boldsymbol{\theta}_{\varphi p}}|\boldsymbol{0}, s_z'\rangle] \\
&= N(\boldsymbol{p})N^{-1}(\boldsymbol{\varphi}\boldsymbol{p})\sum_{s_z'=-s}^{s}\mathscr{D}^s_{s_z's_z}(\boldsymbol{\varphi})|\boldsymbol{\varphi}\boldsymbol{p}, s_z'\rangle \\
&= \sum_{s_z'=-s}^{s}\mathscr{D}^s_{s_z's_z}(\boldsymbol{\varphi})|\boldsymbol{\varphi}\boldsymbol{p}, s_z'\rangle.
\end{aligned} \tag{5.10}$$

This means that both momentum and spin of the particle rotate through the angle $\boldsymbol{\varphi}$, as expected.

5.1.6 Action of boosts on momentum eigenvectors

The action of boosts on the basis vectors $|\boldsymbol{p}, s_z\rangle$ is slightly more complicated. Boosts of the spin components s_z will be considered in Subsection 5.1.7, and here we will be interested in the effect of boost on the momentum eigenvalue $\boldsymbol{p}$. So, for brevity, we omit spin indices.

It is easy to see that boost transforms momentum eigenvectors into each other. Indeed, using formula (4.3), one can verify that for any $|\boldsymbol{p}\rangle$ the boosted state vector $|\boldsymbol{p}'\rangle = e^{-\frac{ic}{\hbar}\boldsymbol{K}\cdot\boldsymbol{\theta}}|\boldsymbol{p}\rangle$ is also an eigenvector of the momentum operator $\boldsymbol{P}$:

$$\boldsymbol{P}|\boldsymbol{p}'\rangle = \boldsymbol{P}e^{-\frac{ic}{\hbar}\boldsymbol{K}\cdot\boldsymbol{\theta}}|\boldsymbol{p}\rangle = e^{-\frac{ic}{\hbar}\boldsymbol{K}\cdot\boldsymbol{\theta}}e^{\frac{ic}{\hbar}\boldsymbol{K}\cdot\boldsymbol{\theta}}\boldsymbol{P}e^{-\frac{ic}{\hbar}\boldsymbol{K}\cdot\boldsymbol{\theta}}|\boldsymbol{p}\rangle$$

15 The action of a rotation on the momentum vector $(\boldsymbol{\varphi}\boldsymbol{p})$ should be taken from (D.22).

$$= e^{-\frac{ic}{\hbar}\boldsymbol{K}\cdot\boldsymbol{\theta}}\left(\boldsymbol{P} + \frac{\boldsymbol{\theta}}{\theta}\left[\left(\boldsymbol{P}\cdot\frac{\boldsymbol{\theta}}{\theta}\right)(\cosh\theta - 1) + \frac{H}{c}\sinh\theta\right]\right)|\boldsymbol{p}\rangle$$
$$= (\boldsymbol{\theta p})|\boldsymbol{p}'\rangle,$$

where we denote by

$$\boldsymbol{\theta p} \equiv \boldsymbol{p} + \frac{\boldsymbol{\theta}}{\theta}\left[\left(\boldsymbol{p}\cdot\frac{\boldsymbol{\theta}}{\theta}\right)(\cosh\theta - 1) + \frac{\omega_{\boldsymbol{p}}}{c}\sinh\theta\right] \tag{5.11}$$

the active boost transformation in the momentum space.

5.1.7 Action of boosts on spin components

We have just established that a boost $\boldsymbol{\theta}$ transforms an eigenvector of momentum $|\boldsymbol{p}\rangle$ into another eigenvector $|\boldsymbol{\theta p}\rangle$. In the case of nonzero spin, this means that basis vectors $|\boldsymbol{p}, s_z\rangle$ in the subspace $\mathscr{H}_{\boldsymbol{p}}$ will be transformed by $e^{-\frac{ic}{\hbar}\boldsymbol{K}\cdot\boldsymbol{\theta}}$ into a linear combination of basis vectors in the subspace $\mathscr{H}_{\boldsymbol{\theta p}}$. In other words,

$$e^{-\frac{ic}{\hbar}\boldsymbol{K}\cdot\boldsymbol{\theta}}|\boldsymbol{p}, s_z\rangle = \sum_{s_z'=-s}^{s} G_{s_z' s_z}(\boldsymbol{p}, \boldsymbol{\theta})|\boldsymbol{\theta p}, s_z'\rangle, \tag{5.12}$$

where the unitary $(2s+1)\times(2s+1)$ matrix $G_{s_z' s_z}(\boldsymbol{p}, \boldsymbol{\theta})$ depends on the momentum $\boldsymbol{p}$ and on the applied boost $\boldsymbol{\theta}$. Here we would like to determine the nature of this dependence.

Using equation (5.5), we obtain

$$e^{-\frac{ic}{\hbar}\boldsymbol{K}\cdot\boldsymbol{\theta}}|\boldsymbol{p}, s_z\rangle = N(\boldsymbol{p})e^{-\frac{ic}{\hbar}\boldsymbol{K}\cdot\boldsymbol{\theta}}e^{-\frac{ic}{\hbar}\boldsymbol{K}\cdot\boldsymbol{\theta}_{\boldsymbol{p}}}|\boldsymbol{0}, s_z\rangle. \tag{5.13}$$

On the right-hand side we see the product of two boosts. In general, these boosts are not collinear, and their product is a transformation from the Lorentz subgroup, which is representable in the canonical form (2.6) (boost) × (rotation),

$$\boldsymbol{\theta} \circ \boldsymbol{\theta}_{\boldsymbol{p}} = \boldsymbol{\theta}_? \circ \boldsymbol{\varphi}_?. \tag{5.14}$$

Here $\boldsymbol{\theta}_?$ and $\boldsymbol{\varphi}_?$ are as yet undefined boost and rotation vectors, and now we will try to learn more about them. Multiplying both sides of equation (5.14) from the left by $\boldsymbol{\theta}_?^{-1}$, we get

$$\boldsymbol{\theta}_?^{-1} \circ \boldsymbol{\theta} \circ \boldsymbol{\theta}_{\boldsymbol{p}} = \boldsymbol{\varphi}_? \tag{5.15}$$

or in terms of unitary representatives

$$U(\boldsymbol{\theta}_?; \boldsymbol{0}; \boldsymbol{0}; 0)^{-1}U(\boldsymbol{\theta}; \boldsymbol{0}; \boldsymbol{0}; 0)U(\boldsymbol{\theta}_{\boldsymbol{p}}; \boldsymbol{0}; \boldsymbol{0}; 0) = U(\boldsymbol{0}; \boldsymbol{\varphi}_?; \boldsymbol{0}; 0). \tag{5.16}$$

This means that the sequence of boosts on the left-hand side of (5.16) is a pure rotation. Acting on a state vector with zero momentum $|\boldsymbol{0}, s_z\rangle$, this product will return this vector back to the zero-momentum subspace $\mathscr{H}_{\boldsymbol{0}}$, as shown in Figure 5.4: the

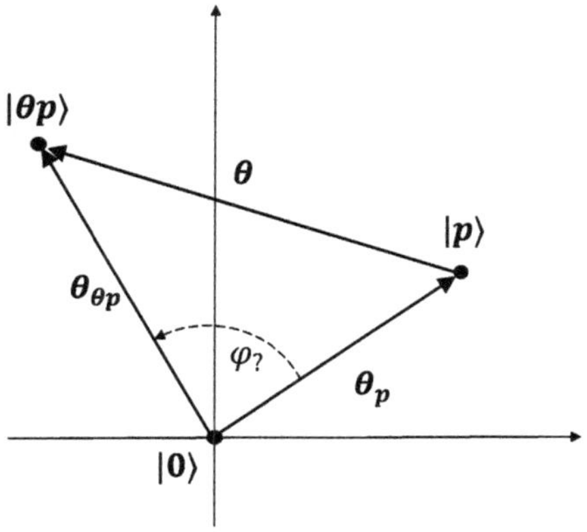

Figure 5.4: To the derivation of formula (5.18) for the Wigner angle.

zero-momentum vector $|\mathbf{0}, s_z\rangle$ is first mapped into a vector with momentum $\boldsymbol{p}$ by the boost $\boldsymbol{\theta}_{\boldsymbol{p}}$; the subsequent application of $\boldsymbol{\theta}$ transforms this vector into another momentum eigenstate with an eigenvalue $\boldsymbol{p}' = \boldsymbol{\theta p}$. Then from Figure 5.4 it follows that the final boost $\boldsymbol{\theta}_?^{-1}$ should bring this vector back to the zero-momentum subspace $\mathscr{H}_0$ and the inverse $\boldsymbol{\theta}_?$ translates $\mathbf{0} \to \boldsymbol{\theta p}$. In other words, $\boldsymbol{\theta}_? = \boldsymbol{\theta}_{\boldsymbol{\theta p}}$ and

$$U(\boldsymbol{\theta}_?; \mathbf{0}; \mathbf{0}; 0) = U(\boldsymbol{\theta}_{\boldsymbol{\theta p}}; \mathbf{0}; \mathbf{0}; 0). \tag{5.17}$$

5.1.8 Wigner angle

For the rotation angle on the right-hand side of equation (5.15) we will use the special symbol $\boldsymbol{\varphi}_W(\boldsymbol{p}, \boldsymbol{\theta})$ and call it the *Wigner angle*. We have[16]

$$\boldsymbol{\varphi}_? \equiv \boldsymbol{\varphi}_W(\boldsymbol{p}, \boldsymbol{\theta}) = \boldsymbol{\theta}_{\boldsymbol{\theta p}}^{-1} \circ \boldsymbol{\theta} \circ \boldsymbol{\theta}_{\boldsymbol{p}}. \tag{5.18}$$

Substituting these results in (5.14) and (5.13), we get[17]

$$\begin{aligned}
e^{-\frac{ic}{\hbar}\boldsymbol{K}\cdot\boldsymbol{\theta}}|\boldsymbol{p}, s_z\rangle &= N(\boldsymbol{p}) e^{-\frac{ic}{\hbar}\boldsymbol{K}\cdot\boldsymbol{\theta}} e^{-\frac{ic}{\hbar}\boldsymbol{K}\cdot\boldsymbol{\theta}_{\boldsymbol{p}}}|\mathbf{0}, s_z\rangle \\
&= N(\boldsymbol{p}) e^{-\frac{ic}{\hbar}\boldsymbol{K}\cdot\boldsymbol{\theta}_{\boldsymbol{\theta p}}} e^{-\frac{i}{\hbar}\boldsymbol{J}\cdot\boldsymbol{\varphi}_W(\boldsymbol{p},\boldsymbol{\theta})}|\mathbf{0}, s_z\rangle \\
&= N(\boldsymbol{p}) e^{-\frac{ic}{\hbar}\boldsymbol{K}\cdot\boldsymbol{\theta}_{\boldsymbol{\theta p}}} \sum_{s_z'=-s}^{s} \mathscr{D}^s_{s_z' s_z}(\boldsymbol{\varphi}_W(\boldsymbol{p}, \boldsymbol{\theta}))|\mathbf{0}, s_z'\rangle \\
&= \frac{N(\boldsymbol{p})}{N(\boldsymbol{\theta p})} \sum_{s_z'=-s}^{s} \mathscr{D}^s_{s_z' s_z}(\boldsymbol{\varphi}_W(\boldsymbol{p}, \boldsymbol{\theta}))|\boldsymbol{\theta p}, s_z'\rangle.
\end{aligned} \tag{5.19}$$

16 Explicit formulas for $\boldsymbol{\varphi}_W(\boldsymbol{p}, \boldsymbol{\theta})$ can be found, for example, in [73].

17 This means that the matrix introduced in (5.12) has the form $G_{s_z' s_z}(\boldsymbol{p}, \boldsymbol{\theta}) = N(\boldsymbol{p})N^{-1}(\boldsymbol{\theta p}) \times \mathscr{D}^s_{s_z' s_z}(\boldsymbol{\varphi}_W(\boldsymbol{p}, \boldsymbol{\theta}))$. It is also clear that Wigner's rotation is a purely relativistic effect. In the nonrelativistic case, where boosts are commuting, any composition of boosts is always equal to a pure boost, and Wigner's angle on the left-hand side of (5.18) is zero.

5.1.9 Irreducibility of representation U_g

Equations (5.10) and (5.19) show that rotations and boosts are accompanied by rotations of basis vectors in each subspace $\mathscr{H}_{\boldsymbol{p}}$ by matrices $\mathscr{D}^s$. If the representation $\mathscr{D}^s$ of the rotation group were reducible, then each subspace $\mathscr{H}_{\boldsymbol{p}}$ would be represented as a direct sum of irreducible components $\mathscr{H}^k_{\boldsymbol{p}}$,

$$\mathscr{H}_{\boldsymbol{p}} = \bigoplus_k \mathscr{H}^k_{\boldsymbol{p}},$$

and each subspace

$$\mathscr{H}^k = \bigoplus_{\boldsymbol{p} \in R^3} \mathscr{H}^k_{\boldsymbol{p}}$$

would be reducible with respect to the entire Poincaré group. Therefore, in order to construct an *irreducible* representation of this group in $\mathscr{H}$, the representation $\mathscr{D}^s$ must be an irreducible unitary representation of the rotation group, as we announced in Subsection 5.1.2.

In this book, we will mostly deal with electrons and protons, which are massive particles with spin $s = 1/2$. Therefore, we are especially interested in the two-dimensional irreducible representation $\mathscr{D}^{1/2}$ of the rotation group from Appendix I.5.

5.1.10 Method of induced representations

Let us repeat the basic steps in our construction of the unitary irreducible representation U_g of the Poincaré group for a massive particle. These steps are known as Mackey's method of *induced representation* [51, 87]. This construction was based on the basis $|\boldsymbol{p}\rangle$ of eigenvectors of the momentum operator $\boldsymbol{P}$. We chose a 3-vector of the so-called *standard momentum* $\boldsymbol{\kappa} = \boldsymbol{0}$ and found the corresponding *small group*, i. e., the subgroup of those transformations in the Lorentz subgroup which leave this vector invariant.[18] In our case the small group turned out to be the group of rotations. Then we found that if the subspace of the standard vector $\mathscr{H}_{\boldsymbol{\kappa}}$ carries an irreducible representation V_g of the small group, then the entire Hilbert space $\mathscr{H}$ will carry an irreducible representation U_g of the full Poincaré group.

The desired representation U_g of the Poincaré group was *induced* from the irreducible representation $\mathscr{D}^s$ of the rotation group. Translations multiply state vectors by phase factors (5.8) and (5.9), while rotations and boosts are represented by formulas (5.10) and (5.19), respectively.

18 One can show [51] that any other selection of the standard momentum ($\boldsymbol{\kappa} \neq \boldsymbol{0}$) would lead us to an equivalent representation of the Poincaré group.

5.2 Momentum representation

So far we have discussed the action of inertial transformations on the basis of common eigenvectors $|\boldsymbol{p}, s_z\rangle$ of operators $\boldsymbol{P}$ and S_z. All other state vectors $|\Psi\rangle$ in the Hilbert space $\mathscr{H}$ can be represented as linear combinations of these basis vectors, i. e., state vectors $|\Psi\rangle$ can be written in terms of *wave functions* $\psi(\boldsymbol{p}, s_z)$ in the momentum–spin representation. In this section we will consider in more detail how states are described by wave functions and how these functions change under the action of inertial transformations. For simplicity, we restrict our treatment to spinless particles only.

5.2.1 Resolution of identity

Two basis vectors with different momenta $|\boldsymbol{p}\rangle$ and $|\boldsymbol{p}'\rangle$ are eigenvectors of the Hermitian operator $\boldsymbol{P}$ with different eigenvalues, and hence they are orthogonal, so

$$\langle\boldsymbol{p}|\boldsymbol{p}'\rangle = 0 \quad \text{if } \boldsymbol{p} \neq \boldsymbol{p}'.$$

If the spectrum of eigenvalues $\boldsymbol{p}$ were discrete, we could simply normalize the basis vectors to unity: $\langle\boldsymbol{p}|\boldsymbol{p}\rangle = 1$. However, such a normalization becomes problematic in the continuous momentum space. It turns out to be more convenient to work with *nonnormalizable* (= improper) eigenvectors of momentum $|\boldsymbol{p}\rangle$ and use them for writing normalized state vectors $|\Psi\rangle$ in the form of integrals,

$$|\Psi\rangle = \int d\boldsymbol{p}\psi(\boldsymbol{p})|\boldsymbol{p}\rangle, \tag{5.20}$$

where the set of coefficients $\psi(\boldsymbol{p})$ is called the wave function in the *momentum representation*. By analogy with (1.21), it is usually required that the normalized wave functions $\psi(\boldsymbol{p})$ are expressed through the inner product, so

$$\psi(\boldsymbol{p}) = \langle\boldsymbol{p}|\Psi\rangle = \int d\boldsymbol{p}'\psi(\boldsymbol{p}')\langle\boldsymbol{p}|\boldsymbol{p}'\rangle.$$

It then follows that the inner product of two improper basis vectors is equal to the Dirac delta function (see Appendix A)

$$\langle\boldsymbol{p}|\boldsymbol{p}'\rangle = \delta(\boldsymbol{p} - \boldsymbol{p}'). \tag{5.21}$$

Then, by analogy with equation (G.3), we can define the following spectral resolution of the identity operator:

$$I = \int d\boldsymbol{p}|\boldsymbol{p}\rangle\langle\boldsymbol{p}|. \tag{5.22}$$

As expected, the action of this operator on any normalized state vector $|\Psi\rangle$ is trivial. We have

$$I|\Psi\rangle = \int d\boldsymbol{p}|\boldsymbol{p}\rangle\langle\boldsymbol{p}|\Psi\rangle = \int d\boldsymbol{p}|\boldsymbol{p}\rangle\psi(\boldsymbol{p}) = |\Psi\rangle.$$

Of course, the identity operator must be invariant with respect to Poincaré transformations, that is, we expect that

$$I = U_g I U_g^{-1}$$

for all g. The invariance of I with respect to translations follows directly from equations (5.8) and (5.9). Invariance with respect to rotations can be proved as follows:

$$\begin{aligned}
I' &= e^{-\frac{i}{\hbar}\boldsymbol{J}\cdot\boldsymbol{\varphi}} I e^{\frac{i}{\hbar}\boldsymbol{J}\cdot\boldsymbol{\varphi}} = e^{-\frac{i}{\hbar}\boldsymbol{J}\cdot\boldsymbol{\varphi}}\left(\int d\boldsymbol{p}|\boldsymbol{p}\rangle\langle\boldsymbol{p}|\right)e^{\frac{i}{\hbar}\boldsymbol{J}\cdot\boldsymbol{\varphi}} \\
&= \int d\boldsymbol{p}|\boldsymbol{\varphi p}\rangle\langle\boldsymbol{\varphi p}| = \int d(\boldsymbol{\varphi p})\left|\det\left[\frac{d(\boldsymbol{\varphi p})}{d\boldsymbol{p}}\right]\right|^{-1}|\boldsymbol{\varphi p}\rangle\langle\boldsymbol{\varphi p}| \\
&= \int d\boldsymbol{q}|\boldsymbol{q}\rangle\langle\boldsymbol{q}| = I,
\end{aligned}$$

where we used (5.10) and the fact that the Jacobian for the change of variables $\boldsymbol{p} \to \boldsymbol{q} = \boldsymbol{\varphi p}$ is equal to $|\det[d(\boldsymbol{\varphi p})/d\boldsymbol{p}]|^{-1} = |\det[R_{\boldsymbol{\varphi}}]|^{-1} = 1$.

5.2.2 Boost transformation

We now consider in more detail the invariance of I with respect to boosts. From equation (5.19) we get

$$\begin{aligned}
I' &= e^{-\frac{ic}{\hbar}\boldsymbol{K}\cdot\boldsymbol{\theta}} I e^{\frac{ic}{\hbar}\boldsymbol{K}\cdot\boldsymbol{\theta}} = e^{-\frac{ic}{\hbar}\boldsymbol{K}\cdot\boldsymbol{\theta}}\left(\int d\boldsymbol{q}|\boldsymbol{q}\rangle\langle\boldsymbol{q}|\right)e^{\frac{ic}{\hbar}\boldsymbol{K}\cdot\boldsymbol{\theta}} \\
&= \int d\boldsymbol{q}\frac{N^2(\boldsymbol{q})}{N^2(\boldsymbol{\theta q})}|\boldsymbol{\theta q}\rangle\langle\boldsymbol{\theta q}| = \int d\boldsymbol{p}\left|\det\left[\frac{d(\boldsymbol{\theta}^{-1}\boldsymbol{p})}{d\boldsymbol{p}}\right]\right|\frac{N^2(\boldsymbol{\theta}^{-1}\boldsymbol{p})}{N(\boldsymbol{p})}|\boldsymbol{p}\rangle\langle\boldsymbol{p}|,
\end{aligned} \tag{5.23}$$

where $N(\boldsymbol{p})$ is the normalization factor introduced in (5.5) and $|\det[d(\boldsymbol{\theta}^{-1}\boldsymbol{q})/d\boldsymbol{q}]|$ is the Jacobian for the change of variables $\boldsymbol{q} \to \boldsymbol{p} = \boldsymbol{\theta q}$. This Jacobian cannot depend on the direction of $\boldsymbol{\theta}$, so without loss of generality and to simplify our calculations, we choose this direction along the z-axis. Then, replacing $\boldsymbol{\theta} \to -\boldsymbol{\theta}$ in (5.11), we obtain

$$\boldsymbol{\theta}^{-1}p_x = p_x, \tag{5.24}$$

$$\boldsymbol{\theta}^{-1}p_y = p_y, \tag{5.25}$$

$$\boldsymbol{\theta}^{-1}p_z = p_z\cosh\theta - \frac{1}{c}\sqrt{m^2c^4 + p^2c^2}\sinh\theta, \tag{5.26}$$

$$\omega_{\boldsymbol{\theta}^{-1}\boldsymbol{p}} = \sqrt{m^2c^4 + p^2c^2}\cosh\theta - cp_z\sinh\theta,$$

$$\det\left[\frac{d(\boldsymbol{\theta}^{-1}\boldsymbol{p})}{d\boldsymbol{p}}\right] = \det\begin{bmatrix} 1 & 0 & 0 \\ 0 & 1 & 0 \\ \frac{-cp_x \sinh\theta}{\sqrt{m^2c^4+p^2c^2}} & \frac{-cp_y \sinh\theta}{\sqrt{m^2c^4+p^2c^2}} & \cosh\theta - \frac{cp_z \sinh\theta}{\sqrt{m^2c^4+p^2c^2}} \end{bmatrix}$$
$$= \cosh\theta - \frac{cp_z \sinh\theta}{\sqrt{m^2c^4+p^2c^2}} = \frac{\omega_{\boldsymbol{\theta}^{-1}\boldsymbol{p}}}{\omega_{\boldsymbol{p}}}. \tag{5.27}$$

Substituting (5.27) in (5.23), we get

$$I' = \int d\boldsymbol{p} \frac{\omega_{\boldsymbol{\theta}^{-1}\boldsymbol{p}}}{\omega_{\boldsymbol{p}}} \frac{N^2(\boldsymbol{\theta}^{-1}\boldsymbol{p})}{N^2(\boldsymbol{p})} |\boldsymbol{p}\rangle\langle\boldsymbol{p}|. \tag{5.28}$$

Therefore, in order to ensure the invariance of I, we must define our normalization factor as

$$N(\boldsymbol{p}) = \sqrt{\frac{mc^2}{\omega_{\boldsymbol{p}}}}. \tag{5.29}$$

Indeed, in this case the product of numerical factors under the integral sign in (5.28) is equal to unity. In particular,

$$\frac{\omega_{\boldsymbol{\theta}^{-1}\boldsymbol{p}}}{\omega_{\boldsymbol{p}}} \cdot \frac{N^2(\boldsymbol{\theta}^{-1}\boldsymbol{p})}{N^2(\boldsymbol{p})} = \frac{\omega_{\boldsymbol{\theta}^{-1}\boldsymbol{p}}}{\omega_{\boldsymbol{p}}} \cdot \frac{mc^2}{\omega_{\boldsymbol{\theta}^{-1}\boldsymbol{p}}} \cdot \frac{\omega_{\boldsymbol{p}}}{mc^2} = 1,$$

and we obtain the desired relation $I' = I$.

Taking into account (5.29), formula (5.19) for boost transformations can be written in its final form

$$e^{-\frac{ic}{\hbar}\boldsymbol{K}\cdot\boldsymbol{\theta}}|\boldsymbol{p}, s_z\rangle = \sqrt{\frac{\omega_{\boldsymbol{\theta p}}}{\omega_{\boldsymbol{p}}}} \sum_{s'_z=-s}^{s} \mathscr{D}^s_{s'_z s_z}(\boldsymbol{\varphi}_W(\boldsymbol{p}, \boldsymbol{\theta}))|\boldsymbol{\theta p}, s'_z\rangle. \tag{5.30}$$

Note that the result (5.27) means, in particular, that inside three-dimensional momentum integrals we are allowed to make the following change of variables:

$$\boldsymbol{p} \to \boldsymbol{\theta p},$$
$$d\boldsymbol{p} \to d(\boldsymbol{\theta p})\frac{\omega_{\boldsymbol{p}}}{\omega_{\boldsymbol{\theta p}}}.$$

In other words, $d\boldsymbol{p}/\omega_{\boldsymbol{p}}$ is the "Lorentz-invariant measure"

$$\frac{d\boldsymbol{p}}{\omega_{\boldsymbol{p}}} = \frac{d(\boldsymbol{\theta p})}{\omega_{\boldsymbol{\theta p}}}. \tag{5.31}$$

We will use this property quite often in our calculations.

5.2.3 Wave function in momentum representation

The inner product of two normalized vectors $|\Psi\rangle = \int d\boldsymbol{p}\psi(\boldsymbol{p})|\boldsymbol{p}\rangle$ and $|\Phi\rangle = \int d\boldsymbol{p}\phi(\boldsymbol{p})|\boldsymbol{p}\rangle$ can be written in terms of their wave functions

$$\begin{aligned}\langle\Psi|\Phi\rangle &= \int d\boldsymbol{p}d\boldsymbol{p}'\psi^*(\boldsymbol{p})\phi(\boldsymbol{p}')\langle\boldsymbol{p}|\boldsymbol{p}'\rangle = \int d\boldsymbol{p}d\boldsymbol{p}'\psi^*(\boldsymbol{p})\phi(\boldsymbol{p}')\delta(\boldsymbol{p}-\boldsymbol{p}') \\ &= \int d\boldsymbol{p}\psi^*(\boldsymbol{p})\phi(\boldsymbol{p}).\end{aligned} \tag{5.32}$$

Therefore, if the state vector $|\Psi\rangle$ is normalized to unity, its wave function $\psi(\boldsymbol{p})$ must satisfy the normalization condition

$$1 = \langle\Psi|\Psi\rangle = \int d\boldsymbol{p}|\psi(\boldsymbol{p})|^2.$$

This wave function has a direct probabilistic interpretation: if Ω is a region of the momentum space, then the integral $\int_\Omega d\boldsymbol{p}|\psi(\boldsymbol{p})|^2$ expresses the probability of finding the particle inside this region.

Inertial transformations of the state vector $|\Psi\rangle$ can be regarded as transformations of the corresponding wave function. For example, by Hermitian conjugation of equation (5.30), we get

$$\langle\boldsymbol{p}|e^{-\frac{ic}{\hbar}\boldsymbol{K}\cdot\boldsymbol{\theta}} = [e^{\frac{ic}{\hbar}\boldsymbol{K}\cdot\boldsymbol{\theta}}|\boldsymbol{p}\rangle]^\dagger = \langle\boldsymbol{\theta}^{-1}\boldsymbol{p}|\sqrt{\frac{\omega_{\boldsymbol{\theta}^{-1}\boldsymbol{p}}}{\omega_{\boldsymbol{p}}}}$$

and the transformed wave function is

$$\begin{aligned}\psi'(\boldsymbol{p}) &\equiv e^{-\frac{ic}{\hbar}\hat{\boldsymbol{K}}\cdot\boldsymbol{\theta}}\psi(\boldsymbol{p}) \equiv \langle\boldsymbol{p}|e^{-\frac{ic}{\hbar}\boldsymbol{K}\cdot\boldsymbol{\theta}}|\Psi\rangle \\ &= \sqrt{\frac{\omega_{\boldsymbol{\theta}^{-1}\boldsymbol{p}}}{\omega_{\boldsymbol{p}}}}\langle\boldsymbol{\theta}^{-1}\boldsymbol{p}|\Psi\rangle = \sqrt{\frac{\omega_{\boldsymbol{\theta}^{-1}\boldsymbol{p}}}{\omega_{\boldsymbol{p}}}}\psi(\boldsymbol{\theta}^{-1}\boldsymbol{p}).\end{aligned} \tag{5.33}$$

Then the invariance of the inner product (5.32) with respect to boosts can be easily deduced from the property (5.31). We have

$$\begin{aligned}\langle\Phi'|\Psi'\rangle &= \int d\boldsymbol{p}(\phi')^*(\boldsymbol{p})\psi'(\boldsymbol{p}) = \int d\boldsymbol{p}\sqrt{\frac{\omega_{\boldsymbol{\theta}^{-1}\boldsymbol{p}}}{\omega_{\boldsymbol{p}}}}\phi^*(\boldsymbol{\theta}^{-1}\boldsymbol{p})\sqrt{\frac{\omega_{\boldsymbol{\theta}^{-1}\boldsymbol{p}}}{\omega_{\boldsymbol{p}}}}\psi(\boldsymbol{\theta}^{-1}\boldsymbol{p}) \\ &= \int \frac{d\boldsymbol{p}}{\omega_{\boldsymbol{p}}}\omega_{\boldsymbol{\theta}^{-1}\boldsymbol{p}}\phi^*(\boldsymbol{\theta}^{-1}\boldsymbol{p})\psi(\boldsymbol{\theta}^{-1}\boldsymbol{p}) = \int \frac{d(\boldsymbol{\theta}^{-1}\boldsymbol{p})}{\omega_{\boldsymbol{\theta}^{-1}\boldsymbol{p}}}\omega_{\boldsymbol{\theta}^{-1}\boldsymbol{p}}\phi^*(\boldsymbol{\theta}^{-1}\boldsymbol{p})\psi(\boldsymbol{\theta}^{-1}\boldsymbol{p}) \\ &= \int d\boldsymbol{p}\phi^*(\boldsymbol{p})\psi(\boldsymbol{p}) = \langle\Phi|\Psi\rangle.\end{aligned} \tag{5.34}$$

The actions of Poincaré group generators on momentum–space wave functions are derived from formulas (5.8)–(5.10) and (5.30). We have

$$\hat{P}_x\psi(\boldsymbol{p}) = i\hbar\lim_{a\to 0}\frac{d}{da}e^{-\frac{i}{\hbar}\hat{P}_x a}\psi(\boldsymbol{p}) = p_x\psi(\boldsymbol{p}), \tag{5.35}$$

$$\hat{H}\psi(\boldsymbol{p}) = -i\hbar \lim_{t\to 0} \frac{d}{dt} e^{\frac{i}{\hbar}\hat{H}t}\psi(\boldsymbol{p}) = \omega_{\boldsymbol{p}}\psi(\boldsymbol{p}), \tag{5.36}$$

$$\begin{aligned}
\hat{K}_x\psi(\boldsymbol{p}) &= \frac{i\hbar}{c} \lim_{\theta\to 0} \frac{d}{d\theta} e^{-\frac{ic}{\hbar}\hat{K}_x\theta}\psi(\boldsymbol{p}) \\
&= \frac{i\hbar}{c} \lim_{\theta\to 0} \frac{d}{d\theta} \sqrt{\frac{\sqrt{m^2c^4 + p^2c^2}\cosh\theta - cp_x \sinh\theta}{\sqrt{m^2c^4+p^2c^2}}} \\
&\quad \times \psi\left(p_x\cosh\theta - \frac{1}{c}\sqrt{m^2c^4+p^2c^2}\sinh\theta, p_y, p_z\right) \\
&= i\hbar\left(-\frac{\omega_{\boldsymbol{p}}}{c^2}\frac{d}{dp_x} - \frac{p_x}{2\omega_{\boldsymbol{p}}}\right)\psi(\boldsymbol{p}).
\end{aligned} \tag{5.37}$$

We can also obtain expressions for the position and angular momentum operators,

$$\begin{aligned}
\hat{R}_x\psi(\boldsymbol{p}) &= -\frac{c^2}{2}(\hat{H}^{-1}\hat{K}_x + \hat{K}_x\hat{H}^{-1})\psi(\boldsymbol{p}) \\
&= -\frac{i\hbar}{2}\left(-\omega_{\boldsymbol{p}}^{-1}\omega_{\boldsymbol{p}}\frac{d}{dp_x} - \omega_{\boldsymbol{p}}\frac{d}{dp_x}\omega_{\boldsymbol{p}}^{-1} - \frac{p_x c^2}{\omega_{\boldsymbol{p}}^2}\right)\psi(\boldsymbol{p}) = i\hbar\frac{d}{dp_x}\psi(\boldsymbol{p}),
\end{aligned} \tag{5.38}$$

$$\hat{J}_x\psi(\boldsymbol{p}) = (\hat{R}_y\hat{P}_z - \hat{R}_z\hat{P}_y)\psi(\boldsymbol{p}) = i\hbar\left(p_z\frac{d}{dp_y} - p_y\frac{d}{dp_z}\right)\psi(\boldsymbol{p}). \tag{5.39}$$

5.3 Position representation

In the previous section, we discussed wave functions in the momentum representation, i. e., with respect to common eigenvectors of the three commuting components (P_x, P_y, P_z) of the momentum operator. According to Theorem 4.1, the three components of position (R_x, R_y, R_z) also commute with each other. So, their common eigenvectors $|\boldsymbol{r}\rangle$ also form a basis in the Hilbert space $\mathscr{H}$ of a single massive particle. In this section we will consider particle wave functions with respect to this basis, i. e., in the *position representation*. It is remarkable that our formulas for relativistic wave functions will be very similar to their analogs in nonrelativistic quantum mechanics.

5.3.1 Basis of localized functions

First, we expand the eigenvector $|\boldsymbol{r}\rangle$ in the momentum basis

$$|\boldsymbol{r}\rangle = \int d\boldsymbol{p}\psi_{\boldsymbol{r}}(\boldsymbol{p})|\boldsymbol{p}\rangle \tag{5.40}$$

and obtain eigenfunctions of the position operator in the momentum representation,

$$\psi_{\boldsymbol{r}}(\boldsymbol{p}) = \langle\boldsymbol{p}|\boldsymbol{r}\rangle = \frac{e^{-\frac{i}{\hbar}\boldsymbol{p}\cdot\boldsymbol{r}}}{(2\pi\hbar)^{3/2}}. \tag{5.41}$$

This can be verified by substituting (5.38) and (5.41) in the eigenvalue equation to obtain

$$\begin{aligned}\hat{\boldsymbol{R}}\psi_{\boldsymbol{r}}(\boldsymbol{p}) &= \frac{1}{(2\pi\hbar)^{3/2}}\hat{\boldsymbol{R}}e^{-\frac{i}{\hbar}\boldsymbol{p}\cdot\boldsymbol{r}} = \frac{i\hbar}{(2\pi\hbar)^{3/2}}\frac{d}{d\boldsymbol{p}}e^{-\frac{i}{\hbar}\boldsymbol{p}\cdot\boldsymbol{r}} = \boldsymbol{r}\left(\frac{e^{-\frac{i}{\hbar}\boldsymbol{p}\cdot\boldsymbol{r}}}{(2\pi\hbar)^{3/2}}\right)\\ &= \boldsymbol{r}\psi_{\boldsymbol{r}}(\boldsymbol{p}).\end{aligned}$$

Since the operator $\boldsymbol{R}$ is Hermitian, its eigenvectors with different eigenvalues $\boldsymbol{r}$ and $\boldsymbol{r}'$ must be orthogonal. Indeed, using equation (A.1), we get the inner product in the form of a delta function,

$$\begin{aligned}\langle\boldsymbol{r}'|\boldsymbol{r}\rangle &= \frac{1}{(2\pi\hbar)^3}\int d\boldsymbol{p}d\boldsymbol{p}'e^{-\frac{i}{\hbar}\boldsymbol{p}\cdot\boldsymbol{r}+\frac{i}{\hbar}\boldsymbol{p}'\cdot\boldsymbol{r}'}\langle\boldsymbol{p}'|\boldsymbol{p}\rangle\\ &= \frac{1}{(2\pi\hbar)^3}\int d\boldsymbol{p}d\boldsymbol{p}'e^{-\frac{i}{\hbar}\boldsymbol{p}\cdot\boldsymbol{r}+\frac{i}{\hbar}\boldsymbol{p}'\cdot\boldsymbol{r}'}\delta(\boldsymbol{p}-\boldsymbol{p}')\\ &= \frac{1}{(2\pi\hbar)^3}\int d\boldsymbol{p}e^{-\frac{i}{\hbar}\boldsymbol{p}\cdot(\boldsymbol{r}-\boldsymbol{r}')} = \delta(\boldsymbol{r}-\boldsymbol{r}').\end{aligned} \tag{5.42}$$

This means that $|\boldsymbol{r}\rangle$ are nonnormalizable states, just like $|\boldsymbol{p}\rangle$.[19]

By analogy with (5.20), any normalized state vector $|\Psi\rangle$ can be represented as an integral over the position space, so

$$|\Psi\rangle = \int d\boldsymbol{r}\psi(\boldsymbol{r})|\boldsymbol{r}\rangle,$$

where $\psi(\boldsymbol{r}) = \langle\boldsymbol{r}|\Psi\rangle$ is the position–space wave function of the state $|\Psi\rangle$. The square of the absolute value $|\psi(\boldsymbol{r})|^2$ of the wave function expresses the probability density of finding the particle at a given space point $\boldsymbol{r}$. The inner product of two state vectors $|\Psi\rangle$ and $|\Phi\rangle$ can be written in terms of their position–space wave functions as follows:

$$\begin{aligned}\langle\Phi|\Psi\rangle &= \int d\boldsymbol{r}d\boldsymbol{r}'\phi^*(\boldsymbol{r})\psi(\boldsymbol{r}')\langle\boldsymbol{r}|\boldsymbol{r}'\rangle = \int d\boldsymbol{r}d\boldsymbol{r}'\phi^*(\boldsymbol{r})\psi(\boldsymbol{r}')\delta(\boldsymbol{r}-\boldsymbol{r}')\\ &= \int d\boldsymbol{r}\phi^*(\boldsymbol{r})\psi(\boldsymbol{r}).\end{aligned} \tag{5.43}$$

Using equations (5.40) and (5.41), we find that the position–space wave function of the momentum eigenstate $|\boldsymbol{p}\rangle$ is an ordinary plane wave,

$$\psi_{\boldsymbol{p}}(\boldsymbol{r}) = \langle\boldsymbol{r}|\boldsymbol{p}\rangle = \frac{e^{\frac{i}{\hbar}\boldsymbol{p}\cdot\boldsymbol{r}}}{(2\pi\hbar)^{3/2}}, \tag{5.44}$$

$$\hat{\boldsymbol{P}}\psi_{\boldsymbol{p}}(\boldsymbol{r}) = \boldsymbol{p}\psi_{\boldsymbol{p}}(\boldsymbol{r}).$$

19 As in nonrelativistic quantum mechanics, eigenvectors of the position operator in the position representation are expressed by delta functions (5.42). Note that eigenfunctions introduced in [62] did not satisfy this important requirement.

This is also an eigenstate of the energy $H = \sqrt{P^2c^2 + m^2c^4}$ and velocity $\boldsymbol{V} = \boldsymbol{P}c^2/H$ operators. So, in this state measurements of the velocity

$$\boldsymbol{v} = \frac{\boldsymbol{p}c^2}{\omega_{\boldsymbol{p}}} \tag{5.45}$$

are unambiguous and predictable, while according to the Heisenberg inequality (6.95), the uncertainty of the position is infinitely high.

From (5.22) we can also obtain a position representation of the identity operator

$$\begin{aligned}\int d\boldsymbol{r}|\boldsymbol{r}\rangle\langle\boldsymbol{r}| &= \frac{1}{(2\pi\hbar)^3}\int d\boldsymbol{r}d\boldsymbol{p}d\boldsymbol{p}' e^{-\frac{i}{\hbar}\boldsymbol{p}'\cdot\boldsymbol{r}}|\boldsymbol{p}\rangle\langle\boldsymbol{p}'|e^{\frac{i}{\hbar}\boldsymbol{p}\cdot\boldsymbol{r}} \\ &= \int d\boldsymbol{p}d\boldsymbol{p}'|\boldsymbol{p}\rangle\langle\boldsymbol{p}'|\delta(\boldsymbol{p}-\boldsymbol{p}') = \int d\boldsymbol{p}|\boldsymbol{p}\rangle\langle\boldsymbol{p}| = I.\end{aligned}$$

5.3.2 Operators of observables in position representation

Similarly to the momentum–space formulas (5.35)–(5.39), we can write actions of Poincaré generators on position–space wave functions. For example, from (5.8), (5.40) and (5.41) it follows that

$$\begin{aligned}e^{-\frac{i}{\hbar}\boldsymbol{P}\cdot\boldsymbol{a}}\int d\boldsymbol{r}\psi(\boldsymbol{r})|\boldsymbol{r}\rangle &= e^{-\frac{i}{\hbar}\boldsymbol{P}\cdot\boldsymbol{a}}\int d\boldsymbol{r}\psi(\boldsymbol{r})\int d\boldsymbol{p}\frac{e^{-\frac{i}{\hbar}\boldsymbol{p}\cdot\boldsymbol{r}}}{(2\pi\hbar)^{3/2}}|\boldsymbol{p}\rangle \\ &= \int d\boldsymbol{r}\psi(\boldsymbol{r})\int d\boldsymbol{p}\frac{e^{-\frac{i}{\hbar}\boldsymbol{p}\cdot(\boldsymbol{r}+\boldsymbol{a})}}{(2\pi\hbar)^{3/2}}|\boldsymbol{p}\rangle = \int d\boldsymbol{r}\psi(\boldsymbol{r})|\boldsymbol{r}+\boldsymbol{a}\rangle = \int d\boldsymbol{r}\psi(\boldsymbol{r}-\boldsymbol{a})|\boldsymbol{r}\rangle.\end{aligned}$$

Hence translations and their generators act on position–space wave functions as follows:

$$\begin{aligned}e^{-\frac{i}{\hbar}\hat{\boldsymbol{P}}\cdot\boldsymbol{a}}\psi(\boldsymbol{r}) &= \psi(\boldsymbol{r}-\boldsymbol{a}), \\ \hat{\boldsymbol{P}}\psi(\boldsymbol{r}) &= i\hbar\lim_{\boldsymbol{a}\to 0}\frac{d}{d\boldsymbol{a}}e^{-\frac{i}{\hbar}\hat{\boldsymbol{P}}\cdot\boldsymbol{a}}\psi(\boldsymbol{r}) = i\hbar\lim_{\boldsymbol{a}\to 0}\frac{d}{d\boldsymbol{a}}\psi(\boldsymbol{r}-\boldsymbol{a}) = -i\hbar\frac{d}{d\boldsymbol{r}}\psi(\boldsymbol{r}).\end{aligned} \tag{5.46}$$

Other important operators have the following position–space representations:

$$\begin{aligned}\hat{H}\psi(\boldsymbol{r}) &= \sqrt{m^2c^4 - \hbar^2c^2\frac{d^2}{d\boldsymbol{r}^2}}\psi(\boldsymbol{r}), \\ \boldsymbol{K}\psi(\boldsymbol{r}) &= \frac{1}{2}\left(\sqrt{m^2c^4 - \hbar^2c^2\frac{d^2}{d\boldsymbol{r}^2}}\boldsymbol{r} + \boldsymbol{r}\sqrt{m^2c^4 - \hbar^2c^2\frac{d^2}{d\boldsymbol{r}^2}}\right)\psi(\boldsymbol{r}), \\ \hat{\boldsymbol{R}}\psi(\boldsymbol{r}) &= \boldsymbol{r}\psi(\boldsymbol{r}), \\ \hat{J}_x\psi(\boldsymbol{r}) &= -i\hbar\left(y\frac{d}{dz} - z\frac{d}{dy}\right)\psi(\boldsymbol{r}).\end{aligned} \tag{5.47}$$

The actions of operators (5.46) and (5.47) are characteristic of the so-called *Schrödinger representation* of the Heisenberg algebra $\mathfrak{h}_3$ (see Appendix I.3).

If necessary, one can switch between momentum and position wave functions of the same state by means of the Fourier transform. For example, suppose that the state $|\Psi\rangle$ has the position wave function $\psi(\boldsymbol{r})$. Then from (5.40) and (5.41) we get

$$\begin{aligned}|\Psi\rangle &= \int d\boldsymbol{r}\psi(\boldsymbol{r})|\boldsymbol{r}\rangle = \frac{1}{(2\pi\hbar)^{3/2}}\int d\boldsymbol{r}\psi(\boldsymbol{r})\int d\boldsymbol{p}e^{-\frac{i}{\hbar}\boldsymbol{p}\cdot\boldsymbol{r}}|\boldsymbol{p}\rangle \\ &= \int d\boldsymbol{p}\left(\frac{1}{(2\pi\hbar)^{3/2}}\int d\boldsymbol{r}\psi(\boldsymbol{r})e^{-\frac{i}{\hbar}\boldsymbol{p}\cdot\boldsymbol{r}}\right)|\boldsymbol{p}\rangle.\end{aligned}$$

Comparing with (5.20), we see that the corresponding momentum–space wave function is

$$\psi(\boldsymbol{p}) = \frac{1}{(2\pi\hbar)^{3/2}}\int d\boldsymbol{r}\psi(\boldsymbol{r})e^{-\frac{i}{\hbar}\boldsymbol{p}\cdot\boldsymbol{r}}. \tag{5.48}$$

Conversely, for the momentum wave function $\psi(\boldsymbol{p})$, its position–space counterpart is

$$\psi(\boldsymbol{r}) = \frac{1}{(2\pi\hbar)^{3/2}}\int d\boldsymbol{p}\psi(\boldsymbol{p})e^{\frac{i}{\hbar}\boldsymbol{p}\cdot\boldsymbol{r}}. \tag{5.49}$$

5.3.3 Inertial transformations of observables and states

In this subsection we would like to discuss in more detail how observables and states change under inertial transformations. We have already touched on this issue in several places in the book, but it will be useful to put these pieces together and clarify the physical meaning of Poincaré transformations with a simple example. What do we mean when we express observables and states in the (primed) frame O' through observables and states in the (non-primed) frame O by the standard formulas

$$F' = U_g F U_g^{-1}, \tag{5.50}$$

$$|\Psi'\rangle = U_g|\Psi\rangle? \tag{5.51}$$

Here

$$U_g = e^{-\frac{ic}{\hbar}\boldsymbol{K}\cdot\boldsymbol{\theta}}e^{-\frac{i}{\hbar}\boldsymbol{J}\cdot\boldsymbol{\varphi}}e^{-\frac{i}{\hbar}\boldsymbol{P}\cdot\boldsymbol{a}}e^{\frac{i}{\hbar}Ht}$$

is a unitary representative of a general inertial transformation g in the Hilbert space of the system.

Let us start with transformations of observables (5.50) in the Heisenberg picture. For definiteness, suppose that the observable F is the x-component of position ($F = R_x$). This means that the operator F is a mathematical representative of the usual

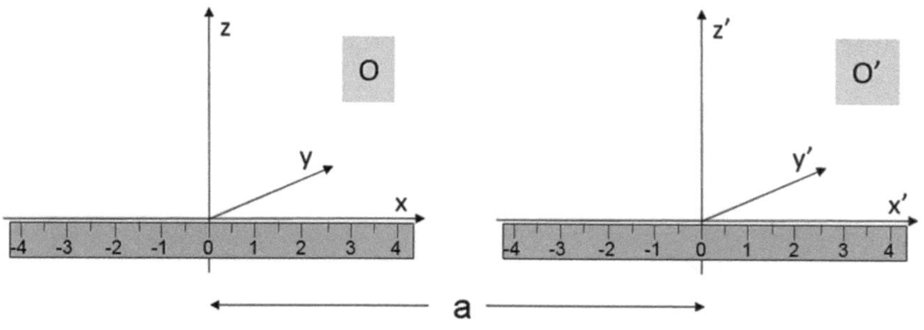

Figure 5.5: Rulers for measuring the x-component of position in the reference frames O and O' shifted by the distance a.

ruler X oriented along the x-axis at rest in the reference frame O, as shown in Figure 5.5. The zero mark on the ruler coincides with the origin of the frame O. For the element g of the Poincaré group we take the shift by the distance a along the x-axis

$$U_g = e^{-\frac{i}{\hbar}P_x a}.$$

Then the transformed observable

$$R'_x = e^{-\frac{i}{\hbar}P_x a} R_x e^{\frac{i}{\hbar}P_x a} \tag{5.52}$$

is the operator describing measurements of positions in the shifted frame of reference O'. These measurements are made with the ruler X' shifted by the distance a with respect to the ruler X. The zero mark on X' coincides with the origin in the frame O'. Of course, position measurements performed by the rulers X and X' on the same particle will give different results. For example, if the particle is at the origin of the frame O, then measurements by the rulers X and X' yield $R_x = 0$ and $R_{x'} = -a$, respectively. For this reason, we say that observables R_x and R'_x are related to each other by

$$R'_x = R_x - a.$$

Of course, the same result can be obtained by formal application of equations (5.52) and (E.14), which gives

$$R'_x = R_x - \frac{i}{\hbar}[P_x, R_x]a = R_x - a.$$

The operator of position can also be represented by its spectral resolution

$$R_x = \int_{-\infty}^{\infty} dx x|x\rangle\langle x|,$$

where $|x\rangle$ are eigenvectors (eigenstates) with positions x. Then equation (5.52) can be rewritten as

$$R'_x = e^{-\frac{i}{\hbar}P_x a}\left(\int_{-\infty}^{\infty} dx x|x\rangle\langle x|\right)e^{\frac{i}{\hbar}P_x a}$$
$$= \int_{-\infty}^{\infty} dx x|x+a\rangle\langle x+a| = \int_{-\infty}^{\infty} dx(x-a)|x\rangle\langle x| = R_x - a.$$

Hence we see that the action of U_g on state vectors

$$e^{-\frac{i}{\hbar}P_x a}|x\rangle = |x+a\rangle$$

should be interpreted as an *active* shift of states, i. e., a translation by the distance a in our case.

However, in practice we are more often interested in how the state of the system looks from the point of view of an inertially transformed observer, i. e., we are interested in passive transformations of states.[20] Obviously, such passive transformations should be expressed by inverse operators U_g^{-1}, so

$$|\Psi'\rangle = U_g^{-1}|\Psi\rangle. \tag{5.53}$$

In particular, this means that if the vector $|\Psi\rangle = |x\rangle$ describes a state localized at the point x from the point of view of O (the value of x is measured by the ruler X), then the *same* state is described by the vector

$$|\Psi'\rangle = U_g^{-1}|\Psi\rangle = e^{\frac{i}{\hbar}P_x a}|x\rangle = |x-a\rangle \tag{5.54}$$

from the point of view of the observer O' (position measured by the ruler X').

Instead of state vectors, we can also apply inertial transformations to their wave functions. For example, the state vector

$$|\Psi\rangle = \int d\boldsymbol{r}\psi(\boldsymbol{r})|\boldsymbol{r}\rangle$$

has the wave function $\psi(x,y,z)$ in the position representation. When we shift the observer by the distance a in the positive x-direction, we must apply a passive transformation (5.54) to the state vector,

$$|\Psi'\rangle = e^{\frac{i}{\hbar}P_x a}|\Psi\rangle = \int d\boldsymbol{r}\psi(x,y,z)|x-a,y,z\rangle = \int d\boldsymbol{r}\psi(x+a,y,z)|x,y,z\rangle,$$

which means that the passive transformation of the wave function has the form

$$e^{\frac{i}{\hbar}\hat{P}_x a}\psi(x,y,z) = \psi(x+a,y,z).$$

---

20 Here we switch to the Schrödinger picture.

5.3.4 Time translations of observables and states

The above reasoning is especially useful in the case when the inertial transformation U_g is a time translation. As we saw in (4.36), the position operator in the time-shifted reference frame O'' has the form

$$R''_x = e^{\frac{i}{\hbar}Ht} R_x e^{-\frac{i}{\hbar}Ht} = R_x + V_x t.$$

If we are interested in how the state vector $|\Psi\rangle$ looks in the frame O'', we have to apply a passive transformation (5.53). We have

$$|\Psi''\rangle = e^{-\frac{i}{\hbar}Ht}|\Psi\rangle. \tag{5.55}$$

It is customary to consider a continuous sequence of such shifts parameterized by time t and talk about the *time evolution* of the state vector $|\Psi(t)\rangle$. Then equation (5.55) can be regarded as a solution of the *time-dependent* Schrödinger equation

$$i\hbar\frac{d}{dt}|\Psi(t)\rangle = H|\Psi(t)\rangle. \tag{5.56}$$

In actual calculations it is more convenient to deal with numerical functions (wave functions in a certain basis/representation) than with abstract state vectors. To obtain the Schrödinger equation for wave functions, equation (5.56) should be multiplied on the left by basis bra-vectors $\langle \boldsymbol{r}|$, so

$$\begin{aligned} i\hbar\frac{d}{dt}\langle \boldsymbol{r}|\Psi(t)\rangle &= \langle \boldsymbol{r}|H|\Psi(t)\rangle, \\ i\hbar\frac{d}{dt}\Psi(\boldsymbol{r},t) &= \hat{H}\Psi(\boldsymbol{r},t), \end{aligned} \tag{5.57}$$

where $\Psi(\boldsymbol{r},t) \equiv \langle \boldsymbol{r}|\Psi(t)\rangle$ is the position–space wave function.

5.4 Massless particles

5.4.1 Spectra of momentum, energy and velocity

For massless ($m = 0$) particles, such as photons, the method for constructing irreducible unitary representations of the Poincaré group, described in Section 5.1, needs to be slightly modified. Indeed, the arguments given at the beginning of Subsection5.1.1 are no longer valid, because for massless particles the Newton–Wigner operator is not defined; and we cannot apply Corollary I.5 to find the spectrum of the momentum operator $\boldsymbol{P}$. To determine this spectrum, we will use other arguments.

Let us choose a state of the massless particle with some nonzero momentum $\boldsymbol{p} \neq 0$.[21] There are two types of inertial transformations, which can change this value: rotations and boosts. Every vector $\boldsymbol{p}'$ obtained from $\boldsymbol{p}$ by means of rotations or boosts is also an eigenvector of $\boldsymbol{P}$.[22] So, we can use these transformations to study the spectrum $\boldsymbol{P}$. Rotations, in general, change the direction of the momentum vector, but retain its length p, so that rotational images of $\boldsymbol{p}$ form a sphere of radius p with its center at zero momentum $\boldsymbol{0}$. Boosts along the vector $\boldsymbol{p}$ do not change its direction, but change its length. To reduce the length of the momentum vector, we can use the boost $\boldsymbol{\theta}$, which is directed opposite to $\boldsymbol{p}$, i. e., $\boldsymbol{\theta}/\theta = -\boldsymbol{p}/p$. Then, from (5.11) and the equality[23]

$$\omega_{\boldsymbol{p}} = cp \tag{5.58}$$

we obtain

$$\boldsymbol{p}' = \boldsymbol{\theta}\boldsymbol{p} = \boldsymbol{p} - \frac{\boldsymbol{p}}{p}[-p(\cosh\theta - 1) + p\sinh\theta] = \boldsymbol{p}[\cosh\theta - \sinh\theta] = \boldsymbol{p}e^{-\theta}, \tag{5.59}$$

so that the transformed momentum reaches zero only in the limit $\theta \to \infty$. This means that it is impossible to reach the point $\boldsymbol{0}$ from $\boldsymbol{p}$ using rotations and finite boosts. So, this point does not belong to the momentum spectrum of massless particles.[24]

On the other hand, the length of the vector $\boldsymbol{p}$ can be increased without limit, using boosts in the collinear direction. Thus, for massless particles, the hyperboloid (5.6) degenerates into a cone (5.58) with its apex ($\omega = 0, \boldsymbol{p} = \boldsymbol{0}$) removed (see Figure 5.3). So, in the massless case, $H = cP$ and the spectrum of the velocity operator $\boldsymbol{V} \equiv \boldsymbol{P}c^2/H$ is a 3-sphere $|\boldsymbol{v}| = c$. It then follows that massless particles can move only at the speed of light in all reference systems. This is the famous second postulate of Einstein's special theory of relativity.

Statement 5.1 (invariance of the speed of light). *The speed of massless particles (e. g., photons) is equal to the speed of light* (c) *regardless of the speed of the source and the observer.*

5.4.2 Representations of small groups

Our next goal is to characterize massless elementary particles by constructing corresponding unitary irreducible representations of the Poincaré group. For this we will apply the induced representation method from Subsection 5.1.10.

---

21 We assume that at least one such value exists in the spectrum of $\boldsymbol{P}$.

22 The proof of this statement is the same as in (5.2) and (5.7).

23 It follows from the general formula (5.6) for $m = 0$.

24 The physical meaning of this result is that there can be no photons with zero momentum and energy.

We already know that the vector $\boldsymbol{p} = (0, 0, 0)$ does not belong to the momentum spectrum of a massless particle. Therefore, by analogy with (5.1), the Hilbert space of our representation should have the form of a direct sum,

$$\mathcal{H} = \bigoplus_{\boldsymbol{p} \in \mathbb{R}^3 \setminus \{\mathbf{0}\}} \mathcal{H}_{\boldsymbol{p}}, \tag{5.60}$$

in which the summand $\mathcal{H}_{\mathbf{0}}$ is absent. Hence, unlike in the massive case, we cannot choose the vector $\boldsymbol{p} = \mathbf{0}$ as the standard momentum for constructing the induced representation. However, this is not a problem, since we know that the choice of the standard momentum is, in fact, arbitrary, and representations constructed on different standard momenta are unitarily equivalent. Therefore, without loss of generality, we choose another standard momentum,

$$\boldsymbol{\kappa} = (0, 0, 1). \tag{5.61}$$

The next step is to find the small group corresponding to the vector $\boldsymbol{\kappa}$, i. e., the subgroup of the Lorentz group, which leaves this vector invariant. The energy–momentum 4-vector associated with our standard vector (5.61) is $(c\kappa, c\boldsymbol{\kappa}) = (c, 0, 0, c)$. Therefore, in the four-dimensional notation from Appendix J the matrices $\tilde{\Sigma}$ of the small group must satisfy the equation

$$\tilde{\Sigma} \begin{bmatrix} c \\ 0 \\ 0 \\ c \end{bmatrix} = \begin{bmatrix} c \\ 0 \\ 0 \\ c \end{bmatrix}.$$

$\tilde{\Sigma}$ is also an element of the Lorentz group. Therefore, the condition (J.5)

$$\tilde{\Sigma}^T \eta \tilde{\Sigma} = \eta$$

must be satisfied as well. One can verify that the most general 4×4 matrix with these properties has the form [94]

$$\tilde{\Sigma}(X_1, X_2, \theta) = \begin{bmatrix} 1 + (X_1^2 + X_2^2)/2 & X_1 & X_2 & -(X_1^2 + X_2^2)/2 \\ X_1 \cos\theta + X_2 \sin\theta & \cos\theta & \sin\theta & -X_1 \cos\theta - X_2 \sin\theta \\ -X_1 \sin\theta + X_2 \cos\theta & -\sin\theta & \cos\theta & X_1 \sin\theta - X_2 \cos\theta \\ (X_1^2 + X_2^2)/2 & X_1 & X_2 & 1 - (X_1^2 + X_2^2)/2 \end{bmatrix}. \tag{5.62}$$

It depends on three real parameters X_1, X_2 and θ. The three corresponding generators can be obtained by differentiation.[25] We have

$$T_1 = \lim_{X_1, X_2, \theta \to 0} \frac{\partial}{\partial X_1} \tilde{\Sigma}(X_1, X_2, \theta) = \begin{bmatrix} 0 & 1 & 0 & 0 \\ 1 & 0 & 0 & -1 \\ 0 & 0 & 0 & 0 \\ 0 & 1 & 0 & 0 \end{bmatrix} = \mathcal{J}_y + c\mathcal{K}_x,$$

25 Here $\mathcal{J}$ and $\mathcal{K}$ are the familiar generators of the Lorentz group (J.16) and (J.17)–(J.19).

$$T_2 = \lim_{X_1,X_2,\theta\to 0} \frac{\partial}{\partial X_2}\tilde{\Sigma}(X_1,X_2,\theta) = \begin{bmatrix} 0 & 0 & 1 & 0 \\ 0 & 0 & 0 & 0 \\ 1 & 0 & 0 & -1 \\ 0 & 0 & 1 & 0 \end{bmatrix} = -\mathcal{J}_x + c\mathcal{K}_y,$$

$$R = \lim_{X_1,X_2,\theta\to 0} \frac{\partial}{\partial \theta}\tilde{\Sigma}(X_1,X_2,\theta) = \begin{bmatrix} 0 & 0 & 0 & 0 \\ 0 & 0 & 1 & 0 \\ 0 & -1 & 0 & 0 \\ 0 & 0 & 0 & 0 \end{bmatrix} = \mathcal{J}_z.$$

Their commutators

$$[T_1, T_2] = 0,$$
$$[R, T_2] = T_1,$$
$$[R, T_1] = -T_2$$

are characteristic of the Lie algebra of the group of "translations" (T_1 and T_2) and rotations (R) in the usual two-dimensional plane, which we will call the κ-space.

The next step is to find the complete set of unitary irreducible representations of the small group constructed above. We will do this by the same prescription of the "induced representation," this time applying it to the two-dimensional κ-space. First, we introduce three Hermitian operators, $\mathbf{\Pi} = (\Pi_1, \Pi_2)$ and $\Theta \equiv J_z$, which represent Lie algebra generators $\boldsymbol{T} = (T_1, T_2)$ and R, respectively.[26] Hence, the small group of κ-translations and κ-rotations is represented in the subspace of the standard momentum $\mathscr{H}_{\boldsymbol{\kappa}}$ by unitary operators $e^{-\frac{i}{\hbar}\Pi_1 x}$, $e^{-\frac{i}{\hbar}\Pi_2 y}$ and $e^{-\frac{i}{\hbar}\Theta\phi}$.

Next, let us clarify the structure of the subspace $\mathscr{H}_{\boldsymbol{\kappa}}$, bearing in mind that this subspace must contain an irreducible representation of the small group. First we assume that the subspace $\mathscr{H}_{\boldsymbol{\kappa}}$ contains the state vector $|\boldsymbol{\pi}\rangle$ with nonzero κ-momentum, $\boldsymbol{\pi} \neq 0$, so

$$\mathbf{\Pi}|\boldsymbol{\pi}\rangle = (\pi_1, \pi_2)|\boldsymbol{\pi}\rangle.$$

Then the rotated vector

$$e^{-\frac{i}{\hbar}\Theta\varphi}|\pi_1, \pi_2\rangle = |\pi_1 \cos\varphi + \pi_2 \sin\varphi, \pi_1 \sin\varphi - \pi_2 \cos\varphi\rangle \tag{5.63}$$

also belongs to the subspace $\mathscr{H}_{\boldsymbol{\kappa}}$. Vectors (5.63) form a circle $\pi_1^2 + \pi_2^2 = \text{const}$ in the plane of κ-momenta (see Figure 5.6). The linear span $\mathscr{H}_{\boldsymbol{\kappa}}$ of all these vectors forms an infinite-dimensional Hilbert space. If we used this representation of the small group to build a unitary irreducible representation of the full Poincaré group, we would obtain massless particles with an infinite number of internal (spin) degrees of freedom or, in other words, with "continuous" spin. Such particles do not exist in nature, so we will not discuss this option further.

---

26 One can notice a formal analogy of the operators $\mathbf{\Pi}$ and Θ with two-dimensional "momentum" and "angular momentum", respectively.

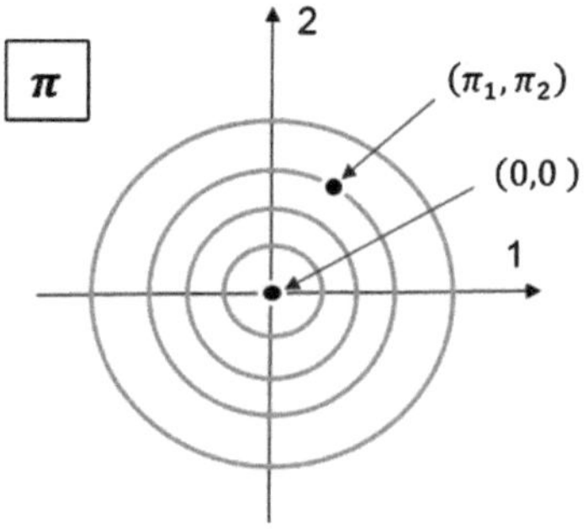

Figure 5.6: Orbits (5.63) of the small group representation in the subspace $\mathscr{H}_\kappa$ of the standard vector $\boldsymbol{\kappa}$.

The only possibility relevant to physics is the circle of "zero radius," $\boldsymbol{\pi} = (0,0)$. It corresponds to a one-dimensional irreducible subspace $\mathscr{H}_\kappa$, where κ-translations are represented trivially,

$$e^{-\frac{i}{\hbar}\boldsymbol{\Pi}\cdot\boldsymbol{r}}|\boldsymbol{\pi} = \mathbf{0}\rangle = |\boldsymbol{\pi} = \mathbf{0}\rangle \tag{5.64}$$

and rotations about the z-axis are represented by unimodular factors, so

$$e^{-\frac{i}{\hbar}\Theta\varphi}|\boldsymbol{\pi} = \mathbf{0}\rangle \equiv e^{-\frac{i}{\hbar}J_z\varphi}|\boldsymbol{\pi} = \mathbf{0}\rangle = e^{i\tau\varphi}|\boldsymbol{\pi} = \mathbf{0}\rangle. \tag{5.65}$$

The allowed values of the parameter τ can be obtained from the fact that our representation must be either single-valued or double-valued (see Statement 3.2). Hence, a rotation over the angle $\varphi = 2\pi$ should be represented by either 1 or −1. Therefore, τ is either an integer or a half-integer

$$\tau = \ldots, -1, -1/2, 0, 1/2, 1, \ldots. \tag{5.66}$$

The parameter τ is called *helicity*. It marks different unitary irreducible massless representations of the Poincaré group, i. e., different possible types of massless elementary particles.

5.4.3 Basis in Hilbert space of massless particle

In the previous subsection we constructed unitary irreducible representations of the small group in the one-dimensional subspace $\mathscr{H}_\kappa$ of the standard momentum $\boldsymbol{\kappa} = (0,0,1)$. In this subsection we will make the next step and construct a basis in the entire Hilbert space (5.60) of a massless particle with a given helicity τ.

Let us choose an arbitrary normalized vector $|\boldsymbol{\kappa}\rangle^\tau$ in the one-dimensional subspace $\mathscr{H}_\kappa$. In the same way as we did in the massive case, we are going to propagate this basis vector to other momentum values $\boldsymbol{p} \neq \boldsymbol{\kappa}$, using transformations from the

Lorentz subgroup. To do this, we need to define canonical elements $\lambda_{\boldsymbol{p}}$ of the Lorentz subgroup that connect the standard vector $\boldsymbol{\kappa}$ with other momenta $\boldsymbol{p}$, written

$$\lambda_{\boldsymbol{p}}\boldsymbol{\kappa} = \boldsymbol{p}.$$

As in the massive case, the choice of such canonical transformations $\lambda_{\boldsymbol{p}}$ is not unique. However, it can be shown that representations constructed using different $\lambda_{\boldsymbol{p}}$ are unitarily equivalent. Hence, we are free to choose a set of $\lambda_{\boldsymbol{p}}$ transformations that is more convenient for computations. Our solution is to define $\lambda_{\boldsymbol{p}}$ as a rotation $\boldsymbol{\varphi}_{\boldsymbol{p}}$, which brings the vector $\boldsymbol{\kappa} = (0, 0, 1)$ to the direction of $\boldsymbol{p}/p$, followed by a Lorentz boost $\boldsymbol{\theta}_{\boldsymbol{p}}$, which takes the momentum $\boldsymbol{p}/p$ to $\boldsymbol{p}$.[27] We have

$$\lambda_{\boldsymbol{p}} = \boldsymbol{\theta}_{\boldsymbol{p}} \circ \boldsymbol{\varphi}_{\boldsymbol{p}}; \tag{5.67}$$

see Figure 5.7.

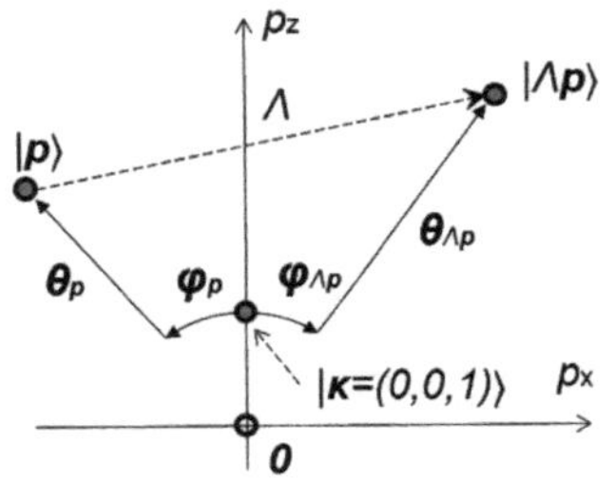

Figure 5.7: Any point $\boldsymbol{p}$ (except $\boldsymbol{p} = \boldsymbol{0}$) in the momentum space of a massless particle can be reached from the standard momentum $\boldsymbol{\kappa} = (0, 0, 1)$ by combined application of a rotation $\boldsymbol{\varphi}_{\boldsymbol{p}}$ followed by a boost $\boldsymbol{\theta}_{\boldsymbol{p}}$.

The complete basis $|\boldsymbol{p}\rangle^\tau$ in $\mathscr{H}$ is now obtained by translating the initial basis vector $|\boldsymbol{\kappa}\rangle^\tau$ to all other fixed momentum subspaces $\mathscr{H}_{\boldsymbol{p}}$, so[28]

$$|\boldsymbol{p}\rangle^\tau \equiv \frac{1}{\sqrt{p}} e^{-\frac{ic}{\hbar}\boldsymbol{K}\cdot\boldsymbol{\theta}_p} e^{-\frac{i}{\hbar}\boldsymbol{J}\cdot\boldsymbol{\varphi}_p} |\boldsymbol{\kappa}\rangle^\tau \equiv \frac{1}{\sqrt{p}} U(\lambda_{\boldsymbol{p}}; \boldsymbol{0}; 0)|\boldsymbol{\kappa}\rangle^\tau. \tag{5.68}$$

5.4.4 Massless representations of Poincaré group

The next step is to see how transformations from the Poincaré group act on the basis vectors (5.68). First, we apply a transformation $U(\Lambda; \boldsymbol{0}; 0)$ from the Lorentz subgroup to an arbitrary basis vector $|\boldsymbol{p}\rangle^\tau$ and obtain

$$U(\Lambda; \boldsymbol{0}; 0)|\boldsymbol{p}\rangle^\tau = \frac{1}{\sqrt{p}} U(\Lambda; \boldsymbol{0}; 0) U(\lambda_{\boldsymbol{p}}; \boldsymbol{0}; 0)|\boldsymbol{\kappa}\rangle^\tau$$

27 Note that the definitions of the boost $\boldsymbol{\theta}_{\boldsymbol{p}}$ for massive and massless particles are different.

28 Compare with equalities (5.5) and (5.29). Here we used the notation $U(\lambda_{\boldsymbol{p}}; \boldsymbol{0}; 0)$ introduced in (3.57) for unitary representatives of Lorentz group elements $\lambda_{\boldsymbol{p}}$.

$$= \frac{1}{\sqrt{p}} U(\lambda_{\Lambda p}; \mathbf{0}; 0) U(\lambda_{\Lambda p}^{-1}; \mathbf{0}; 0) U(\Lambda; \mathbf{0}; 0) U(\lambda_{p}; \mathbf{0}; 0) |\boldsymbol{\kappa}\rangle^{\tau}$$
$$= \frac{1}{\sqrt{p}} U(\lambda_{\Lambda p}; \mathbf{0}; 0) U(\lambda_{\Lambda p}^{-1} \Lambda \lambda_{p}; \mathbf{0}; 0) |\boldsymbol{\kappa}\rangle^{\tau}$$
$$= \frac{1}{\sqrt{p}} U(\lambda_{\Lambda p}; \mathbf{0}; 0) U((\boldsymbol{\varphi}_{\Lambda p}^{-1} \circ \boldsymbol{\theta}_{\Lambda p}^{-1}) \Lambda (\boldsymbol{\theta}_{p} \circ \boldsymbol{\varphi}_{p}); \mathbf{0}; 0) |\boldsymbol{\kappa}\rangle^{\tau}.$$

The product of (unitary representatives of) Lorentz elements

$$\lambda_{\Lambda p}^{-1} \Lambda \lambda_{p} = (\boldsymbol{\varphi}_{\Lambda p}^{-1} \circ \boldsymbol{\theta}_{\Lambda p}^{-1}) \Lambda (\boldsymbol{\theta}_{p} \circ \boldsymbol{\varphi}_{p})$$

on the right-hand side transforms the standard vector $\boldsymbol{\kappa}$ back to the subspace $\mathscr{H}_{\boldsymbol{\kappa}}$ (see Figure 5.7), which means that this product is an element of the small group, composed of κ-translation and κ-rotation factors. The κ-translation part of this element is not important to us, because of equation (5.64). The important part is the angle of κ-rotation about the z-axis. It is called the *Wigner angle* $\varphi_W(\boldsymbol{p}, \Lambda)$.[29] According to (5.65), in the Hilbert space this rotation is represented by a unimodular factor,

$$U(\lambda_{\Lambda p}^{-1} \Lambda \lambda_{p}; \mathbf{0}; 0) |\boldsymbol{\kappa}\rangle^{\tau} = e^{i\tau\varphi_W(\boldsymbol{p},\Lambda)} |\boldsymbol{\kappa}\rangle^{\tau}.$$

Therefore, taking into account (5.68), we obtain the following unitary representation of the Lorentz subgroup:

$$U(\Lambda; \mathbf{0}; 0) |\boldsymbol{p}\rangle^{\tau} = \frac{1}{\sqrt{p}} U(\lambda_{\Lambda p}; \mathbf{0}; 0) e^{i\tau\varphi_W(\boldsymbol{p},\Lambda)} |\boldsymbol{\kappa}\rangle^{\tau} = \sqrt{\frac{|\Lambda \boldsymbol{p}|}{p}} e^{i\tau\varphi_W(\boldsymbol{p},\Lambda)} |\Lambda \boldsymbol{p}\rangle^{\tau}.$$

As usual, in the momentum representation, translations are represented by exponential phase factors. Therefore, for the general element of the Poincaré group, we obtain our final formula,

$$U(\Lambda; \boldsymbol{r}; t) |\boldsymbol{p}\rangle^{\tau} = \sqrt{\frac{|\Lambda \boldsymbol{p}|}{p}} e^{-\frac{i}{\hbar}\boldsymbol{p}\cdot\boldsymbol{r} + \frac{ic}{\hbar}pt} e^{i\tau\varphi_W(\boldsymbol{p},\Lambda)} |\Lambda \boldsymbol{p}\rangle^{\tau}, \tag{5.69}$$

which can be compared with its massive counterparts (5.8), (5.9), (5.10) and (5.30).

As we noted in the beginning of this chapter, massless photons, in fact, are not elementary particles. They are described by a reducible representation of the Poincaré group, which is the direct sum of two irreducible representations with helicities $\tau = 1$ and $\tau = -1$. In the language of classical physics, these two irreducible components correspond to the left and right circularly polarized light.

It is well known [42] that in the photon state space it is impossible to define a position operator with commuting components. In our opinion, this does not in the least complicate our interpretation of the photon as a particle, but simply says that

29 Explicit expressions for the Wigner angle can be found in [73, 14].

a photon cannot be prepared in a strictly localized state. Note that even for massive particles the statement of their localizability requires a nontrivial proof in Theorem 4.1. Therefore, we do not consider pointlike localizability to be an indispensable attribute of a particle.

5.4.5 Doppler effect and aberration

To illustrate results obtained in this chapter, let us derive known formulas for the Doppler effect and the aberration of light. These formulas connect the energies and directions of motion of photons in frames of reference moving relative to each other.

Let $H(0)$ be the energy of the photon and $\boldsymbol{P}(0)$ its momentum in the frame of reference O at rest. Then $H(\theta)$ and $\boldsymbol{P}(\theta)$ are the energy and momentum of the same photon in the frame O' moving with velocity $\boldsymbol{v} = c(\boldsymbol{\theta}/\theta)\tanh\theta$. Using (4.4) and (5.58), we immediately obtain the usual formula for the *Doppler effect*,

$$\begin{aligned} H(\theta) &= H(0)\cosh\theta - c\boldsymbol{P}(0)\cdot\frac{\boldsymbol{\theta}}{\theta}\sinh\theta = H(0)\cosh\theta\left(1 - \frac{cP(0)\boldsymbol{P}(0)}{H(0)P(0)}\cdot\frac{\boldsymbol{\theta}}{\theta}\tanh\theta\right) \\ &= H(0)\cosh\theta\left(1 - \frac{v}{c}\cos\varphi\right), \end{aligned} \tag{5.70}$$

where we denote by φ the angle between the photon's direction (as it is perceived by the observer O) and the direction of O' motion with respect to O, so that

$$\cos\varphi \equiv \frac{\boldsymbol{P}(0)}{P(0)}\cdot\frac{\boldsymbol{\theta}}{\theta}. \tag{5.71}$$

Sometimes the formula for the photon energy shift is written in another form, where the angle φ' between the photon momentum and the direction of motion of the reference frame is measured from the point of view of O'. Then we have

$$\cos\varphi' \equiv \frac{\boldsymbol{P}(\theta)}{P(\theta)}\cdot\frac{\boldsymbol{\theta}}{\theta}. \tag{5.72}$$

From (4.4) we can write

$$\begin{aligned} H(0) &= H(\theta)\cosh\theta + c\boldsymbol{P}(\theta)\cdot\frac{\boldsymbol{\theta}}{\theta}\sinh\theta = H(\theta)\cosh\theta\left(1 + \frac{cP(\theta)\boldsymbol{P}(\theta)}{H(\theta)P(\theta)}\cdot\frac{\boldsymbol{\theta}}{\theta}\tanh\theta\right) \\ &= H(\theta)\cosh\theta\left(1 + \frac{v}{c}\cos\varphi'\right). \end{aligned}$$

Therefore

$$H(\theta) = \frac{H(0)}{\cosh\theta(1 + \frac{v}{c}\cos\varphi')}. \tag{5.73}$$

In these formulas, $\boldsymbol{v}$ is the speed of the observer relative to the light source. More common are situations, in which the observer considers himself stationary and the source is moving. Then simply replace the sign of v in equation (5.73). In addition, when dealing with light, physicists prefer to talk not about the energies of quanta, but about the light frequency that is proportional to the energy ($H = \hbar\omega$). So the frequency of light emitted by a moving source is

$$\omega(\theta) = \frac{\omega(0)}{\cosh\theta(1 - \frac{v}{c}\cos\varphi')}. \tag{5.74}$$

The difference between the angles φ and φ', i. e., the observer dependence of the light propagation direction, is known as *aberration*. To see the same star in the sky, two astronomers O and O' must point their telescopes in different directions. These directions form the angles φ and φ', respectively, with the direction $\boldsymbol{\theta}/\theta$ of the relative velocity of the two observers. The relationship between these angles can be found by taking the scalar product of both sides of (4.3) with $\boldsymbol{\theta}/\theta$ and taking into account equations (5.70)–(5.72) and $cP(\theta) = H(\theta)$. Then we have

$$\cos\varphi' = \frac{P(0)}{P(\theta)}(\cosh\theta\cos\varphi - \sinh\theta) = \frac{\cosh\theta\cos\varphi - \sinh\theta}{\cosh\theta(1 - \frac{v}{c}\cos\varphi)} = \frac{\cos\varphi - v/c}{1 - \frac{v}{c}\cos\varphi}.$$

The Doppler effect will be discussed from a different point of view in Subsection 6.5.3.

6 Interaction

I myself, a professional mathematician, on re-reading my own work find it strains my mental powers to recall to mind from the figures the meanings of the demonstrations, meanings which I myself originally put into the figures and the text from my mind. But when I attempt to remedy the obscurity of the material by putting in extra words, I see myself falling into the opposite fault of becoming chatty in something mathematical.
Johannes Kepler

In the previous chapter, we were interested in isolated elementary particles moving freely in space. Starting with the present chapter, we turn our attention to composite systems consisting of two or more particles. In addition, we allow energy and momentum to be redistributed between different parts of the system. In other words, we assume the presence of *interaction*. In this chapter, our analysis will be limited to systems of several massive spinless particles, whose numbers cannot change. Starting with the second volume, we will remove this restriction and consider interacting systems in complete generality.

6.1 Hilbert space of multiparticle system

In this section we will construct the Hilbert space of a composite system with a fixed number of particles. In textbooks on quantum mechanics it is assumed without explanation that this space has to be constructed as a tensor product of Hilbert spaces of the components. Here we will follow the example of works [54, 2] and show how to derive this statement from postulates of quantum logic. For simplicity, let us start with a two-particle system.

6.1.1 Tensor product theorem

Let $\mathcal{L}_1$, $\mathcal{L}_2$ and $\mathcal{L}_{1+2}$ be quantum-propositional systems (logics) related to the particles 1, 2 and the composite system 1 + 2, respectively. It seems reasonable to assume that every proposition about subsystem 1 (or 2) also has meaning in the composite system. Hence, propositions in $\mathcal{L}_1$ and $\mathcal{L}_2$ should be also represented by propositions in $\mathcal{L}_{1+2}$. Let us formulate this idea as a new postulate.

Postulate 6.1 (properties of compound systems). If $\mathcal{L}_1$ and $\mathcal{L}_2$ are quantum logics describing two physical systems and $\mathcal{L}_{1+2}$ is the quantum logic of the composite system 1 + 2, then there exist two mappings,

$$\mathbb{K}_1 : \mathcal{L}_1 \to \mathcal{L}_{1+2},$$
$$\mathbb{K}_2 : \mathcal{L}_2 \to \mathcal{L}_{1+2},$$

https://doi.org/10.1515/9783110492132-006

satisfying the following conditions:

(I) The mappings $\mathbb{K}_1$, $\mathbb{K}_2$ preserve all logical relations between propositions, so that for $j = 1, 2$

$$\mathbb{K}_j(\emptyset_{\mathcal{L}_j}) = \emptyset_{\mathcal{L}_{1+2}},$$
$$\mathbb{K}_j(\mathcal{I}_{\mathcal{L}_j}) = \mathcal{I}_{\mathcal{L}_{1+2}}$$

and for any two $\mathcal{X}, \mathcal{Y} \in \mathcal{L}_j$

$$\mathcal{X} \leq \mathcal{Y} \quad \Leftrightarrow \quad \mathbb{K}_j(\mathcal{X}) \leq \mathbb{K}_j(\mathcal{Y}),$$
$$\mathbb{K}_j(\mathcal{X} \wedge \mathcal{Y}) = \mathbb{K}_j(\mathcal{X}) \wedge \mathbb{K}_j(\mathcal{Y}),$$
$$\mathbb{K}_j(\mathcal{X} \vee \mathcal{Y}) = \mathbb{K}_j(\mathcal{X}) \vee \mathbb{K}_j(\mathcal{Y}),$$
$$\mathbb{K}_j(\mathcal{X}^{\perp}) = (\mathbb{K}_j(\mathcal{X}))^{\perp}.$$

(II) Measurements on the two subsystems can be performed simultaneously and independently. This means that in the composite system all statements about subsystem 1 are compatible with statements about subsystem 2, so

$$\mathbb{K}_1(\mathcal{X}_1) \leftrightarrow \mathbb{K}_2(\mathcal{X}_2),$$

where $\mathcal{X}_1 \in \mathcal{L}_1$, $\mathcal{X}_2 \in \mathcal{L}_1$.

(III) When we have complete information about subsystems 1 and 2, we also have complete information about the combined system 1+2. This means that if $\mathcal{X}_1 \in \mathcal{L}_1$ and $\mathcal{X}_2 \in \mathcal{L}_2$ are atoms, then the meet of their images $\mathbb{K}_1(\mathcal{X}_1) \wedge \mathbb{K}_2(\mathcal{X}_2)$ is also an atomic proposition in $\mathcal{L}_{1+2}$.

The next theorem [54, 2] allows us to translate the above properties from the language of quantum logic to the more familiar language of Hilbert spaces.

Theorem 6.2 (Matolcsi). *Suppose that $\mathscr{H}_1$, $\mathscr{H}_2$ and $\mathscr{H}_{1+2}$ are three complex Hilbert spaces corresponding to the propositional lattices $\mathcal{L}_1$, $\mathcal{L}_2$ and $\mathcal{L}_{1+2}$ from Postulate 6.1. Suppose, also, that $\mathbb{K}_1$ and $\mathbb{K}_2$ are two mappings whose existence is required by the Postulate. Then the Hilbert space $\mathscr{H}_{1+2}$ of the composite system 1 + 2 is equal to one of the following four possible tensor products[1]: $\mathscr{H}_{1+2} = \mathscr{H}_1 \otimes \mathscr{H}_2$, $\mathscr{H}_{1+2} = \mathscr{H}_1^* \otimes \mathscr{H}_2$, or $\mathscr{H}_{1+2} = \mathscr{H}_1 \otimes \mathscr{H}_2^*$ or $\mathscr{H}_{1+2} = \mathscr{H}_1^* \otimes \mathscr{H}_2^*$.*

The proof of this theorem is beyond the scope of our book.

So, we have four ways to connect two one-particle Hilbert spaces into one two-particle space. In quantum mechanics, only the first possibility is used: $\mathscr{H}_{1+2} = \mathscr{H}_1 \otimes$

1 A definition of the tensor product of two Hilbert spaces can be found in Appendix F.4. The asterisk marks dual Hilbert spaces, introduced in Appendix F.3.

$\mathscr{H}_2$.[2] In particular, this means that if particle 1 is in the state $|1\rangle \in \mathscr{H}_1$ and particle 2 is in the state $|2\rangle \in \mathscr{H}_2$, then the state of the composite system is described by the vector $|1\rangle \otimes |2\rangle \in \mathscr{H}_1 \otimes \mathscr{H}_2$.

6.1.2 Particle observables in multiparticle systems

The functions $\mathbb{K}_1$ and $\mathbb{K}_2$ from Postulate 6.1 map propositions (projections) from Hilbert spaces $\mathscr{H}_1$ and $\mathscr{H}_2$ of individual particles into the Hilbert space $\mathscr{H}_{1+2} = \mathscr{H}_1 \otimes \mathscr{H}_2$ of the composite system. Therefore, they also map observables of particles from $\mathscr{H}_1$ and $\mathscr{H}_2$ to $\mathscr{H}_{1+2}$. For example, consider an observable of particle 1, which is represented in the Hilbert space $\mathscr{H}_1$ by a Hermitian operator with spectral resolution (1.9),

$$F_1 = \sum_f f P_{1f}.$$

According to the proof of the Matolcsi theorem, the mapping $\mathbb{K}_1 : \mathcal{L}_1 \to \mathcal{L}_{1+2}$ has a unitary representative $\Bbbk_1 : \mathscr{H}_1 \to \mathscr{H}_{1+2}$.[3] Then the unitary mapping $\Bbbk_1$ transforms F_1 into the Hermitian operator $\Bbbk_1(F_1)$ in the compound Hilbert space $\mathscr{H}_{1+2}$, so

$$\Bbbk_1(F_1) = \sum_f f \Bbbk_1(P_{1f}).$$

This operator has the same spectrum f as F_1, and $\Bbbk_1(P_{1f})$ are its spectral projections.

Thus, all observables of individual particles have a well-defined meaning also in the Hilbert space $\mathscr{H}_{1+2}$ of the composite system. In the following, we will use lowercase letters to indicate such one-particle observables in $\mathscr{H}_{1+2}$.[4] For example, the images of position and momentum operators of particle 1 in $\mathscr{H}_{1+2}$ will be denoted by $\boldsymbol{r}_1$ and $\boldsymbol{p}_1$, respectively. We will write the energy operator of particle 1 as $h_1 = \sqrt{m_1^2 c^4 + p_1^2 c^2}$, etc. Similarly, observables of particle 2 in $\mathscr{H}_{1+2}$ are denoted by $\boldsymbol{r}_2$, $\boldsymbol{p}_2$ and h_2. In accordance with Postulate 6.1 (II), spectral projections of observables of different particles commute with each other in $\mathscr{H}_{1+2}$. Hence, observables of different particles also commute.

As in the one-particle case considered in Chapter 5, two-particle states can also be described in terms of wave functions. From the properties of the tensor product of Hilbert spaces it follows that if $\psi_1(\boldsymbol{r}_1 s_{1z})$ is a wave function of particle 1 in the position–spin representation and $\psi_2(\boldsymbol{r}_2 s_{2z})$ is a wave function of particle 2, then the wave function of the unified system is expressed by the product

$$\psi(\boldsymbol{r}_1 s_{1z}, \boldsymbol{r}_2 s_{2z}) = \psi_1(\boldsymbol{r}_1 s_{1z}) \psi_2(\boldsymbol{r}_2 s_{2z}). \tag{6.1}$$

2 It is not yet clear what is the physical meaning of the other three options.

3 The relationship between $\mathbb{K}_1$ and $\Bbbk_1$ is the same as that between mappings $\mathbb{K}_g$ and $\Bbbk_g$ in Wigner's Theorem 3.1.

4 We continue to use uppercase letters for total observables $H, \boldsymbol{P}, \boldsymbol{J}, \boldsymbol{R}, \ldots$ of the composite system.

In this case, both particles 1 and 2 and the composite system are in pure quantum states. However, the most general pure two-particle state in $\mathscr{H}_1 \otimes \mathscr{H}_2$ is described by a general function of two variables, $\psi(\boldsymbol{r}_1 s_{1z}, \boldsymbol{r}_2 s_{2z})$, which is not necessarily representable as the product (6.1). In this case, the individual particles are in mixed states: the results of measurements made over particle 1 correlate with the results of particle 2 measurements, even if the particles do not interact with each other. The existence of such *entangled states* is a special feature of quantum mechanics, which has no analog in the classical world.

6.1.3 Statistics

The above construction of the two-particle Hilbert space $\mathscr{H}_{1+2} = \mathscr{H}_1 \otimes \mathscr{H}_2$ is appropriate when the two particles are of different types. If particles 1 and 2 are identical, then in $\mathscr{H}_1 \otimes \mathscr{H}_2$ there are vectors that do not correspond to any physically realizable state and the Hilbert space of two particles is actually "smaller" than $\mathscr{H}_1 \otimes \mathscr{H}_2$. Indeed, if the two particles are identical, then no measurable quantity can change when these particles change places. Therefore, after such a permutation, the wave function can, at most, acquire an innocuous unimodular phase factor β:

$$\psi(\boldsymbol{r}_2 s_{2z}, \boldsymbol{r}_1 s_{1z}) = \beta \psi(\boldsymbol{r}_1 s_{1z}, \boldsymbol{r}_2 s_{2z}). \tag{6.2}$$

If, however, we swap the particles again, we must return to the original wave function

$$\psi(\boldsymbol{r}_1 s_{1z}, \boldsymbol{r}_2 s_{2z}) = \beta \psi(\boldsymbol{r}_2 s_{2z}, \boldsymbol{r}_1 s_{1z}) = \beta^2 \psi(\boldsymbol{r}_1 s_{1z}, \boldsymbol{r}_2 s_{2z}).$$

Thus $\beta^2 = 1$, which implies that the factor β for any physical state $\psi(\boldsymbol{r}_1 s_{1z}, \boldsymbol{r}_2 s_{2z})$ in $\mathscr{H}_{1+2}$ can be either 1 or -1.

Is it possible that one state $\phi_1(\boldsymbol{r}_1 s_{1z}, \boldsymbol{r}_2 s_{2z})$ has the permutation factor β equal to 1, so that

$$\phi_1(\boldsymbol{r}_1 s_{1z}, \boldsymbol{r}_2 s_{2z}) = \phi_1(\boldsymbol{r}_2 s_{2z}, \boldsymbol{r}_1 s_{1z}), \tag{6.3}$$

while another state $\phi_2(\boldsymbol{r}_1 s_{1z}, \boldsymbol{r}_2 s_{2z})$ have this factor equal to -1, so that

$$\phi_2(\boldsymbol{r}_1 s_{1z}, \boldsymbol{r}_2 s_{2z}) = -\phi_2(\boldsymbol{r}_2 s_{2z}, \boldsymbol{r}_1 s_{1z})\,? \tag{6.4}$$

If equations (6.3) and (6.4) were satisfied, then the nontrivial linear combination of states ϕ_1 and ϕ_2

$$\psi(\boldsymbol{r}_1 s_{1z}, \boldsymbol{r}_2 s_{2z}) = a\phi_1(\boldsymbol{r}_1 s_{1z}, \boldsymbol{r}_2 s_{2z}) + b\phi_2(\boldsymbol{r}_1 s_{1z}, \boldsymbol{r}_2 s_{2z})$$

would not have the simple transformation (6.2) with respect to the permutations

$$\begin{aligned}\psi(\boldsymbol{r}_2 s_{2z}, \boldsymbol{r}_1 s_{1z}) &= a\phi_1(\boldsymbol{r}_2 s_{2z}, \boldsymbol{r}_1 s_{1z}) + b\phi_2(\boldsymbol{r}_2 s_{2z}, \boldsymbol{r}_1 s_{1z}) \\ &= a\phi_1(\boldsymbol{r}_1 s_{1z}, \boldsymbol{r}_2 s_{2z}) - b\phi_2(\boldsymbol{r}_1 s_{1z}, \boldsymbol{r}_2 s_{2z}) \neq \pm\psi(\boldsymbol{r}_1 s_{1z}, \boldsymbol{r}_2 s_{2z}).\end{aligned}$$

Such states are unphysical. Hence, the factor β must be the same for all states in the Hilbert space $\mathcal{H}_{1+2}$ of two identical particles. It then follows that all particles in nature fall into two categories: *bosons* and *fermions*.

For bosons $\beta = 1$, and two-particle wave functions are *symmetric* with respect to permutations. Wave functions of the bosons form a "symmetrized" linear subspace $\mathcal{H}_1 \otimes_{sym} \mathcal{H}_2 \subseteq \mathcal{H}_1 \otimes \mathcal{H}_2$. In particular, this means that two identical bosons can occupy the same quantum state, i. e., wave functions like $\psi(\boldsymbol{r}_1 s_{1z})\psi(\boldsymbol{r}_2 s_{2z})$ are permissible and belong to the bosonic subspace $\mathcal{H}_1 \otimes_{sym} \mathcal{H}_2$.

For fermions $\beta = -1$ and two-particle wave functions are *antisymmetric* with respect to permutations of arguments. The Hilbert space of two identical fermions is the subspace of antisymmetric functions $\mathcal{H}_1 \otimes_{asym} \mathcal{H}_2 \subseteq \mathcal{H}_1 \otimes \mathcal{H}_2$. This, in particular, means that two identical fermions cannot occupy the same quantum state.[5] In particular, the wave function $\psi(\boldsymbol{r}_1 s_{1z})\psi(\boldsymbol{r}_2 s_{2z})$ does not belong to the antisymmetric fermionic subspace $\mathcal{H}_1 \otimes_{asym} \mathcal{H}_2$.

All results obtained in this section for two particles can be immediately generalized to n-particle systems where $n > 2$. For example, the Hilbert space of n identical bosons is a symmetrized tensor product $\mathcal{H}_{sym} = \mathcal{H}_1 \otimes_{sym} \mathcal{H}_2 \otimes_{sym} \cdots \otimes_{sym} \mathcal{H}_n$, and the Hilbert space of n identical fermions is an antisymmetrized tensor product $\mathcal{H}_{asym} = \mathcal{H}_1 \otimes_{asym} \mathcal{H}_2 \otimes_{asym} \cdots \otimes_{asym} \mathcal{H}_n$.

A remarkable theorem on the *connection between spin and statistics* was proved in the framework of QFT. It states (in full agreement with the experiment) that all particles with integer spin (for example, photons) are bosons and that all particles with half-integer spin (neutrinos, electrons, protons, etc.) are fermions.

6.2 Relativistic Hamiltonian dynamics

To complete our description of two-particle systems initiated in the previous section, we need to specify a unitary representation U_g of the Poincaré group in the Hilbert space $\mathcal{H}_{1+2} = \mathcal{H}_1 \otimes \mathcal{H}_2$.[6] From Chapter 4 we already know that such a construction is equivalent to finding 10 generators $\{H, \boldsymbol{P}, \boldsymbol{J}, \boldsymbol{K}\}$, which also play the role of total observables in our system. From Subsection 6.1.2 we also know how to find observables $\{\boldsymbol{p}_1, \boldsymbol{r}_1, \boldsymbol{p}_2, \boldsymbol{r}_2\}$ of individual particles in $\mathcal{H}_{1+2}$. It is reasonable to assume that the total observables can be expressed as functions of these one-particle observables. Then the construction of U_g is reduced to finding 10 Hermitian operator functions,

$$H(\boldsymbol{p}_1, \boldsymbol{r}_1; \boldsymbol{p}_2, \boldsymbol{r}_2), \tag{6.5}$$

$$\boldsymbol{P}(\boldsymbol{p}_1, \boldsymbol{r}_1; \boldsymbol{p}_2, \boldsymbol{r}_2), \tag{6.6}$$

5 This property is called the *Pauli exclusion principle*.

6 For simplicity, we assume that particles 1 and 2 are massive, spinless and distinguishable.

$$\boldsymbol{J}(\boldsymbol{p}_1, \boldsymbol{r}_1; \boldsymbol{p}_2, \boldsymbol{r}_2), \tag{6.7}$$

$$\boldsymbol{K}(\boldsymbol{p}_1, \boldsymbol{r}_1; \boldsymbol{p}_2, \boldsymbol{r}_2), \tag{6.8}$$

which satisfy commutation relations (3.49)–(3.55) of the Poincaré Lie algebra.

Even in the simplest two-particle case, this problem has an infinite number of solutions. So, additional physical considerations need to be invoked to find the unique representation (6.5)–(6.8) of the Poincaré Lie algebra that agrees with observations. For multiparticle systems, the construction of the representation U_g, consistent with experiments, is the most difficult part of relativistic quantum theory. It is important to understand that once this task is fulfilled, we get everything we need for a full theoretical description of the physical system. Further calculations of measurable properties become only a technical matter.

6.2.1 Noninteracting representation of Poincaré group

As we already mentioned, there are infinitely many ways to define a representation U_g of the Poincaré group in the Hilbert space of two particles $\mathscr{H}_{1+2} = \mathscr{H}_1 \otimes \mathscr{H}_2$. Let us begin our analysis with one completely legitimate choice, which has a very clear physical meaning. We know from Chapter 5 that one-particle Hilbert spaces $\mathscr{H}_1$ and $\mathscr{H}_2$ carry irreducible unitary representations U_g^1 and U_g^2 of the Poincaré group. The unitary mappings $\mathbb{k}_1$ and $\mathbb{k}_2$, defined in Subsection 6.1.2, allow us to map these representations to the Hilbert space $\mathscr{H}_{1+2}$ of the combined system, i. e., they define two representations $\mathbb{k}_1(U_g^1)$ and $\mathbb{k}_2(U_g^2)$ of the Poincaré group in $\mathscr{H}_{1+2}$. We can also construct a third representation U_g^0 in $\mathscr{H}_{1+2}$ by combining representations $\mathbb{k}_1(U_g^1)$ and $\mathbb{k}_2(U_g^2)$ in a tensor product. In more detail, for any vector of the form $|1\rangle \otimes |2\rangle \in \mathscr{H}_{1+2}$ we define the action of U_g^0 as

$$U_g^0(|1\rangle \otimes |2\rangle) = \mathbb{k}_1(U_g^1)|1\rangle \otimes \mathbb{k}_2(U_g^2)|2\rangle \tag{6.9}$$

and its action on other vectors in $\mathscr{H}_{1+2}$ extends by the principle of linearity. The newly constructed representation (6.9) is called the *tensor product* of unitary representations of U_g^1 and U_g^2 and is written as $U_g^0 = U_g^1 \otimes U_g^2$. Its generators are expressed as sums of single-particle generators:

$$H_0 = h_1 + h_2, \tag{6.10}$$

$$\boldsymbol{P}_0 = \boldsymbol{p}_1 + \boldsymbol{p}_2, \tag{6.11}$$

$$\boldsymbol{J}_0 = \boldsymbol{j}_1 + \boldsymbol{j}_2, \tag{6.12}$$

$$\boldsymbol{K}_0 = \boldsymbol{k}_1 + \boldsymbol{k}_2. \tag{6.13}$$

Poincaré commutators for these generators follow directly from the facts that generators of the same particle have usual Poincaré commutators (3.49)–(3.55) and that generators of different particles commute with each other. For example,

$$[H_0, \boldsymbol{K}_0] = [(h_1 + h_2), (\boldsymbol{k}_1 + \boldsymbol{k}_2)] = [h_1, \boldsymbol{k}_1] + [h_2, \boldsymbol{k}_2] = i\hbar\boldsymbol{p}_1 + i\hbar\boldsymbol{p}_2 = i\hbar\boldsymbol{P}_0.$$

From formulas (6.10)–(6.13) it is not difficult to find transformations of single-particle observables under the action of the group representation U_g^0. For example, positions of particles 1 and 2 change with time independently of each other, as if these particles were alone:

$$\boldsymbol{r}_1(t) = e^{\frac{i}{\hbar}H_0 t}\boldsymbol{r}_1 e^{-\frac{i}{\hbar}H_0 t} = e^{\frac{i}{\hbar}h_1 t}\boldsymbol{r}_1 e^{-\frac{i}{\hbar}h_1 t} = \boldsymbol{r}_1 + \boldsymbol{v}_1 t,$$
$$\boldsymbol{r}_2(t) = \boldsymbol{r}_2 + \boldsymbol{v}_2 t.$$

Therefore, representation (6.9)–(6.13) corresponds to the absence of interaction. It is called the *noninteracting* representation of the Poincaré group. Generators (6.10)–(6.13) of the noninteracting representation will play a special role in our constructions, and we will always label them with the subscript "0."

6.2.2 Dirac's forms of dynamics

Obviously, the simple choice of generators (6.10)–(6.13) is unrealistic, because particles in nature do interact with each other. Therefore, we have to use a certain *interacting representation* U_g of the Poincaré group in $\mathscr{H}_{1+2}$, which differs from U_g^0. First, we write generators $\{H, \boldsymbol{P}, \boldsymbol{J}, \boldsymbol{K}\}$ of this desired representation U_g in the most general form, where all of the generators are different from their noninteracting counterparts by the presence of interaction terms, which we denote V, $\boldsymbol{U}$, $\boldsymbol{Y}$, $\boldsymbol{Z}$. We have[7]

$$H = H_0 + V(\boldsymbol{p}_1, \boldsymbol{r}_1; \boldsymbol{p}_2, \boldsymbol{r}_2), \tag{6.14}$$
$$\boldsymbol{P} = \boldsymbol{P}_0 + \boldsymbol{U}(\boldsymbol{p}_1, \boldsymbol{r}_1; \boldsymbol{p}_2, \boldsymbol{r}_2), \tag{6.15}$$
$$\boldsymbol{J} = \boldsymbol{J}_0 + \boldsymbol{Y}(\boldsymbol{p}_1, \boldsymbol{r}_1; \boldsymbol{p}_2, \boldsymbol{r}_2), \tag{6.16}$$
$$\boldsymbol{K} = \boldsymbol{K}_0 + \boldsymbol{Z}(\boldsymbol{p}_1, \boldsymbol{r}_1; \boldsymbol{p}_2, \boldsymbol{r}_2). \tag{6.17}$$

It may happen that some of the interaction operators on the right-hand sides of equations (6.14)–(6.17) are zero. Then these generators and the corresponding finite transformations coincide with those in the noninteracting representation U_g^0. Such generators and transformations will be called *kinematical*. Generators and transformations containing interaction terms will be called *dynamical*.

Our description of the interaction by formulas (6.14)–(6.17) generalizes traditional classical *nonrelativistic Hamiltonian dynamics*, in which the only dynamical generator is the Hamiltonian H. We want our relativistic theory to reduce to the familiar nonrelativistic approach in the $c \to \infty$ limit. Therefore, we postulate that time translations

7 Our approach to describing interactions in this book is based on equalities (6.14)–(6.17) and their generalizations for many-particle systems and systems with a variable number of particles. This approach is called *relativistic Hamiltonian dynamics* [46]. There exist many other (non-Hamiltonian) approaches for describing interactions. Their surveys and further references can be found in [45, 92, 68].

are generated by a dynamical Hamiltonian, $H = H_0 + V$, with a nontrivial interaction (= potential energy) V. At this point we do not apply any restrictions on the choice of other generators, except Poincaré commutation relations (3.49)–(3.55). This restriction implies that kinematical generators form a subalgebra of the full Lie algebra.[8] Unlike in the Galilei Lie algebra, in the Poincaré algebra the set of generators $\{\boldsymbol{P}, \boldsymbol{J}, \boldsymbol{K}\}$ does not form a subalgebra.[9] This explains why in the relativistic case we have no right to add interaction *only* to the Hamiltonian.

Table 6.1: Comparison of three Dirac forms of relativistic dynamics.

Instant form	**Point form**	**Front form**
Kinematical generators		
P_{0x}	K_{0x}	P_{0x}
P_{0y}	K_{0y}	P_{0y}
P_{0z}	K_{0z}	$\frac{1}{\sqrt{2}}(H_0 + P_{0z})$
J_{0x}	J_{0x}	$\frac{1}{\sqrt{2}}(K_{0x} + J_{0y})$
J_{0y}	J_{0y}	$\frac{1}{\sqrt{2}}(K_{0y} - J_{0x})$
J_{0z}	J_{0z}	J_{0z}
		K_{0z}
Dynamical generators		
H	H	$\frac{1}{\sqrt{2}}(H - P_z)$
K_x	P_x	$\frac{1}{\sqrt{2}}(K_x - J_y)$
K_y	P_y	$\frac{1}{\sqrt{2}}(K_y + J_x)$
K_z	P_z	

So, in order to remain in harmony with the principle of relativity, we must add interaction operators to at least some of the generators $\boldsymbol{P}, \boldsymbol{J}$ or $\boldsymbol{K}$. We shall say that interacting representations having different kinematical subgroups belong to different *forms of dynamics*. In his famous article [23], Dirac developed a classification of interactions based on this principle. In Table 6.1, we present the three Dirac forms of dynamics that are most commonly used in the literature. In the *instant form*, the kinematical subgroup is the subgroup of spatial translations and rotations. In the *point form* [90], the Lorentz subgroup is kinematical. In both these cases, the dimension of the kinematical subgroup is six. The *front form* dynamics has the largest number (seven) of kinematical generators.

---

8 Indeed, if two generators A and B do not contain interaction terms, then their commutator $[A, B]$ must be also kinematical.

9 For example, the commutator (3.54) is outside the subset $\{\boldsymbol{P}, \boldsymbol{J}, \boldsymbol{K}\}$.

6.2.3 Total observables in multiparticle systems

As soon as the interacting representation of the Poincaré group is defined and its generators $\{H, \boldsymbol{P}, \boldsymbol{J}, \boldsymbol{K}\}$ are written out, we immediately get expressions for the *total* observables of the physical system. These are the operators of *total energy* H, *total momentum* $\boldsymbol{P}$ and *total angular momentum* $\boldsymbol{J}$. Other total observables (mass M, spin $\boldsymbol{S}$, center-of-energy position $\boldsymbol{R}$, etc.) are obtained as functions of the basic generators by formulas from Chapter 4.

Our insistence on preserving Poincaré commutators between interacting generators $\{H, \boldsymbol{P}, \boldsymbol{J}, \boldsymbol{K}\}$ guarantees that inertial transformations of the total observables $\{H, \boldsymbol{P}, \boldsymbol{J}, \boldsymbol{K}\}$ coincide with those in Chapter 4, regardless of the form and strength of the interaction. For example, the total energy H and the total momentum $\boldsymbol{P}$ form a 4-vector whose boost transformations *always* have the form (4.3)–(4.4). In any physical system, time translations lead to a uniform motion of the center-of-energy position $\boldsymbol{R}$ with constant velocity along a straight line (4.36). Universal formulas for boosts of the position operator were derived in Subsection 4.3.10.

However, it should also be clear that the presence of interaction may affect inertial transformations of observables of *individual particles* in the compound system. This important point will be analyzed in Section 8.3 of Volume 3.

6.3 Instant form of dynamics

So, which form of dynamics should be chosen to describe the physics of interacting systems? On this issue, there are different opinions, up to the claim that this choice is absolutely unimportant. We do not agree with this point of view. In the third volume of our book, we will see that the instant form of dynamics agrees with observations better than other forms. This is one of our postulates.

Postulate 6.3 (instant form of dynamics). The unitary representation of the Poincaré group acting in the Hilbert space of any interacting physical system belongs to the instant form of dynamics.

6.3.1 General instant-form interaction

In the instant form, equations (6.14)–(6.17) simplify to

$$H = H_0 + V, \tag{6.18}$$

$$\boldsymbol{P} = \boldsymbol{P}_0, \tag{6.19}$$

$$\boldsymbol{J} = \boldsymbol{J}_0, \tag{6.20}$$

$$\boldsymbol{K} = \boldsymbol{K}_0 + \boldsymbol{Z}. \tag{6.21}$$

In particular, in the two-particle system we have

$$H = h_1 + h_2 + V, \tag{6.22}$$

$$\boldsymbol{P} = \boldsymbol{p}_1 + \boldsymbol{p}_2, \tag{6.23}$$

$$\boldsymbol{J} = \boldsymbol{j}_1 + \boldsymbol{j}_2, \tag{6.24}$$

$$\boldsymbol{K} = \boldsymbol{k}_1 + \boldsymbol{k}_2 + \boldsymbol{Z}. \tag{6.25}$$

The interaction operator V in the Hamiltonian H is usually called the *potential energy*. By analogy, the operator $\boldsymbol{Z}$ will be referred to as the *potential boost*.

In accordance with the principle of relativity, ten operators (6.18)–(6.21) must have commutation relations (3.49)–(3.55). They are equivalent to the following set of commutators for V and $\boldsymbol{Z}$:

$$[\boldsymbol{J}_0, V] = [\boldsymbol{P}_0, V] = 0, \tag{6.26}$$

$$[Z_i, P_{0j}] = -\frac{i\hbar\delta_{ij}}{c^2}V, \tag{6.27}$$

$$[J_{0i}, Z_j] = i\hbar\sum_{k=1}^{3}\epsilon_{ijk}Z_k, \tag{6.28}$$

$$[K_{0i}, Z_j] + [Z_i, K_{0j}] + [Z_i, Z_j] = 0, \tag{6.29}$$

$$[\boldsymbol{Z}, H_0] + [\boldsymbol{K}_0, V] + [\boldsymbol{Z}, V] = 0. \tag{6.30}$$

Thus, we reduced the problem of constructing a Poincaré-invariant theory of interacting particles to finding a nontrivial solution (or solutions) for the system of equations (6.26)–(6.30) with respect to the yet unknown operators V and $\boldsymbol{Z}$. These equations are necessary and sufficient conditions for the relativistic invariance of our theory.

6.3.2 Bakamjian–Thomas construction

The system of equations (6.26)–(6.30) is rather complex. We will see later that it has an infinite number of solutions. The first nontrivial group of solutions was obtained by Bakamjian and Thomas [6]. Their idea was as follows. Instead of struggling with ten generators $\{H, \boldsymbol{P}, \boldsymbol{J}, \boldsymbol{K}\}$ and their complex commutators, it is more convenient to use the alternative set of operators $\{M, \boldsymbol{P}, \boldsymbol{R}, \boldsymbol{S}\}$ introduced in Subsection 4.3.5. Let us denote by $\{M_0, \boldsymbol{P}_0, \boldsymbol{R}_0, \boldsymbol{S}_0\}$ and $\{M, \boldsymbol{P}_0, \boldsymbol{R}, \boldsymbol{S}\}$ such alternative sets obtained by formulas (4.43)–(4.45) from the noninteracting $\{H_0, \boldsymbol{P}_0, \boldsymbol{J}_0, \boldsymbol{K}_0\}$ and interacting $\{H, \boldsymbol{P}_0, \boldsymbol{J}_0, \boldsymbol{K}\}$ sets of generators, respectively. In a general instant form dynamics, all three operators $\boldsymbol{R}$, $\boldsymbol{S}$ and M can contain interaction terms. However, Bakamjian and Thomas tried to find a simpler solution in which the center-of-energy position remained kinematical,

$$\boldsymbol{R} \equiv -\frac{c^2}{2}(\boldsymbol{K}H^{-1} + H^{-1}\boldsymbol{K}) - \frac{c[\boldsymbol{P} \times \boldsymbol{W}]}{MH(Mc^2 + H)}$$

$$= \boldsymbol{R}_0 \equiv -\frac{c^2}{2}(\boldsymbol{K}_0 H_0^{-1} + H_0^{-1}\boldsymbol{K}_0) - \frac{c[\boldsymbol{P}_0 \times \boldsymbol{W}_0]}{M_0 H_0 (M_0 c^2 + H_0)}. \tag{6.31}$$

This immediately implied that

$$\boldsymbol{S} = \boldsymbol{J} - [\boldsymbol{R} \times \boldsymbol{P}] = \boldsymbol{J}_0 - [\boldsymbol{R}_0 \times \boldsymbol{P}_0] = \boldsymbol{S}_0$$

(i. e., spin is kinematical as well), and that interaction was present only in the operator of mass,

$$M = M_0 + N. \tag{6.32}$$

The desired operator of "potential mass" N is related to the total Hamiltonian by the following formulas:

$$H = +\sqrt{(M_0 + N)^2 c^4 + P_0^2 c^2}, \tag{6.33}$$

$$N = \frac{1}{c^2}\sqrt{H^2 - P_0^2 c^2} - M_0.$$

From commutators of the Poincaré algebra, we conclude that this operator must meet the following requirements:

$$[\boldsymbol{P}_0, N] = \left[\boldsymbol{P}_0, c^{-2}\sqrt{H^2 - P_0^2 c^2}\right] = 0, \tag{6.34}$$

$$[\boldsymbol{J}_0, N] = \left[\boldsymbol{J}_0, c^{-2}\sqrt{H^2 - P_0^2 c^2}\right] = 0, \tag{6.35}$$

$$[\boldsymbol{R}_0, N] = \left[\boldsymbol{R}_0, c^{-2}\sqrt{H^2 - P_0^2 c^2}\right] = \frac{H[\boldsymbol{R}_0, H] - c^2 \boldsymbol{P}_0}{c^2\sqrt{H^2 - P_0^2 c^2}} = 0. \tag{6.36}$$

Thus, we succeeded in reducing the system of equations (6.26)–(6.30) to the simpler problem of finding a single operator N with properties (6.34)–(6.36). Indeed, knowing this operator and the noninteracting operators $\{M_0, \boldsymbol{P}_0, \boldsymbol{R}_0, \boldsymbol{S}_0\}$, we can recover not only the Hamiltonian (6.33), but also other generators by using formulas (4.47)–(4.48). In particular,

$$\boldsymbol{K} = -\frac{1}{2c^2}(\boldsymbol{R}_0 H + H\boldsymbol{R}_0) - \frac{[\boldsymbol{P}_0 \times \boldsymbol{S}_0]}{Mc^2 + H}, \tag{6.37}$$

$$\boldsymbol{J} = \boldsymbol{J}_0 = [\boldsymbol{R}_0 \times \boldsymbol{P}_0] + \boldsymbol{S}_0. \tag{6.38}$$

6.3.3 Example: two-particle system

Now we are going to construct the operator N in the case of a system consisting of two massive spinless particles. In this case, the Newton–Wigner operator simplifies to[10]

$$\boldsymbol{R}_0 = -\frac{c^2 \boldsymbol{K}_0}{H_0} = \frac{h_1 \boldsymbol{r}_1 + h_2 \boldsymbol{r}_2}{h_1 + h_2},$$

10 Here, for brevity, we ignore the noncommutativity of operators and write, for example, $\boldsymbol{K}_0/H_0$ instead of the more correct $(1/2)(\boldsymbol{K}_0 H_0^{-1} + H_0^{-1}\boldsymbol{K}_0)$.

which justifies our interpretation of this quantity as the *center-of-energy* position. Suppose that we were able to find two vector operators $\{\boldsymbol{\rho}, \boldsymbol{\pi}\}$ such that they form the Heisenberg algebra $\mathfrak{h}_3$,

$$[\rho_i, \pi_j] = i\hbar\delta_{ij}, \tag{6.39}$$

$$[\pi_i, \pi_j] = [\rho_i, \rho_j] = 0, \tag{6.40}$$

commuting with total observables $\{\boldsymbol{R}_0, \boldsymbol{P}_0\}$, so that

$$[\boldsymbol{\pi}, \boldsymbol{P}_0] = [\boldsymbol{\pi}, \boldsymbol{R}_0] = [\boldsymbol{\rho}, \boldsymbol{P}_0] = [\boldsymbol{\rho}, \boldsymbol{R}_0] = 0. \tag{6.41}$$

The existence of the operators $\{\boldsymbol{\rho}, \boldsymbol{\pi}\}$ can be guessed from the following heuristic considerations [63]. In the two-particle Hilbert space we already have representations of two mutually commuting Heisenberg algebras with generators $\{\boldsymbol{r}_1, \boldsymbol{p}_1\}$ and $\{\boldsymbol{r}_2, \boldsymbol{p}_2\}$, respectively. The total observables $\{\boldsymbol{R}_0, \boldsymbol{P}_0\}$ also form the Heisenberg algebra. Then, according to Corollary I.4, there exists a unitary operator $\mathfrak{W}$ that connects the two algebras, so $\mathfrak{W}\{\boldsymbol{r}_1, \boldsymbol{p}_1\}\mathfrak{W}^{-1} = \{\boldsymbol{R}_0, \boldsymbol{P}_0\}$. Acting by the same transformation on $\{\boldsymbol{r}_2, \boldsymbol{p}_2\}$, we can obtain the desired algebra $\{\boldsymbol{\rho}, \boldsymbol{\pi}\} = \mathfrak{W}\{\boldsymbol{r}_2, \boldsymbol{p}_2\}\mathfrak{W}^{-1}$, which satisfies all necessary commutators (6.39)–(6.41).

Note that in the nonrelativistic ($c \to \infty$) limit our total operators reduce to the well-known expressions

$$\boldsymbol{P}_0^{NR} = \boldsymbol{p}_1 + \boldsymbol{p}_2,$$

$$\boldsymbol{R}_0^{NR} = \frac{m_1\boldsymbol{r}_1 + m_2\boldsymbol{r}_2}{m_1 + m_2}.$$

We would like to interpret the new observables as the *relative momentum* $\boldsymbol{\pi}$ and the *relative position* $\boldsymbol{\rho}$ of the two-particle system, so that in the nonrelativistic limit

$$\boldsymbol{\pi}^{NR} = \frac{m_2\boldsymbol{p}_1 - m_1\boldsymbol{p}_2}{m_1 + m_2}, \tag{6.42}$$

$$\boldsymbol{\rho}^{NR} = \boldsymbol{r}_1 - \boldsymbol{r}_2 \tag{6.43}$$

and the required commutators (6.39)–(6.41) are easily verified. For example,

$$[\rho_i^{NR}, \pi_j^{NR}] = \frac{m_2[r_{1i}, p_{1j}] + m_1[r_{2i}, p_{2j}]}{m_1 + m_2} = i\hbar\delta_{ij}.$$

We already said that observables in the Hilbert space $\mathscr{H}_{1+2}$ depend on particle observables $\{\boldsymbol{p}_1, \boldsymbol{r}_1; \boldsymbol{p}_2, \boldsymbol{r}_2\}$. Since this set is unitarily connected with $\{\boldsymbol{P}_0, \boldsymbol{R}_0, \boldsymbol{\pi}, \boldsymbol{\rho}\}$, any observable can be also expressed as a function of $\{\boldsymbol{P}_0, \boldsymbol{R}_0, \boldsymbol{\pi}, \boldsymbol{\rho}\}$. In particular, the interaction operator N, satisfying conditions $[N, \boldsymbol{P}_0] = [N, \boldsymbol{R}_0] = 0$, can depend only on the relative operators $\boldsymbol{\pi}$ and $\boldsymbol{\rho}$. To satisfy the last condition (6.34), $[N, \boldsymbol{J}_0] = 0$, it is sufficient to require that N is an arbitrary function of rotationally invariant combinations, so

$$N = N(\pi^2, \rho^2, (\boldsymbol{\pi} \cdot \boldsymbol{\rho})). \tag{6.44}$$

Thus, in our formulation the problem of constructing a relativistically invariant interaction has been reduced to the search for operators of the relative position $\boldsymbol{\rho}$ and momentum $\boldsymbol{\pi}$ satisfying equations (6.39)–(6.41) and approaching the limits (6.42)–(6.43) as $c \to \infty$. This (rather difficult) mathematical problem was solved in a number of works [6, 8, 63, 29, 17]. For our purposes, explicit expressions for the relative operators are not required, so we will not reproduce them here.

For systems of n massive spinless particles ($n > 2$), similar arguments are applicable, but instead of one pair of relative observables $\boldsymbol{\pi}$ and $\boldsymbol{\rho}$, one has to define $n-1$ such pairs,

$$\boldsymbol{\pi}_r, \boldsymbol{\rho}_r, \quad r = 1, 2, \ldots, n-1. \tag{6.45}$$

These operators must form the Heisenberg algebra $\mathfrak{h}_{3(n-1)}$ commuting with total observables $\boldsymbol{P}_0$ and $\boldsymbol{R}_0$. Once these formulas are found, one can construct relativistic interactions in the n-particle system by the Bakamjian–Thomas method, defining N as a function of the rotationally invariant combinations of relative operators (6.45). Then we have

$$N = N(\pi_1^2, \rho_1^2, (\boldsymbol{\pi}_1 \cdot \boldsymbol{\rho}_1), \pi_2^2, \rho_2^2, (\boldsymbol{\pi}_2 \cdot \boldsymbol{\rho}_2), (\boldsymbol{\pi}_1 \cdot \boldsymbol{\rho}_2), (\boldsymbol{\pi}_2 \cdot \boldsymbol{\rho}_1), \ldots). \tag{6.46}$$

An extension of this formalism to particles with spin is also possible.

6.3.4 Other variants of instant-form dynamics

In the method of Bakamjian and Thomas, it was assumed that $\boldsymbol{R} = \boldsymbol{R}_0$, but this restriction was rather artificial; pretty soon we will see that it is violated in real interactions. So, we need to consider interactions in non-Bakamjian–Thomas forms as well, where $\boldsymbol{R} \neq \boldsymbol{R}_0$. We are going to show that any such representation can be connected by a unitary transformation with a Bakamjian–Thomas one.

Theorem 6.4 (Coester–Polyzou, Theorem 3.4 in [18]). *A representation U_g of the Poincaré group with a strictly positive mass operator M belongs to the instant form of dynamics if and only if there is a unitary operator $\mathfrak{W}$ commuting with $\boldsymbol{P}_0$ and $\boldsymbol{J}_0$ and transforming U_g into some Bakamjian–Thomas representation V_g of the Poincaré group. Then we have*

$$U_g = \mathfrak{W} V_g \mathfrak{W}^{-1}. \tag{6.47}$$

Proof. The direct statement of the theorem is easy to prove. Condition (6.47) implies that U_g is a unitary representation of the Poincaré group, and the property $[\mathfrak{W}, \boldsymbol{P}_0] = [\mathfrak{W}, \boldsymbol{J}_0] = 0$ guarantees the instant form of this representation.

To prove the reverse claim, note that the positivity of the mass operator M guarantees the existence of the Newton–Wigner position operator[11]

$$\boldsymbol{R} = -\frac{c^2}{2}(\boldsymbol{K}H^{-1} + H^{-1}\boldsymbol{K}) - \frac{c[\boldsymbol{P}_0 \times \boldsymbol{W}_0]}{MH(Mc^2 + H)}$$

in the representation U_g. Consequently, U_g generates a reducible representation of the Heisenberg algebra $\{\boldsymbol{R}, \boldsymbol{P}_0\}$. Then, according to Corollary I.4, there is a unitary transformation $\mathfrak{W}^{-1}$ that takes operators $\{\boldsymbol{R}, \boldsymbol{P}_0\}$ to the noninteracting pair $\{\boldsymbol{R}_0, \boldsymbol{P}_0\} = \mathfrak{W}^{-1}\{\boldsymbol{R}, \boldsymbol{P}_0\}\mathfrak{W}$. Obviously, this transformation takes U_g to some kind of Bakamjian–Thomas representation V_g. □

6.4 Cluster separability

As we saw above, relativistic invariance imposes rather weak restrictions on the choice of interactions. For example, in the Bakamjian–Thomas method, the interaction operator (6.46) could be chosen as any function of its arguments. However, not all such functions are acceptable from the physical point of view. In particular, the interaction operator must satisfy additional requirements of *cluster separability*, which will play an important role in our studies.

6.4.1 Definition of cluster separability

From experience, we know that all interactions between particles vanish when the particles are separated by large distances.[12] So, if we remove particle 2 to infinity,[13] then interaction (6.44) should turn to zero, so

$$\lim_{\boldsymbol{a}\to\infty} e^{-\frac{i}{\hbar}\boldsymbol{p}_2\cdot\boldsymbol{a}} N(\pi^2, \rho^2, (\boldsymbol{\pi}\cdot\boldsymbol{\rho})) e^{\frac{i}{\hbar}\boldsymbol{p}_2\cdot\boldsymbol{a}} = 0, \tag{6.48}$$

or, using (6.33),

$$\begin{aligned}
&\lim_{\boldsymbol{a}\to\infty} e^{-\frac{i}{\hbar}\boldsymbol{p}_2\cdot\boldsymbol{a}} V e^{\frac{i}{\hbar}\boldsymbol{p}_2\cdot\boldsymbol{a}} \\
&= \lim_{\boldsymbol{a}\to\infty} e^{-\frac{i}{\hbar}\boldsymbol{p}_2\cdot\boldsymbol{a}} \left(\sqrt{(M_0 + N)^2c^4 + P_0^2c^2} - \sqrt{M_0^2c^4 + P_0^2c^2}\right) e^{\frac{i}{\hbar}\boldsymbol{p}_2\cdot\boldsymbol{a}} = 0.
\end{aligned} \tag{6.49}$$

This condition is not difficult to satisfy in the two-particle case.

---

11 Here we take into account that in the instant form, operators $\boldsymbol{P} = \boldsymbol{P}_0$ and $W \equiv (\boldsymbol{P}\cdot\boldsymbol{J}) = (\boldsymbol{P}_0\cdot\boldsymbol{J}_0) = W_0$ are interaction-free.

12 Here we do not consider hypothetical potentials between quarks, which supposedly grow as linear functions of distance and lead to the "confinement" of quarks inside hadrons.

13 This can be achieved by applying the operator $\exp(\frac{i}{\hbar}\boldsymbol{p}_2 \cdot \boldsymbol{a})$, which performs the following space translation of particle 2: $\exp(\frac{i}{\hbar}\boldsymbol{p}_2 \cdot \boldsymbol{a})\boldsymbol{r}_2 \exp(-\frac{i}{\hbar}\boldsymbol{p}_2 \cdot \boldsymbol{a}) = \boldsymbol{r}_2 + \boldsymbol{a}$.

A cluster-separable n-particle interaction (6.46) should be constructed in such a way that every spatially isolated m-particle group ($m < n$) behaves as if it were alone. This, in particular, means that we cannot independently choose interactions in systems with different numbers of particles. Interaction in the n-particle sector of the theory must be consistent with interactions in all m-particle sectors, where $m < n$.

We will postulate that only cluster-separable interactions can exist in nature.

Postulate 6.5 (cluster separability of interactions). Physical interactions are cluster-separable. This means that for any partitioning of an n-particle system ($n \geq 2$) into two spatially remote groups (or *clusters*), composed of l and m particles ($l + m = n$),
(1) the interaction also separates, i. e., the clusters move independently of each other;
(2) the interaction in each cluster is the same as in separate l-particle and m-particle systems, respectively.

We require the property of cluster separability to be fulfilled for both potential energy and potential boosts. For example, in the relativistic system of three massive particles with interacting generators

$$H = H_0 + V_{123}(\boldsymbol{p}_1, \boldsymbol{r}_1; \boldsymbol{p}_2, \boldsymbol{r}_2; \boldsymbol{p}_3, \boldsymbol{r}_3),$$
$$\boldsymbol{K} = \boldsymbol{K}_0 + \boldsymbol{Z}_{123}(\boldsymbol{p}_1, \boldsymbol{r}_1; \boldsymbol{p}_2, \boldsymbol{r}_2; \boldsymbol{p}_3, \boldsymbol{r}_3),$$

the cluster separability implies, in particular, that

$$\lim_{\boldsymbol{a}\to\infty} e^{\frac{i}{\hbar}\boldsymbol{p}_3\cdot\boldsymbol{a}} V_{123}(\boldsymbol{p}_1, \boldsymbol{r}_1; \boldsymbol{p}_2, \boldsymbol{r}_2; \boldsymbol{p}_3, \boldsymbol{r}_3) e^{-\frac{i}{\hbar}\boldsymbol{p}_3\cdot\boldsymbol{a}} = V_{12}(\boldsymbol{p}_1, \boldsymbol{r}_1; \boldsymbol{p}_2, \boldsymbol{r}_2), \tag{6.50}$$

$$\lim_{\boldsymbol{a}\to\infty} e^{\frac{i}{\hbar}\boldsymbol{p}_3\cdot\boldsymbol{a}} \boldsymbol{Z}_{123}(\boldsymbol{p}_1, \boldsymbol{r}_1; \boldsymbol{p}_2, \boldsymbol{r}_2; \boldsymbol{p}_3, \boldsymbol{r}_3) e^{-\frac{i}{\hbar}\boldsymbol{p}_3\cdot\boldsymbol{a}} = \boldsymbol{Z}_{12}(\boldsymbol{p}_1, \boldsymbol{r}_1; \boldsymbol{p}_2, \boldsymbol{r}_2), \tag{6.51}$$

where V_{12} and $\boldsymbol{Z}_{12}$ are interaction operators in the two-particle (1 + 2) system.

6.4.2 Examples of interaction potentials

Familiar nonrelativistic interactions from classical physics (Coulomb, gravitational, etc.) certainly satisfy the postulate of cluster separability. For example, the nonrelativistic Hamiltonian of the system of two charges is equal to

$$H = h_1 + h_2 + V_{12},$$

where the Coulomb interaction potential

$$V_{12} = \frac{q_1 q_2}{4\pi|\boldsymbol{r}_1 - \boldsymbol{r}_2|} \equiv \frac{q_1 q_2}{4\pi\rho} \tag{6.52}$$

satisfies condition (6.49):

$$\lim_{\boldsymbol{a}\to\infty} e^{\frac{i}{\hbar}\boldsymbol{p}_2\cdot\boldsymbol{a}} V_{12} e^{-\frac{i}{\hbar}\boldsymbol{p}_2\cdot\boldsymbol{a}} = \lim_{\boldsymbol{a}\to\infty} \frac{q_1 q_2}{4\pi|\boldsymbol{r}_1 - \boldsymbol{r}_2 - \boldsymbol{a}|} = 0.$$

In the system of three charged particles shown in Figure 6.1 (a), the potential energy is obtained by simply adding together two-particle terms. We have

$$V_{123} = V_{12} + V_{13} + V_{23} = \frac{q_1 q_2}{4\pi|\boldsymbol{r}_1 - \boldsymbol{r}_2|} + \frac{q_2 q_3}{4\pi|\boldsymbol{r}_2 - \boldsymbol{r}_3|} + \frac{q_1 q_3}{4\pi|\boldsymbol{r}_1 - \boldsymbol{r}_3|}. \tag{6.53}$$

The distance between particle 3 and the cluster 1 + 2 can be increased by applying spatial translation to particle 3, as shown in Figure 6.1 (b). In accordance with Postulate 6.5, this translation effectively destroys the interaction between clusters 3 and 1 + 2, i. e.,

$$\begin{aligned}
&\lim_{\boldsymbol{a}\to\infty} e^{\frac{i}{\hbar}\boldsymbol{p}_3\cdot\boldsymbol{a}}(V_{12} + V_{13} + V_{23})e^{-\frac{i}{\hbar}\boldsymbol{p}_3\cdot\boldsymbol{a}} \\
&\quad = \lim_{\boldsymbol{a}\to\infty}\left(\frac{q_1 q_2}{4\pi|\boldsymbol{r}_1 - \boldsymbol{r}_2|} + \frac{q_2 q_3}{4\pi|\boldsymbol{r}_2 - \boldsymbol{r}_3 - \boldsymbol{a}|} + \frac{q_1 q_3}{4\pi|\boldsymbol{r}_1 - \boldsymbol{r}_3 - \boldsymbol{a}|}\right) \\
&\quad = \frac{q_1 q_2}{4\pi|\boldsymbol{r}_1 - \boldsymbol{r}_2|} = V_{12}.
\end{aligned}$$

As expected, this is the same potential (6.52) as in the isolated two-particle system.

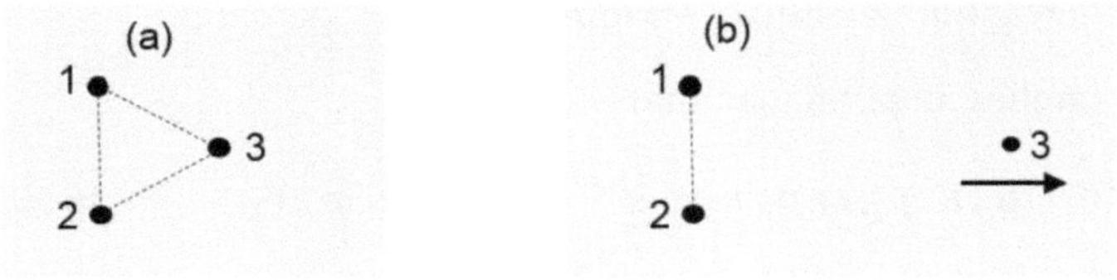

Figure 6.1: Cluster-separable interaction in a system of three particles. (a) Three particles interact at close distances. (b) When particle 3 is removed to infinity, its interaction with 1 + 2 vanishes, and the dynamics in 1 + 2 acquires a form characteristic of a two-particle system.

In the nonrelativistic theory, the potential boost vanishes, $\boldsymbol{Z} = \boldsymbol{0}$, so the separability condition for boosts is trivial. Thus, both conditions (1) and (2) of Postulate 6.5 are satisfied, and interaction (6.53) is cluster-separable. As we will see in Subsections 6.4.4–6.4.6, the construction of a relativistic cluster-separable interaction is a much more difficult task.

A simple counterexample of a nonseparable interaction can be constructed in the case of four particles. The interaction Hamiltonian

$$V \propto \frac{1}{|\boldsymbol{r}_1 - \boldsymbol{r}_2||\boldsymbol{r}_3 - \boldsymbol{r}_4|} \tag{6.54}$$

is such that no matter how far apart the pairs of particles (1 + 2 and 3 + 4) are, the distance in the pair 3 + 4 affects the force acting between particles 1 and 2 and vice versa. Such interactions with an unlimited range of action are unknown in nature.

6.4.3 Smooth potentials

Let us now introduce several definitions that will be useful in our discussions of cluster separability. A *smooth m*-particle potential $V^{(m)}$ is an operator that depends on dynamical variables[14] of m particles and tends to zero when any subgroup of these particles is removed to infinity.[15] For example, potential (6.52) is smooth, while (6.54) is not.

In the general case, a cluster-separable interaction in an n-particle system can be written as the sum

$$V = \sum_{\{2\}} V^{(2)} + \sum_{\{3\}} V^{(3)} + \cdots + V^{(n)}, \tag{6.55}$$

where $\sum_{\{2\}} V^{(2)}$ is the sum of smooth *two-particle potentials* over all pairs of particles; $\sum_{\{3\}} V^{(3)}$ is the sum of smooth *three-particle potentials* over all triplets of particles, etc. The example in equation (6.53) is the sum of smooth two-particle potentials $V^{(2)}$. So, in this case, we just set the smooth three-particle part of the potential equal to zero, so $V^{(3)} = 0$.

6.4.4 Nonseparability of Bakamjian–Thomas dynamics

Is it possible to observe the principles of cluster separability in the Bakamjian–Thomas relativistic theory? As we know, in this case the interaction is added only to the mass operator (6.32), and the potential energy of a three-particle system takes the form

$$\begin{aligned} V_{123} = H - H_0 &= \sqrt{(\boldsymbol{p}_1 + \boldsymbol{p}_2 + \boldsymbol{p}_3)^2 c^2 + (M_0 + N(\boldsymbol{p}_1, \boldsymbol{r}_1; \boldsymbol{p}_2, \boldsymbol{r}_2; \boldsymbol{p}_3, \boldsymbol{r}_3))^2 c^4} \\ &\quad - \sqrt{(\boldsymbol{p}_1 + \boldsymbol{p}_2 + \boldsymbol{p}_3)^2 c^2 + M_0^2 c^4}. \end{aligned}$$

Removing particle 3 to infinity, we obtain

$$\begin{aligned} &\lim_{\boldsymbol{a}\to\infty} e^{\frac{i}{\hbar}\boldsymbol{p}_3\cdot\boldsymbol{a}} V_{123}(\boldsymbol{p}_1, \boldsymbol{r}_1; \boldsymbol{p}_2, \boldsymbol{r}_2; \boldsymbol{p}_3, \boldsymbol{r}_3) e^{-\frac{i}{\hbar}\boldsymbol{p}_3\cdot\boldsymbol{a}} \\ &\quad = \sqrt{(\boldsymbol{p}_1 + \boldsymbol{p}_2 + \boldsymbol{p}_3)^2 c^2 + (M_0 + N(\boldsymbol{p}_1, \boldsymbol{r}_1; \boldsymbol{p}_2, \boldsymbol{r}_2; \boldsymbol{p}_3, \infty))^2 c^4} \\ &\quad\quad - \sqrt{(\boldsymbol{p}_1 + \boldsymbol{p}_2 + \boldsymbol{p}_3)^2 c^2 + M_0^2 c^4}. \end{aligned}$$

In accordance with the condition (6.50), the right-hand side should depend only on observables of particles 1 and 2. However, this is only possible in the case when

$$N(\boldsymbol{p}_1, \boldsymbol{r}_1; \boldsymbol{p}_2, \boldsymbol{r}_2; \boldsymbol{p}_3, \infty) = 0,$$

14 Positions, momenta and spins.

15 In Section 2.4 of Volume 2, it will become clear why we call such potentials smooth.

i. e., when

$$\lim_{\boldsymbol{a}\to\infty} e^{\frac{i}{\hbar}\boldsymbol{p}_3\cdot\boldsymbol{a}} V_{123}(\boldsymbol{p}_1,\boldsymbol{r}_1;\boldsymbol{p}_2,\boldsymbol{r}_2;\boldsymbol{p}_3,\boldsymbol{r}_3) e^{-\frac{i}{\hbar}\boldsymbol{p}_3\cdot\boldsymbol{a}} = V_{12}(\boldsymbol{p}_1,\boldsymbol{r}_1;\boldsymbol{p}_2,\boldsymbol{r}_2) = 0,$$

and the interaction in the two-particle sector $1+2$ vanishes. Similarly, it can be shown that interaction V_{123} should tend to zero, when either particle 1 or particle 2 is removed to infinity. Hence, we conclude that V_{123} is a smooth three-particle potential. This means that V_{123} is different from zero only when three or more particles are close together, i. e., any interaction in two-particle subsystems is excluded, which is clearly unphysical. Therefore, the Bakamjian–Thomas method cannot be used to construct a nontrivial cluster-separable theory. The rigorous proof of this statement can be found in [59].

6.4.5 Cluster-separable three-particle interaction

So, to build a relativistic cluster-separated theory, it is necessary to use non-Bakamjian–Thomas interactions in the instant form of dynamics. Our goal in this and the following subsection is to present a nontrivial example of such a theory. In particular, we are going to specify the energy and boost operators

$$H = H_0 + V_{123},$$
$$\boldsymbol{K} = \boldsymbol{K}_0 + \boldsymbol{Z}_{123}$$

in the Hilbert space $\mathscr{H} = \mathscr{H}_1 \otimes \mathscr{H}_2 \otimes \mathscr{H}_3$ of a three-particle system.[16] Here, V_{123} and $\boldsymbol{Z}_{123}$ are interaction operators, which we have to define in accordance with the postulate of separability, Postulate 6.5. In this construction, we will follow the work of [18]; see also [60].

Let us assume that we know two-particle potentials V_{ij} and $\boldsymbol{Z}_{ij}$ $(i,j = 1,2,3,\ i \neq j)$, which are obtained by removing the third particle $k \neq i,j$ to infinity. For example, if particle 3 is removed to infinity, the 3-particle operators take the form[17]

$$\lim_{\boldsymbol{a}\to\infty} e^{\frac{i}{\hbar}\boldsymbol{p}_3\cdot\boldsymbol{a}} H e^{-\frac{i}{\hbar}\boldsymbol{p}_3\cdot\boldsymbol{a}} = H_0 + V_{12} \equiv H_{12}, \tag{6.56}$$

$$\lim_{\boldsymbol{a}\to\infty} e^{\frac{i}{\hbar}\boldsymbol{p}_3\cdot\boldsymbol{a}} \boldsymbol{K} e^{-\frac{i}{\hbar}\boldsymbol{p}_3\cdot\boldsymbol{a}} = \boldsymbol{K}_0 + \boldsymbol{Z}_{12} \equiv \boldsymbol{K}_{12}, \tag{6.57}$$

$$\lim_{\boldsymbol{a}\to\infty} e^{\frac{i}{\hbar}\boldsymbol{p}_3\cdot\boldsymbol{a}} M e^{-\frac{i}{\hbar}\boldsymbol{p}_3\cdot\boldsymbol{a}} = \frac{1}{c^2}\sqrt{H_{12}^2 - P_0^2 c^2} \equiv M_{12}, \tag{6.58}$$

16 For simplicity we choose these particles to be massive and distinguishable.

17 Similar equalities are valid when particles 1 or 2 are removed to infinity. They are obtained from (6.56)–(6.59) by cyclic permutation of the indices $(1,2,3)$.

$$\lim_{\boldsymbol{a}\to\infty} e^{\frac{i}{\hbar}\boldsymbol{p}_3\cdot\boldsymbol{a}}\boldsymbol{R}e^{-\frac{i}{\hbar}\boldsymbol{p}_3\cdot\boldsymbol{a}} = -\frac{c^2}{2}(\boldsymbol{K}_{12}H_{12}^{-1} + H_{12}^{-1}\boldsymbol{K}_{12}) - \frac{c[\boldsymbol{P}_0 \times \boldsymbol{W}_{12}]}{M_{12}H_{12}(M_{12}c^2 + H_{12})}$$
$$\equiv \boldsymbol{R}_{12}, \qquad (6.59)$$

where the interacting operators of energy H_{12}, boost $\boldsymbol{K}_{12}$, mass M_{12} and the center-of-energy position $\boldsymbol{R}_{12}$ are assumed to be given in our construction.[18]

According to the cluster separability, when particles 1 and 2 are moved away from each other, the operators V_{12} and $\boldsymbol{Z}_{12}$ should tend to zero. Therefore

$$\lim_{\boldsymbol{a}\to\infty} e^{\frac{i}{\hbar}\boldsymbol{p}_1\cdot\boldsymbol{a}}M_{12}e^{-\frac{i}{\hbar}\boldsymbol{p}_1\cdot\boldsymbol{a}} = M_0, \qquad (6.60)$$

$$\lim_{\boldsymbol{a}\to\infty} e^{\frac{i}{\hbar}\boldsymbol{p}_2\cdot\boldsymbol{a}}M_{12}e^{-\frac{i}{\hbar}\boldsymbol{p}_2\cdot\boldsymbol{a}} = M_0, \qquad (6.61)$$

$$\lim_{\boldsymbol{a}\to\infty} e^{\frac{i}{\hbar}\boldsymbol{p}_3\cdot\boldsymbol{a}}M_{12}e^{-\frac{i}{\hbar}\boldsymbol{p}_3\cdot\boldsymbol{a}} = M_{12}. \qquad (6.62)$$

Now we would like to combine the given two-particle potentials V_{ij} and $\boldsymbol{Z}_{ij}$ in order to obtain a cluster-separable three-particle interaction $(V_{123}, \boldsymbol{Z}_{123})$. By a naïve analogy with the nonrelativistic formula (6.53), we could have tried to define these interactions as simple sums of two-particle potentials. However, as we shall see below, in the relativistic case this would not work: such simple sums violate Poincaré commutators. Therefore, we come to the conclusion that

$$V_{123} \neq V_{12} + V_{23} + V_{13},$$
$$\boldsymbol{Z}_{123} \neq \boldsymbol{Z}_{12} + \boldsymbol{Z}_{23} + \boldsymbol{Z}_{13}$$

and the relativistic *addition of interactions* should be done in a more complex way.

6.4.6 Relativistic addition of interactions

Operators $\{H_{12}, \boldsymbol{K}_{12}, M_{12}, \boldsymbol{R}_{12}\}$, see (6.56)–(6.59), define an instant-form representation U_g^{12} of the Poincaré group in the three-particle Hilbert space $\mathscr{H}$. The corresponding position operator $\boldsymbol{R}_{12}$ is generally different from the noninteracting Newton–Wigner operator $\boldsymbol{R}_0$ (6.31) characteristic of the Bakamjian–Thomas dynamics. However, according to Theorem 6.4, we can unitarily transform the representation U_g^{12} in such a way that it takes the Bakamjian–Thomas form with operators $\{\overline{H}_{12}, \overline{\boldsymbol{K}}_{12}, \overline{M}_{12}, \boldsymbol{R}_0\}$. We denote the unitary operator of this transformation by $\mathfrak{W}_{12}$. The same considerations can be repeated for the other two pairs of particles (1 + 3) and (2 + 3), so we can write for $i, j = 1, 2, 3$, $i \neq j$,

$$\mathfrak{W}_{ij}H_{ij}\mathfrak{W}_{ij}^{-1} = \overline{H}_{ij},$$

18 Of course, these operators must depend only on interaction between the 1st and 2nd particle. The Pauli–Lubanski operator $\boldsymbol{W}_{12}$ is a function (4.14) of $H_{12}, \boldsymbol{K}_{12}$ and noninteracting operators $\boldsymbol{P}_{12} = \boldsymbol{p}_1 + \boldsymbol{p}_2$, $\boldsymbol{J}_{12} = \boldsymbol{j}_1 + \boldsymbol{j}_2$.

$$\mathfrak{W}_{ij}\boldsymbol{K}_{ij}\mathfrak{W}_{ij}^{-1} = \overline{\boldsymbol{K}}_{ij},$$
$$\mathfrak{W}_{ij}M_{ij}\mathfrak{W}_{ij}^{-1} = \overline{M}_{ij},$$
$$\mathfrak{W}_{ij}\boldsymbol{R}_{ij}\mathfrak{W}_{ij}^{-1} = \boldsymbol{R}_0,$$

where operators $\mathfrak{W}_{ij} = \{\mathfrak{W}_{12}, \mathfrak{W}_{13}, \mathfrak{W}_{23}\}$ commute with $\boldsymbol{P}_0$ и $\boldsymbol{J}_0$. Representations U_g^{ij} must become noninteracting when the distance between particles i and j tends to infinity. Therefore, for example,

$$\lim_{\boldsymbol{a}\to\infty} e^{\frac{i}{\hbar}\boldsymbol{p}_3\cdot\boldsymbol{a}}\mathfrak{W}_{13}e^{-\frac{i}{\hbar}\boldsymbol{p}_3\cdot\boldsymbol{a}} = 1, \tag{6.63}$$

$$\lim_{\boldsymbol{a}\to\infty} e^{\frac{i}{\hbar}\boldsymbol{p}_3\cdot\boldsymbol{a}}\mathfrak{W}_{23}e^{-\frac{i}{\hbar}\boldsymbol{p}_3\cdot\boldsymbol{a}} = 1, \tag{6.64}$$

$$\lim_{\boldsymbol{a}\to\infty} e^{\frac{i}{\hbar}\boldsymbol{p}_3\cdot\boldsymbol{a}}\mathfrak{W}_{12}e^{-\frac{i}{\hbar}\boldsymbol{p}_3\cdot\boldsymbol{a}} = \mathfrak{W}_{12}. \tag{6.65}$$

According to Bakamjian and Thomas, the mass operators $\overline{M}_{ij}$ commute with $\boldsymbol{R}_0$. By adding $\overline{M}_{ij}$ it is not difficult to construct a new mass operator, where interactions are present symmetrically for all pairs, so

$$\begin{aligned}\overline{M} &\equiv \overline{M}_{12} + \overline{M}_{13} + \overline{M}_{23} - 2M_0 \\ &= \mathfrak{W}_{12}M_{12}\mathfrak{W}_{12}^{-1} + \mathfrak{W}_{13}M_{13}\mathfrak{W}_{13}^{-1} + \mathfrak{W}_{23}M_{23}\mathfrak{W}_{23}^{-1} - 2M_0.\end{aligned}$$

By construction, this operator also commutes with $\boldsymbol{R}_0$. Hence it can be used to formulate a new Bakamjian–Thomas representation, $\{\overline{H}, \overline{\boldsymbol{K}}, \overline{M}, \boldsymbol{R}_0\}$, where by formulas (4.46) and (4.48)

$$\overline{H} = \sqrt{P_0^2c^2 + \overline{M}^2c^4}, \tag{6.66}$$

$$\overline{\boldsymbol{K}} = -\frac{1}{2c^2}(\boldsymbol{R}_0\overline{H} + \overline{H}\boldsymbol{R}_0) - \frac{[\boldsymbol{P}_0 \times \boldsymbol{S}_0]}{\overline{M}c^2 + \overline{H}} \tag{6.67}$$

and all three particles interact with each other. As we already know from Subsection 6.4.4, in this representation the property of cluster separability does not hold. For example, by removing particle 3 to infinity, we do not obtain the mass operator M_{12}, which is characteristic of the two-particle subsystem $1+2$. Instead, we get an operator that differs from M_{12} by a unitary transformation,[19]

$$\begin{aligned}&\lim_{\boldsymbol{a}\to\infty} e^{\frac{i}{\hbar}\boldsymbol{p}_3\cdot\boldsymbol{a}}\overline{M}e^{-\frac{i}{\hbar}\boldsymbol{p}_3\cdot\boldsymbol{a}} \\ &= \lim_{\boldsymbol{a}\to\infty} e^{\frac{i}{\hbar}\boldsymbol{p}_3\cdot\boldsymbol{a}}(\mathfrak{W}_{12}M_{12}\mathfrak{W}_{12}^{-1} + \mathfrak{W}_{13}M_{13}\mathfrak{W}_{13}^{-1} + \mathfrak{W}_{23}M_{23}\mathfrak{W}_{23}^{-1} - 2M_0)e^{-\frac{i}{\hbar}\boldsymbol{p}_3\cdot\boldsymbol{a}} \\ &= \mathfrak{W}_{12}M_{12}\mathfrak{W}_{12}^{-1} - 2M_0 + \lim_{\boldsymbol{a}\to\infty}(e^{\frac{i}{\hbar}\boldsymbol{p}_3\cdot\boldsymbol{a}}M_{13}e^{-\frac{i}{\hbar}\boldsymbol{p}_3\cdot\boldsymbol{a}} + e^{\frac{i}{\hbar}\boldsymbol{p}_3\cdot\boldsymbol{a}}M_{23}e^{-\frac{i}{\hbar}\boldsymbol{p}_3\cdot\boldsymbol{a}}) \\ &= \mathfrak{W}_{12}M_{12}\mathfrak{W}_{12}^{-1} - 2M_0 + 2M_0 = \mathfrak{W}_{12}M_{12}\mathfrak{W}_{12}^{-1}.\end{aligned} \tag{6.68}$$

19 Here we apply formulas like (6.60)–(6.62) and (6.63)–(6.65).

In order to achieve cluster separability, let us unitarily transform the representation $\{\overline{H}, \overline{\boldsymbol{K}}, \overline{M}, \boldsymbol{R}_0,\}$ into a new representation $\{H, \boldsymbol{K}, M, \boldsymbol{R}\}$, i. e.,

$$H = \mathfrak{W}^{-1}\overline{H}\mathfrak{W}, \tag{6.69}$$

$$\boldsymbol{K} = \mathfrak{W}^{-1}\overline{\boldsymbol{K}}\mathfrak{W}, \tag{6.70}$$

$$M = \mathfrak{W}^{-1}\overline{M}\mathfrak{W}, \tag{6.71}$$

$$\boldsymbol{R} = \mathfrak{W}^{-1}\boldsymbol{R}_0\mathfrak{W} \tag{6.72}$$

by choosing a unitary operator $\mathfrak{W}$, which removes the factors $\mathfrak{W}_{ij}$ and $\mathfrak{W}_{ij}^{-1}$ on the right-hand sides of equalities like (6.68). For the transformation $\mathfrak{W}$, we can take any unitary operator with the following limits[20]:

$$\lim_{\boldsymbol{a}\to\infty} e^{\frac{i}{\hbar}\boldsymbol{p}_3\cdot\boldsymbol{a}}\mathfrak{W}e^{-\frac{i}{\hbar}\boldsymbol{p}_3\cdot\boldsymbol{a}} = \mathfrak{W}_{12}, \tag{6.73}$$

$$\lim_{\boldsymbol{a}\to\infty} e^{\frac{i}{\hbar}\boldsymbol{p}_2\cdot\boldsymbol{a}}\mathfrak{W}e^{-\frac{i}{\hbar}\boldsymbol{p}_2\cdot\boldsymbol{a}} = \mathfrak{W}_{13}, \tag{6.74}$$

$$\lim_{\boldsymbol{a}\to\infty} e^{\frac{i}{\hbar}\boldsymbol{p}_1\cdot\boldsymbol{a}}\mathfrak{W}e^{-\frac{i}{\hbar}\boldsymbol{p}_1\cdot\boldsymbol{a}} = \mathfrak{W}_{23}. \tag{6.75}$$

It is easy to verify that one valid choice is

$$\mathfrak{W} = \exp(\ln\mathfrak{W}_{12} + \ln\mathfrak{W}_{13} + \ln\mathfrak{W}_{23}).$$

Indeed, using equations (6.63)–(6.65), with this choice we get

$$\begin{aligned}
&\lim_{\boldsymbol{a}\to\infty} e^{\frac{i}{\hbar}\boldsymbol{p}_3\cdot\boldsymbol{a}}\mathfrak{W}e^{-\frac{i}{\hbar}\boldsymbol{p}_3\cdot\boldsymbol{a}} \\
&\quad = \lim_{\boldsymbol{a}\to\infty} e^{\frac{i}{\hbar}\boldsymbol{p}_3\cdot\boldsymbol{a}}\exp(\ln\mathfrak{W}_{12} + \ln\mathfrak{W}_{13} + \ln\mathfrak{W}_{23})e^{-\frac{i}{\hbar}\boldsymbol{p}_3\cdot\boldsymbol{a}} = \exp(\ln\mathfrak{W}_{12}) \\
&\quad = \mathfrak{W}_{12}.
\end{aligned}$$

Then one can show that the interacting representation of the Poincaré group, defined by the operators $\{H, \boldsymbol{K}, M, \boldsymbol{R}\}$, satisfies all conditions of cluster separability (6.56)–(6.59). For example,

$$\begin{aligned}
&\lim_{\boldsymbol{a}\to\infty} e^{\frac{i}{\hbar}\boldsymbol{p}_3\cdot\boldsymbol{a}}He^{-\frac{i}{\hbar}\boldsymbol{p}_3\cdot\boldsymbol{a}} \\
&\quad = \lim_{\boldsymbol{a}\to\infty} e^{\frac{i}{\hbar}\boldsymbol{p}_3\cdot\boldsymbol{a}}\mathfrak{W}^{-1}\overline{H}\mathfrak{W}e^{-\frac{i}{\hbar}\boldsymbol{p}_3\cdot\boldsymbol{a}} = \lim_{\boldsymbol{a}\to\infty} \mathfrak{W}_{12}^{-1}e^{\frac{i}{\hbar}\boldsymbol{p}_3\cdot\boldsymbol{a}}\sqrt{P_0^2c^2 + \overline{M}^2c^4}e^{-\frac{i}{\hbar}\boldsymbol{p}_3\cdot\boldsymbol{a}}\mathfrak{W}_{12} \\
&\quad = \mathfrak{W}_{12}^{-1}\sqrt{P_0^2c^2 + (\mathfrak{W}_{12}M_{12}\mathfrak{W}_{12}^{-1})^2c^4}\mathfrak{W}_{12} = \sqrt{P_0^2c^2 + M_{12}^2c^4} = H_{12}.
\end{aligned}$$

---

20 In addition, this operator must commute with $\boldsymbol{P}_0$ and $\boldsymbol{J}_0$ in order to preserve our chosen instant form of dynamics.

Thus, the interaction (6.69)–(6.72) solves the problem of the relativistic addition of two-particle potentials V_{ij} and $\boldsymbol{Z}_{ij}$, formulated in Subsection 6.4.5. The corresponding representation does not belong to the Bakamjian–Thomas form.

Obviously, the above method is extremely cumbersome. Moreover, it is not at all obvious how to apply this construction to realistic systems, where the number of particles can change. In the second volume, we will consider another approach for constructing separable relativistic interactions, which turns out to be more practical. It will be based on the idea of quantum fields.

6.5 Bound states and time evolution

We have already mentioned that the knowledge of the representation U_g of the Poincaré group in the Hilbert space $\mathcal{H}$ of a multiparticle system is sufficient to obtain any desired physical information about this system. Here we are going to illustrate this statement by considering two types of data comparable to experiments: the mass/energy spectrum and the time evolution of observables. In Chapter 7, we will discuss scattering experiments, which are currently the most informative approach to studying microscopic systems.

6.5.1 Spectra of mass and energy operators

The mass operator in a noninteracting two-particle system can be expressed through particle observables as follows:

$$\begin{aligned} M_0 &= +\frac{1}{c^2}\sqrt{H_0^2 - P_0^2 c^2} = +\frac{1}{c^2}\sqrt{(h_1 + h_2)^2 - (\boldsymbol{p}_1 + \boldsymbol{p}_2)^2 c^2} \\ &= +\frac{1}{c^2}\sqrt{\left(\sqrt{m_1^2 c^4 + p_1^2 c^2} + \sqrt{m_2^2 c^4 + p_2^2 c^2}\right)^2 - (\boldsymbol{p}_1 + \boldsymbol{p}_2)^2 c^2}. \end{aligned} \tag{6.76}$$

The particle momenta $\boldsymbol{p}_1$ and $\boldsymbol{p}_2$ are allowed to take any values in the 3D momentum space, thus eigenvalues μ_n of the mass operator form a continuous spectrum in the interval

$$m_1 + m_2 \le \mu_n < \infty, \tag{6.77}$$

where the lower limit $m_1 + m_2$ is obtained from (6.76), when both particles are at rest, so $\boldsymbol{p}_1 = \boldsymbol{p}_2 = 0$. It then follows that the common spectrum of the mutually commuting operators $\boldsymbol{P}_0$ and $H_0 = +\sqrt{M_0^2 c^4 + P_0^2 c^2}$ is a union of mass hyperboloids[21] in the 4D energy–momentum space. This spectrum is shown by the hatched area in Figure 6.2 (a).

21 With masses μ_n from the interval (6.77).

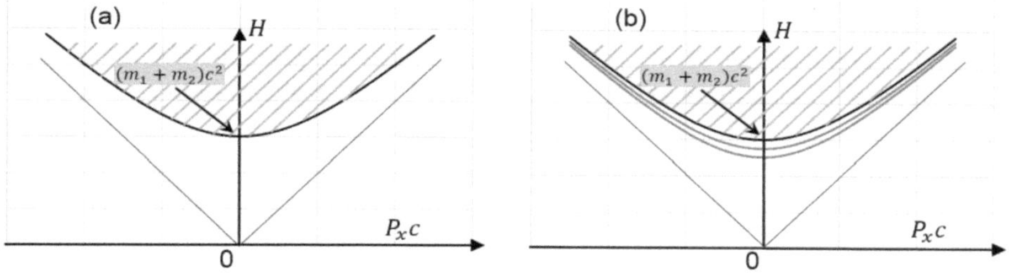

Figure 6.2: Typical energy–momentum spectrum in a two-particle system. (a) Noninteracting. (b) Interacting.

In the presence of interaction, the eigenvalues μ_n of the mass operator $M = M_0 + N$ can be found by solving the *stationary* Schrödinger equation

$$(M_0 + N)|\Psi_n\rangle = \mu_n|\Psi_n\rangle. \tag{6.78}$$

If the interaction N is sufficiently weak, then the spectrum of M remains close to (6.77). For example, if N is an attractive potential (i. e., the characteristic interaction energies are negative), then several discrete levels with masses $\mu_\beta < m_1 + m_2$ and energy hyperboloids $h_\beta = \sqrt{\mu_\beta^2 c^4 + p^2c^2}$ can split off from the continuous noninteracting mass spectrum, as shown in Figure 6.2 (b). The corresponding eigenvectors are called *bound states*, since their wave functions, as a rule, describe two particles that are in close proximity to each other. The simplest realistic example whose mass spectrum contains both continuous and discrete parts is the hydrogen atom, which will be discussed in Section 3.2 of the third volume.

The mass eigenvalues μ_n are strongly degenerate. For example, if $|\Psi_n\rangle$ is an eigenvector corresponding to the mass eigenvalue μ_n, then for any element of the Poincaré group g, the vector $U_g|\Psi_n\rangle$ is also an eigenvector with the same mass. Indeed,[22]

$$M(U_g|\Psi_n\rangle) = U_g M|\Psi_n\rangle = U_g\mu_n|\Psi_n\rangle = \mu_n(U_g|\Psi_n\rangle).$$

The operators $\{\boldsymbol{P}_0, H, M\}$ form a commuting set. Hence, there exists a basis of common eigenvectors $|\Psi_{\boldsymbol{p},n}\rangle$ such that

$$\begin{aligned}
M|\Psi_{\boldsymbol{p},n}\rangle &= \mu_n|\Psi_{\boldsymbol{p},n}\rangle,\\
\boldsymbol{P}_0|\Psi_{\boldsymbol{p},n}\rangle &= \boldsymbol{p}|\Psi_{\boldsymbol{p},n}\rangle,\\
H|\Psi_{\boldsymbol{p},n}\rangle &= \sqrt{M^2c^4 + P^2c^2}|\Psi_{\boldsymbol{p},n}\rangle = \sqrt{\mu_n^2c^4 + p^2c^2}|\Psi_{\boldsymbol{p},n}\rangle.
\end{aligned}$$

22 We used the fact that M is a Casimir operator that commutes with all representatives U_g. This also means that eigensubspaces with fixed masses μ_n are invariant under Poincaré group actions.

In applications, we are more often interested in systems at rest ($\boldsymbol{p} = 0$). For them, there is a simple relationship between mass and energy, $H = Mc^2$, and the Schrödinger equation (6.78) is commonly written for eigenvectors and eigenvalues of the Hamiltonian

$$H|\Psi_{\mathbf{0},n}\rangle = E_n|\Psi_{\mathbf{0},n}\rangle. \tag{6.79}$$

6.5.2 Perturbation theory

Quite often we have to deal with the following problem. Suppose that we know the complete solution for the spectrum of the Hamiltonian H, that is, we know the eigenvalues E_n and the eigenvectors $|\Psi_n\rangle$ that are solutions of the eigenvalue problem (6.79). How would this spectrum change if a small perturbation W is added to the Hamiltonian H? This problem is solved practically in all textbooks on quantum mechanics. Here we simply reproduce the final result, which will be useful in future calculations.

We choose one nondegenerate state $|\Psi\rangle = |\Psi_m\rangle$ in the discrete spectrum with energy $E = E_m$. The new energy E' can be represented as a *perturbation theory* series,

$$E' = E + \Delta E^{(1)} + \Delta E^{(2)} + \cdots,$$

where the first-order correction is simply the matrix element of the perturbation operator

$$\Delta E^{(1)} = \langle\Psi|W|\Psi\rangle \tag{6.80}$$

and the second-order correction

$$\Delta E^{(2)} = \sum_{n \neq m} \frac{\langle\Psi|W|\Psi_n\rangle\langle\Psi_n|W|\Psi\rangle}{E - E_n} \tag{6.81}$$

requires summation over the complete basis of eigenstates of the unperturbed Hamiltonian H.

6.5.3 Once again about the Doppler effect

In Subsection 5.4.5 we discussed the Doppler effect by calculating photon energies either measured by a moving observer or emitted by a moving source. To do this, we applied a boost transformation (5.70) to the energy E of a free massless photon. However, it is instructive to look at this problem also from a different point of view. Photons are usually radiated by compound massive systems (atoms, molecules, nuclei, etc.) in

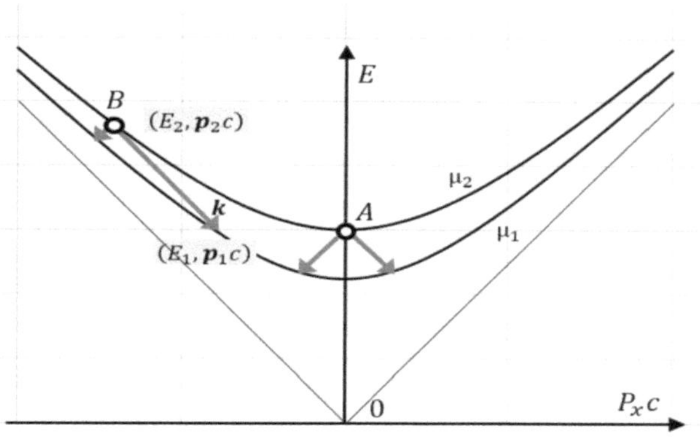

Figure 6.3: Diagram of energy levels in a bound system. The excited state of the system at rest is represented by the point A. Radiative transitions to the ground state are indicated by arrows.

transitions between two stationary energy levels E_2 and E_1, so that the photon energy can be found from the energy conservation law.[23]

When the source moves relative to the observer (or the observer moves relative to the source), the energy levels E_1 and E_2 undergo an inertial transformation according to equation (4.4). Let us check that the Doppler shift calculated by this formula coincides with the result obtained in Subsection 5.4.5. This will give us additional confidence in the consistency of our theory.

Suppose that the composite system emitter has two bound states characterized by mass eigenvalues μ_1 and $\mu_2 > \mu_1$ (see Figure 6.3). Suppose also that initially the system was in the excited state (point B in Figure 6.3) with mass μ_2, total momentum $\boldsymbol{p}_2$ and energy $E_2 = \sqrt{\mu_2^2c^4 + p_2^2c^2}$. In the final state, we have the same system with a smaller mass μ_1. Due to recoil, the final state of the emitter has a different momentum $\boldsymbol{p}_1$ and energy $E_1 = \sqrt{\mu_1^2c^4 + p_1^2c^2}$. In addition, a photon with momentum $\boldsymbol{k}$ and energy ck is created. From the conservation laws, we can write

$$\begin{aligned}
\boldsymbol{p}_2 &= \boldsymbol{p}_1 + \boldsymbol{k}, \\
E_2 &= E_1 + ck, \\
\sqrt{\mu_2^2c^4 + p_2^2c^2} &= \sqrt{\mu_1^2c^4 + p_1^2c^2} + ck = \sqrt{\mu_1^2c^4 + (\boldsymbol{p}_2 - \boldsymbol{k})^2c^2} + ck.
\end{aligned}$$

Squaring both sides of the last equality, we obtain

$$k\sqrt{\mu_1^2c^2 + (\boldsymbol{p}_2 - \boldsymbol{k})^2} = \frac{1}{2}(\Delta\mu)^2c^2 + p_2k\cos\varphi' - k^2,$$

23 Actually, the transition energy $E = E_2 - E_1$ does not have a sharp definition, because the excited state, strictly speaking, is not stationary, i. e., it is not an eigenstate of the energy operator (see Section 4.1 in Volume 3). Therefore, our discussion in this subsection is only valid approximately for long-living levels whose width can be neglected.

where $(\Delta\mu)^2 \equiv \mu_2^2 - \mu_1^2$ and φ' is the angle between vectors $\boldsymbol{p}_2$ and $\boldsymbol{k}$.[24] Again raising both sides into a square, we obtain the following quadratic equation:

$$k^2(\mu_2^2c^2 + p_2^2 - p_2^2\cos^2\varphi') - k(\Delta\mu)^2c^2p_2\cos\varphi' - \frac{1}{4}(\Delta\mu)^4c^4 = 0,$$

with the solution[25]

$$k = \frac{(\Delta\mu)^2c^2}{2\mu_2^2c^2 + 2p_2^2\sin^2\varphi'}\left(p_2\cos\varphi' + \sqrt{\mu_2^2c^2 + p_2^2}\right).$$

Denoting by θ the rapidity of the initial state, we get $p_2 = \mu_2 c\sinh\theta$, $\sqrt{\mu_2^2c^2 + p_2^2} = \mu_2 c\cosh\theta$ and

$$k = \frac{(\Delta\mu)^2c(\sinh\theta\cos\varphi' + \cosh\theta)}{2\mu_2(\cosh^2\theta - \sinh^2\theta\cos^2\varphi')} = \frac{(\Delta\mu)^2c}{2\mu_2\cosh\theta(1 - \frac{v}{c}\cos\varphi')}.$$

Multiplying both parts by c, we get the energy of the emitted photon,

$$E(\theta, \varphi') \equiv ck = \frac{E(0)}{\cosh\theta(1 - \frac{v}{c}\cos\varphi')}, \tag{6.82}$$

where $E(0) = (\Delta\mu)^2c^2/(2\mu_2)$ is the energy of the photon emitted by the source at rest ($\theta = 0$). Formula (6.82) coincides with our previous result (5.74).

6.5.4 Time evolution

In addition to the stationary energy spectra described above, we are often interested in the time evolution of many-particle systems. In quantum theory, the progression from time t_0 (earlier) to time t (later) is described by the *time evolution operator*

$$U(t \leftarrow t_0) = e^{-\frac{i}{\hbar}H(t-t_0)}. \tag{6.83}$$

This operator has the following useful properties:

$$U(t \leftarrow t_0) = e^{-\frac{i}{\hbar}H(t-t')}e^{-\frac{i}{\hbar}H(t'-t_0)} = U(t \leftarrow t')U(t' \leftarrow t_0), \tag{6.84}$$

$$U(t \leftarrow t_0) = U^{-1}(t_0 \leftarrow t) \tag{6.85}$$

for all $t \geq t' \geq t_0$.

---

24 Note also that vector $\boldsymbol{k}$ is directed from the light source to the observer, so angle φ' has the same interpretation as in Subsection 5.4.5.

25 Only the positive sign of the square root leads to the physical solution with a positive k.

In the Schrödinger picture the time evolution of the state vector from time t_0 to time t is (5.55)

$$|\Psi(t)\rangle = U(t \leftarrow t_0)|\Psi(t_0)\rangle = e^{-\frac{i}{\hbar}H(t-t_0)}|\Psi(t_0)\rangle. \tag{6.86}$$

The time-dependent state vector $|\Psi(t)\rangle$ is also a solution of the *time-dependent* Schrödinger equation

$$i\hbar\frac{d}{dt}|\Psi(t)\rangle = i\hbar\frac{d}{dt}e^{-\frac{i}{\hbar}H(t-t_0)}|\Psi(t_0)\rangle = He^{-\frac{i}{\hbar}H(t-t_0)}|\Psi(t_0)\rangle = H|\Psi(t)\rangle. \tag{6.87}$$

Despite a deceptively simple appearance of formula (6.86), the calculation of the exponent $\exp(-\frac{i}{\hbar}Ht)$ is an extremely difficult task. In rare cases when all eigenvalues E_n and eigenvectors $|\Psi\rangle_n$ of H are known,

$$H|\psi_n\rangle = E_n|\psi_n\rangle$$

and the initial state can be represented as the sum (and/or integral) of the basis eigenvectors

$$|\Psi(t_0)\rangle = \sum_n C_n|\psi_n\rangle,$$

the time evolution can be obtained in a closed form

$$|\Psi(t)\rangle = e^{-\frac{i}{\hbar}H(t-t_0)}|\Psi(t_0)\rangle = e^{-\frac{i}{\hbar}H(t-t_0)}\sum_n C_n|\psi_n\rangle = \sum_n C_n e^{-\frac{i}{\hbar}E_n(t-t_0)}|\psi_n\rangle. \tag{6.88}$$

There is another useful formula for the time evolution in a theory with the Hamiltonian $H = H_0 + V$. Introducing the notation

$$V(t) = e^{\frac{i}{\hbar}H_0(t-t_0)}Ve^{-\frac{i}{\hbar}H_0(t-t_0)},$$

it is not difficult to verify that the following time-dependent state vector[26]

$$|\Psi(t)\rangle = e^{-\frac{i}{\hbar}H_0(t-t_0)}\left(1-\frac{i}{\hbar}\int_{t_0}^{t}V(t')dt' - \frac{1}{\hbar^2}\int_{t_0}^{t}V(t')dt'\int_{t_0}^{t'}V(t'')dt'' + \cdots\right)|\Psi(t_0)\rangle \tag{6.89}$$

satisfies the Schrödinger equation (6.87) with the additional condition that for $t = t_0$ this solution coincides with the given initial state $|\Psi(t_0)\rangle$. Indeed,

$$\begin{aligned} i\hbar\frac{d}{dt}|\Psi(t)\rangle &= i\hbar\frac{d}{dt}e^{-\frac{i}{\hbar}H_0(t-t_0)} \\ &\quad\times\left(1-\frac{i}{\hbar}\int_{t_0}^{t}V(t')dt' - \frac{1}{\hbar^2}\int_{t_0}^{t}V(t')dt'\int_{t_0}^{t'}V(t'')dt'' + \cdots\right)|\Psi(t_0)\rangle \end{aligned}$$

26 Note that integration variables are related by the inequalities $t \geq t' \geq t'' \geq \cdots \geq t_0$.

$$
\begin{aligned}
&= H_0 e^{-\frac{i}{\hbar}H_0(t-t_0)} \\
&\quad \times \left(1 - \frac{i}{\hbar}\int_{t_0}^{t} V(t')dt' - \frac{1}{\hbar^2}\int_{t_0}^{t} V(t')dt' \int_{t_0}^{t'} V(t'')dt'' + \cdots\right)|\Psi(t_0)\rangle \\
&\quad + e^{-\frac{i}{\hbar}H_0(t-t_0)} V(t)\left(1 - \frac{i}{\hbar}\int_{t_0}^{t} V(t'')dt'' + \cdots\right)|\Psi(t_0)\rangle \\
&= (H_0 + V)|\Psi(t)\rangle .
\end{aligned}
$$

We will find formula (6.89) useful in our discussion of scattering in Subsection 7.1.4.

Unfortunately, the above methods for calculating the time evolution of quantum systems have very limited practical applications. Complete spectra of eigenvalues and eigenvectors of the interacting Hamiltonian H are known only for the simplest models. The convergence of the perturbation series (6.89) is usually very slow. Therefore, time evolution calculations in quantum mechanics are not an easy task. Fortunately, there are two areas where we can make significant progress in solving this problem. First, there is an important class of *scattering experiments*, which do not require a detailed description of the time evolution of quantum states. The formalism of scattering theory will be discussed in Chapter 7. Second, in many cases, quantum effects are too small to influence observations. Then, it is useful to turn to the limit $\hbar \to 0$ and consider classical trajectories of particles. We will deal with the classical limit of quantum mechanics in the next section.

6.6 Classical Hamiltonian dynamics

There are many studies devoted to the so-called problem of *quantization*. This means that given a classical theory one tries to build a corresponding quantum analog. However, our world is fundamentally quantum, and its classical description is a very rough approximation. The construction of an exact theory, based on its approximate special case, is not a well-posed mathematical problem. It is easier to justify the opposite direction of research, i. e., the construction of an (approximate) classical theory, based on its (exact) quantum analog.

In Section 1.4 we found that distributive (classical) systems of propositions are special cases of orthomodular (quantum) logics. Therefore, we can expect that quantum mechanics includes classical mechanics as a particular case in the limit $\hbar \to 0$. However, it remains unclear how the phase space of classical mechanics is related to the quantum Hilbert space. We are going to analyze this relationship in this section. For simplicity, as an example, we choose one spinless particle with nonzero mass $m > 0$.

6.6.1 Quasiclassical states

In the macroscopic world of classical mechanics, we do not encounter localized eigenstates $|\boldsymbol{r}\rangle$ of the position operator. In accordance with equation (5.41), such states must have infinitely large uncertainty of the momentum, which is rather unusual. Similarly, we do not encounter states $|\boldsymbol{p}\rangle$ with well-defined momentum. Such states must be spread across the entire position space (5.44). The reason why the states like $|\boldsymbol{r}\rangle$ or $|\boldsymbol{p}\rangle$ rarely show up in experiments[27] is not very clear yet. Possibly, the eigenstates of the position or momentum are unstable with respect to small perturbations[28] and quickly transform into more stable *wave packets* or *quasiclassical states* where both position and momentum have good, but not perfect, localization.

Thus, in discussing the classical limit of quantum mechanics, we will not consider general quantum states, but confine ourselves to the class of quasiclassical states of particles $|\Psi_{\boldsymbol{r}_0,\boldsymbol{p}_0}\rangle$ whose wave functions are well localized both in the position and momentum spaces near the points $\boldsymbol{r}_0$ and $\boldsymbol{p}_0$, respectively. Without loss of generality, any such wave function can be written in the position representation as follows:

$$\psi_{\boldsymbol{r}_0,\boldsymbol{p}_0}(\boldsymbol{r}) \equiv \langle \boldsymbol{r}|\Psi_{\boldsymbol{r}_0,\boldsymbol{p}_0}\rangle = \eta(\boldsymbol{r}-\boldsymbol{r}_0)e^{\frac{i}{\hbar}\phi}e^{\frac{i}{\hbar}\boldsymbol{p}_0\cdot(\boldsymbol{r}-\boldsymbol{r}_0)}, \tag{6.90}$$

where $\eta(\boldsymbol{r}-\boldsymbol{r}_0)$ is a real smooth (nonoscillating) function with a peak near the expectation value of the position $\boldsymbol{r}_0$, and ϕ is a real phase;[29] see Figure 6.4. The last factor in (6.90) ensures that the expectation value of the momentum is equal to $\boldsymbol{p}_0$ (compare with (5.44)).

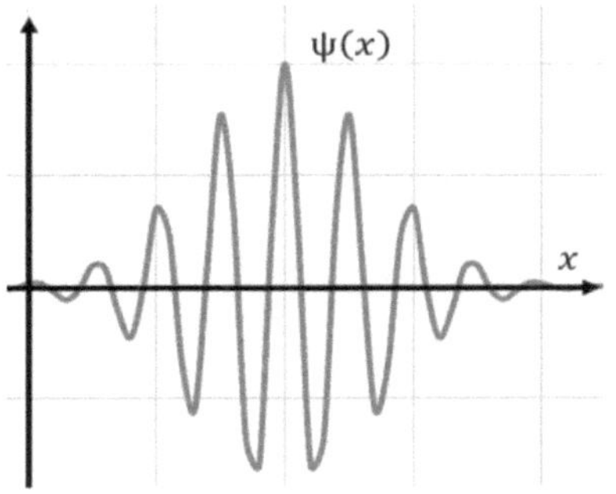

Figure 6.4: A typical quasiclassical wave packet. Only the real part of the complex wave function is shown.

27 Spatially delocalized states of particles play a role in low-temperature effects such as superconductivity and superfluidity and also in lasers.

28 For example, due to thermal fluctuations or external radiation.

29 The introduction of the unimodular phase factor ($|\exp(\frac{i}{\hbar}\phi)| = 1$) may seem superfluous, because any wave function is defined only up to a factor anyway. However, we will see that the factor $\exp(\frac{i}{\hbar}\phi)$ is useful for discussing the interference experiment in Subsection 6.6.6 and also in analyzing the Aharonov–Bohm effect in Volume 3.

As we shall see later, for a qualitative discussion of the classical limit, the choice of the function $\eta(\boldsymbol{r}-\boldsymbol{r}_0)$ is not so important. For example, it is often convenient to select it in the Gaussian form

$$\psi_{\boldsymbol{r}_0,\boldsymbol{p}_0}(\boldsymbol{r}) = Ne^{-(\boldsymbol{r}-\boldsymbol{r}_0)^2/d^2}e^{\frac{i}{\hbar}\boldsymbol{p}_0\cdot(\boldsymbol{r}-\boldsymbol{r}_0)}, \tag{6.91}$$

where parameter d controls the degree of localization in space and the factor N ensures the normalization

$$\int d\boldsymbol{r}|\psi_{\boldsymbol{r}_0,\boldsymbol{p}_0}(\boldsymbol{r})|^2 = 1.$$

The exact value of this factor does not play any role in our discussion, so we will not calculate it here.

6.6.2 Heisenberg uncertainty relation

Wave functions of the type (6.91) cannot have sharp values of both position and momentum simultaneously. They are characterized by uncertainties of both the position $\Delta r > 0$ and the momentum $\Delta p > 0$. These uncertainties are inversely proportional to each other. To understand the nature of this inverse proportionality, let us assume for simplicity that the particle is at rest in the origin; $\boldsymbol{r}_0 = \boldsymbol{p}_0 = 0$. Then the position–space wave function is

$$\psi_{\mathbf{0},\mathbf{0}}(\boldsymbol{r}) = Ne^{-r^2/d^2} \tag{6.92}$$

and its momentum counterpart is[30]

$$\psi_{\mathbf{0},\mathbf{0}}(\boldsymbol{p}) = \frac{N}{(2\pi\hbar)^{3/2}}\int d\boldsymbol{r}e^{-r^2/d^2}e^{-\frac{i}{\hbar}\boldsymbol{p}\cdot\boldsymbol{r}} = \frac{Nd^3}{(2\hbar)^{3/2}}e^{-p^2d^2/(4\hbar^2)}. \tag{6.93}$$

The product of the two uncertainties

$$\Delta p \approx \frac{2\hbar}{d} \tag{6.94}$$

and $\Delta r \approx d$ does not depend on the localization parameter d, as

$$\Delta r\Delta p \approx 2\hbar.$$

This is an example of the *Heisenberg uncertainty relation*, which in its exact form claims that for all quantum states the above deltas obey the famous inequality

$$\Delta r\Delta p \geq \hbar/2. \tag{6.95}$$

---

30 Here we used equation (5.48) and the integral

$$\int d\boldsymbol{r}e^{-ar^2+\boldsymbol{b}\cdot\boldsymbol{r}} = (\pi/a)^{3/2}\exp\big(-b^2/(4a)\big).$$

6.6.3 Spreading of quasiclassical wave packets

Suppose that at the time $t = 0$ the particle was prepared in a state with a well-localized wave function (6.92), and the uncertainty of the position $\Delta r \approx d$ was rather small. The time dependence of the corresponding momentum wave function (6.93) is found fairly easy, as we have

$$\psi(\boldsymbol{p}, t) = e^{-\frac{i}{\hbar}\hat{H}t}\psi_{\mathbf{0},\mathbf{0}}(\boldsymbol{p}, 0) = \frac{Nd^3}{(2\hbar)^{3/2}} e^{-p^2d^2/(4\hbar^2)} e^{-\frac{it}{\hbar}\sqrt{m^2c^4+p^2c^2}}. \tag{6.96}$$

Returning to the position representation, we obtain the time-dependent wave packet[31]

$$\begin{aligned}
\psi(\boldsymbol{r}, t) &= \frac{Nd^3}{(4\pi\hbar^2)^{3/2}} \int d\boldsymbol{p} e^{-p^2d^2/(4\hbar^2)} e^{\frac{i}{\hbar}\boldsymbol{p}\cdot\boldsymbol{r}} e^{-\frac{it}{\hbar}\sqrt{m^2c^4+p^2c^2}} \\
&\approx \frac{Nd^3}{(4\pi\hbar^2)^{3/2}} e^{-\frac{i}{\hbar}mc^2t} \int d\boldsymbol{p} \exp\left(-p^2\left(\frac{d^2}{4\hbar^2} + \frac{it}{2\hbar m}\right) + \frac{i}{\hbar}\boldsymbol{p}\cdot\boldsymbol{r}\right) \\
&= N\left(\frac{d^2m}{d^2m + 2i\hbar t}\right)^{3/2} e^{-\frac{i}{\hbar}mc^2t} \exp\left(-\frac{mr^2}{d^2m + 2i\hbar t}\right).
\end{aligned}$$

The corresponding probability density is

$$\rho(\boldsymbol{r}, t) = |\psi(\boldsymbol{r}, t)|^2 = |N|^2\left(\frac{d^4m^2}{d^4m^2 + 4\hbar^2t^2}\right)^3 \exp\left(-\frac{2r^2d^2m^2}{d^4m^2 + 4\hbar^2t^2}\right).$$

The size of the wave packet at large times $t \to \infty$ can be estimated as

$$\Delta r(t) \approx \sqrt{\frac{d^4m^2 + 4\hbar^2t^2}{d^2m^2}} \approx \frac{2\hbar t}{dm}.$$

So, over time the position wave function spreads out in space, and the velocity v_s of this spreading is directly proportional to the uncertainty of the velocity in the initial state. Using equation (6.94)), we have

$$v_s \approx \frac{2\hbar}{dm} \approx \frac{\Delta p}{m}. \tag{6.97}$$

At large times this velocity does not depend on the shape of the initial wave packet. The only essential parameters are the initial size d of the packet and the mass m of the particle.

A simple estimate shows that for macroscopic objects this "spreading" phenomenon does not matter. For example, for a particle with the mass of $m = 1\,\text{mg}$

31 Due to the factor $\exp(-p^2d^2/(4\hbar^2))$, only small momentum values contribute to the integral, so we have used the nonrelativistic approximation $\sqrt{m^2c^4 + p^2c^2} \approx mc^2 + p^2/(2m)$.

and the initial position uncertainty of $d = 1$ micron, the wave function spreads to a size of 1 cm in 2×10^{11} years. Therefore, in quasiclassical states of macroscopic particles, their positions and momenta are always determined with high precision, and their time evolution is satisfactorily described in the language of classical *trajectories*. In these cases, quantum mechanics can be replaced by the classical one without much error.

6.6.4 Phase space

Let us now clarify how the laws of classical dynamics follow from the quantum Schrödinger equation.

In classical physics, where the resolution of measuring devices is poor,[32] the shape of quasiclassical wave packets $\eta(\boldsymbol{r} - \boldsymbol{r}_0)$ cannot be discerned. In such situations, all quantum states (6.90) with different localized shapes $\eta(\boldsymbol{r} - \boldsymbol{r}_0)$ look indistinguishable, and they are perceived as one and the same classical state. Hence, each (quasi)classical state $|\Psi_{\boldsymbol{r}_0, \boldsymbol{p}_0}\rangle$ is completely characterized by only two numbers: the average position of the packet $\boldsymbol{r}_0$ and its average momentum $\boldsymbol{p}_0$. These states are also approximate eigenstates of both position and momentum operators,

$$\boldsymbol{R}|\Psi_{\boldsymbol{r}_0, \boldsymbol{p}_0}\rangle \approx \boldsymbol{r}_0 |\Psi_{\boldsymbol{r}_0, \boldsymbol{p}_0}\rangle, \tag{6.98}$$

$$\boldsymbol{P}|\Psi_{\boldsymbol{r}_0, \boldsymbol{p}_0}\rangle \approx \boldsymbol{p}_0 |\Psi_{\boldsymbol{r}_0, \boldsymbol{p}_0}\rangle, \tag{6.99}$$

and they can be represented by a single point $(\boldsymbol{r}_0, \boldsymbol{p}_0)$ in the six-dimensional space $\mathbb{R}^6$ with coordinates $\{r_x, r_y, r_z, p_x, p_y, p_z\}$. This is the classical single-particle *phase space*, which we discussed from the logico-probabilistic point of view in Section 1.2.

We know that each one-particle quantum observable F can be expressed as a function of the particle's position $\boldsymbol{r}$ and momentum $\boldsymbol{p}$.[33] Hence, in the classical world, all observables are represented by real functions $f(\boldsymbol{r}, \boldsymbol{p})$ on the phase space, as expected.

6.6.5 Poisson bracket

We can continue the above line of reasoning and translate all quantum concepts into the classical phase space language. It is especially instructive to see what is the classical analog of the quantum commutator.

In quantum mechanics, commutators play two important roles. First, the commutator of two observables determines whether they can be measured simultaneously,

32 Namely, in some sense poorer than the quantum of the action $\hbar$ [48].

33 For example, energy $h = \sqrt{m^2c^4 + p^2c^2}$, angular momentum $\boldsymbol{j} = [\boldsymbol{r} \times \boldsymbol{p}]$, velocity $\boldsymbol{v} = \boldsymbol{p}c^2/h$, etc. Recall that mass m is just a numerical factor in the particle's Hilbert space.

i. e., whether there are states in which both observables have well-defined values. In accordance with (3.49)–(3.55), commutators of basic quantum observables are proportional to $\hbar$, so that in the classical limit $\hbar \to 0$ all operators of observables commute with each other. Therefore, all classical observables are compatible, i. e., simultaneously measurable. This agrees with our Theorem B.14, which establishes the compatibility of all propositions in classical logic. Second, quantum commutators between observables and generators of the Poincaré group determine how these observables are transformed from one reference frame to another. One example of such a transformation is the time translation (3.61). In the classical limit $\hbar \to 0$, all multiple commutators on the right-hand side of (3.61) tend to zero as $\hbar^n$, but they are multiplied by large factors $(-i/\hbar)^n$. So, in the limit $\hbar \to 0$ the right-hand side's dependence on $\hbar$ disappears and we get

$$F(t) = F - [H, F]_P t + \frac{1}{2}[H, [H, F]_P]_P t^2 + O(t^3), \tag{6.100}$$

where the limit

$$[f, g]_P \equiv \lim_{\hbar \to 0} \frac{-i}{\hbar}[f(\boldsymbol{r}, \boldsymbol{p}), g(\boldsymbol{r}, \boldsymbol{p})] \tag{6.101}$$

is called the *Poisson bracket*.

Our next task is to derive a convenient explicit formula for calculating Poisson brackets (6.101). The exact commutator of two quantum-mechanical operators $f(\boldsymbol{r}, \boldsymbol{p})$ and $g(\boldsymbol{r}, \boldsymbol{p})$ can be written as an expansion in powers of $\hbar$,

$$[f, g] = i\hbar k_1 + i\hbar^2 k_2 + i\hbar^3 k_3 + O(\hbar^4),$$

where k_i are some Hermitian operators. From equation (6.101) it should be clear that the Poisson bracket is equal to the coefficient of the dominant first-order term:

$$[f, g]_P = k_1.$$

As a consequence, the classical Poisson bracket $[f, g]_P$ is much easier to compute than the complete quantum commutator $[f, g]$. The following theorem shows that such a calculation can be reduced to simple differentiation.

Theorem 6.6 (Poisson bracket). *If $f(\boldsymbol{r}, \boldsymbol{p})$ and $g(\boldsymbol{r}, \boldsymbol{p})$ are two observables of a massive spinless particle, then*

$$[f(\boldsymbol{r}, \boldsymbol{p}), g(\boldsymbol{r}, \boldsymbol{p})]_P = \frac{\partial f}{\partial \boldsymbol{r}} \cdot \frac{\partial g}{\partial \boldsymbol{p}} - \frac{\partial f}{\partial \boldsymbol{p}} \cdot \frac{\partial g}{\partial \boldsymbol{r}}. \tag{6.102}$$

Proof. For simplicity, we consider the one-dimensional case (the 3D proof is very similar), in which the desired result (6.102) becomes

$$[f(r, p), g(r, p)]_P \equiv \lim_{\hbar \to 0} \frac{-i}{\hbar}[f(r, p), g(r, p)] = \frac{\partial f}{\partial r} \cdot \frac{\partial g}{\partial p} - \frac{\partial f}{\partial p} \cdot \frac{\partial g}{\partial r}. \tag{6.103}$$

Suppose that functions $f(r,p)$ and $g(r,p)$ can be represented by their Taylor expansions near the center of the phase space ($r = 0, p = 0$), i. e.,

$$f(r,p) = C_{00} + C_{10}r + C_{01}p + C_{11}rp + C_{20}r^2 + C_{02}p^2 + C_{21}r^2p + \cdots,$$
$$g(r,p) = D_{00} + D_{10}r + D_{01}p + D_{11}rp + D_{20}r^2 + D_{02}p^2 + D_{21}r^2p + \cdots,$$

where C_{ij} and D_{ij} are numerical multipliers and we agree to write factors r^n to the left from factors p^m. Then it suffices to prove formula (6.103) in the case where f and g are monomials of the form r^np^m. In particular, we would like to prove that

$$\begin{aligned}[][r^np^m, r^qp^s]_P &= \frac{\partial(r^np^m)}{\partial r}\frac{\partial(r^qp^s)}{\partial p} - \frac{\partial(r^np^m)}{\partial p}\frac{\partial(r^qp^s)}{\partial r}\\ &= nsr^{n+q-1}p^{m+s-1} - mqr^{n+q-1}p^{m+s-1}\\ &= (ns - mq)r^{n+q-1}p^{m+s-1}\end{aligned} \tag{6.104}$$

for all integer powers $n, m, q, s \geq 0$. This result is definitely true if f and g are linear functions of r and p. For example, in the case where $n = s = 1$, $m = q = 0$, formula (6.104) yields

$$[r,p]_P = 1,$$

which agrees with our definition (6.101) and with the quantum result (4.25).

To prove (6.104) for higher powers, we use mathematical induction. Suppose that we succeed in proving the validity of (6.104) for the set of powers n, m, q, s and for all lower powers n', m', q', s', where $n' \leq n$, $m' \leq m$, $q' \leq q$ and $s' \leq s$. Then we have to establish the following relations:

$$\begin{aligned}[][r^np^m, r^{q+1}p^s]_P &= (ns - mq - m)r^{n+q}p^{m+s-1},\\ [r^np^m, r^qp^{s+1}]_P &= (ns - mq + n)r^{n+q-1}p^{m+s},\\ [r^{n+1}p^m, r^qp^s]_P &= (ns - mq + s)r^{n+q}p^{m+s-1},\\ [r^np^{m+1}, r^qp^s]_P &= (ns - mq - q)r^{n+q-1}p^{m+s}.\end{aligned}$$

We consider only the first equality (the other three are proved similarly). Using (4.33), (6.104) and (E.12) we, indeed, obtain

$$\begin{aligned}[][r^np^m, r^{q+1}p^s]_P &= -\lim_{\hbar\to 0}\frac{i}{\hbar}[r^np^m, r^{q+1}p^s] = -\lim_{\hbar\to 0}\frac{i}{\hbar}[r^np^m, r]r^qp^s - \lim_{\hbar\to 0}\frac{i}{\hbar}r[r^np^m, r^qp^s]\\ &= [r^np^m, r]_P r^qp^s + r[r^np^m, r^qp^s]_P = -mr^{n+q}p^{m+s-1} + (ns - mq)r^{n+q}p^{m+s-1}\\ &= (ns - mq - m)r^{n+q}p^{m+s-1}.\end{aligned}$$

Therefore, by induction, equation (6.103) holds for all powers $n, m, q, s \geq 0$ and for all smooth functions $f(r,p)$ and $g(r,p)$. □

As a concrete example, let us apply the Poisson bracket formalism to the time evolution. From (6.100) we obtain the classical *Liouville equation*

$$\frac{dF(t)}{dt} = [F, H]_P, \tag{6.105}$$

which is an analog of the quantum Heisenberg equation (3.62). Applying (6.105) to a particle's momentum and position, we obtain the familiar classical *Hamilton equations of motion*

$$\frac{d\boldsymbol{p}(t)}{dt} = [\boldsymbol{p}, H(\boldsymbol{r}, \boldsymbol{p})]_P = -\frac{\partial H(\boldsymbol{r}, \boldsymbol{p})}{\partial \boldsymbol{r}}, \tag{6.106}$$

$$\frac{d\boldsymbol{r}(t)}{dt} = [\boldsymbol{r}, H(\boldsymbol{r}, \boldsymbol{p})]_P = \frac{\partial H(\boldsymbol{r}, \boldsymbol{p})}{\partial \boldsymbol{p}}. \tag{6.107}$$

6.6.6 Time evolution of wave packets

Our results (6.106)–(6.107) imply, in particular, that trajectories of centers of quasi-classical wave packets coincide with predictions of classical Hamiltonian mechanics. In this subsection we would like to demonstrate in more detail how this conclusion follows from solutions of the Schrödinger equation (5.57).

Earlier in this section, we found out that in many cases the spreading of quasi-classical wave packets can be neglected and that the center of the packet moves along a well-defined trajectory $(\boldsymbol{r}_0(t), \boldsymbol{p}_0(t))$. Substituting this (as yet undefined) trajectory in (6.90) and assuming that the phase $\phi(t)$ also depends on time, we arrive at the following *ansatz*:

$$\Psi(\boldsymbol{r}, t) = \eta(\boldsymbol{r} - \boldsymbol{r}_0(t)) \exp\left(\frac{i}{\hbar}\Omega(t)\right), \tag{6.108}$$

where we denote

$$\Omega(t) \equiv \boldsymbol{p}_0(t) \cdot (\boldsymbol{r} - \boldsymbol{r}_0(t)) + \phi(t) \tag{6.109}$$

and $\boldsymbol{r}_0(t), \boldsymbol{p}_0(t), \phi(t)$ are numerical functions that we have to determine. Now we insert (6.108)–(6.109) into the Schrödinger equation (5.57)

$$i\hbar\frac{\partial \Psi(\boldsymbol{r}, t)}{\partial t} + \frac{\hbar^2}{2m}\frac{\partial^2 \Psi(\boldsymbol{r}, t)}{\partial \boldsymbol{r}^2} - V(\boldsymbol{r})\Psi(\boldsymbol{r}, t) = 0, \tag{6.110}$$

which is valid for the position–space wave function $\Psi(\boldsymbol{r}, t)$ of one particle moving in an external potential $V(\boldsymbol{r})$.[34] For brevity, we omit time arguments, use dots to denote

---

34 Here we made several assumptions and approximations to simplify our calculations. First, we consider a particle moving in a fixed potential. This system does not belong to the class of isolated systems,

time derivatives and rewrite the three terms on the left-hand side of (6.110) separately. Then we have

$$i\hbar\frac{\partial\Psi(\boldsymbol{r})}{\partial t} = \left(-i\hbar\left(\frac{\partial\eta}{\partial\boldsymbol{r}}\cdot\dot{\boldsymbol{r}}_0\right) - (\dot{\boldsymbol{p}}_0\cdot\boldsymbol{r})\eta + (\dot{\boldsymbol{p}}_0\cdot\boldsymbol{r}_0)\eta + (\boldsymbol{p}_0\cdot\dot{\boldsymbol{r}}_0)\eta - \dot{\phi}\eta\right)\exp\left(\frac{i}{\hbar}\Omega\right),$$

$$\frac{\hbar^2}{2m}\frac{\partial^2\Psi(\boldsymbol{r})}{\partial\boldsymbol{r}^2} = \frac{\hbar^2}{2m}\left(\frac{\partial^2\eta}{\partial\boldsymbol{r}^2} + \frac{2i}{\hbar}\left(\frac{\partial\eta}{\partial\boldsymbol{r}}\cdot\boldsymbol{p}_0\right) - \frac{p_0^2\eta}{\hbar^2}\right)\exp\left(\frac{i}{\hbar}\Omega\right),$$

$$-V(\boldsymbol{r})\Psi(\boldsymbol{r}) \approx \left(-V(\boldsymbol{r}_0)\eta - \frac{\partial V(\boldsymbol{r})}{\partial\boldsymbol{r}}\Big|_{\boldsymbol{r}=\boldsymbol{r}_0}(\boldsymbol{r}-\boldsymbol{r}_0)\eta\right)\exp\left(\frac{i}{\hbar}\Omega\right).$$

Here we find three types of terms: those proportional to $\hbar^0$, $\hbar^1$ and $\hbar^2$. They should vanish independently. The terms proportional to $\hbar^2$ are too small; they are beyond the accuracy of the quasiclassical approximation and they can be neglected. The terms proportional to $\hbar$ lead to the equation $\dot{\boldsymbol{r}}_0 = \frac{\boldsymbol{p}_0}{m}$, which is the usual velocity–momentum relationship in momentum-independent potentials.[35] The terms $\hbar^0$ lead to the equation

$$\begin{aligned} 0 &= -(\dot{\boldsymbol{p}}_0\cdot\boldsymbol{r}) + (\dot{\boldsymbol{p}}_0\cdot\boldsymbol{r}_0) + (\boldsymbol{p}_0\cdot\dot{\boldsymbol{r}}_0) - \dot{\phi} - \frac{p_0^2}{2m} - V(\boldsymbol{r}_0) - \frac{\partial V(\boldsymbol{r})}{\partial\boldsymbol{r}}\Big|_{\boldsymbol{r}=\boldsymbol{r}_0}(\boldsymbol{r}-\boldsymbol{r}_0) + \cdots \\ &= \frac{p_0^2}{2m} - \dot{\phi} - V(\boldsymbol{r}_0) + \left(-\dot{\boldsymbol{p}}_0 - \frac{\partial V(\boldsymbol{r})}{\partial\boldsymbol{r}}\Big|_{\boldsymbol{r}=\boldsymbol{r}_0}\right)(\boldsymbol{r}-\boldsymbol{r}_0) + \cdots . \end{aligned}$$

The right-hand side can be regarded as the beginning of a Taylor series for some function of $\boldsymbol{r}$ near the initial value $\boldsymbol{r} = \boldsymbol{r}_0$. Hence, both the constant term and the term proportional to $(\boldsymbol{r} - \boldsymbol{r}_0)$ must go to zero independently. The vanishing parenthesis implies

$$\dot{\boldsymbol{p}}_0 = -\frac{\partial V(\boldsymbol{r})}{\partial\boldsymbol{r}}\Big|_{\boldsymbol{r}=\boldsymbol{r}_0},$$

which is the first Hamilton equation (6.106). Setting the first ($\boldsymbol{r}$-independent) term to zero results in the following equation for the phase function $\phi(t)$:

$$\frac{\partial\phi}{\partial t} = \frac{p_0^2(t)}{2m} - V(\boldsymbol{r}_0(t)).$$

which is the main subject of our book. Nevertheless, our approximation is completely justified in the case when the object creating the potential $V(\boldsymbol{r})$ is so heavy that it can be regarded as fixed. Second, the potential $V(\boldsymbol{r})$ is assumed to be independent of the particle momentum $\boldsymbol{p}$. Thus, we work in the nonrelativistic approximation with the Hamiltonian $H = p^2/(2m) + V(\boldsymbol{r}) = -\hbar^2/(2m)\partial^2/\partial\boldsymbol{r}^2 + V(\boldsymbol{r})$.

35 This is also the second Hamilton equation (6.107).

The solution of this equation within the time interval $[t_0, t]$ is given by the so-called *action integral*,[36]

$$\phi(t) = \phi(t_0) + \int_{t_0}^{t} dt' \left[\frac{p_0^2(t')}{2m} - V(\boldsymbol{r}_0(t')) \right]. \tag{6.111}$$

So, we conclude that in our approximation the center of the quasiclassical wave packet really moves along a trajectory determined by the classical Hamilton equations of motion (6.106)–(6.107). In addition, we have a purely quantum effect: a change of the overall phase factor in accordance with equation (6.111).

6.6.7 Once again about experiments with two holes

In this subsection we would like to illustrate the importance of the phase ϕ for describing quantum effects. Consider, for example, particle interference in the two-hole experiment from Section 1.1.

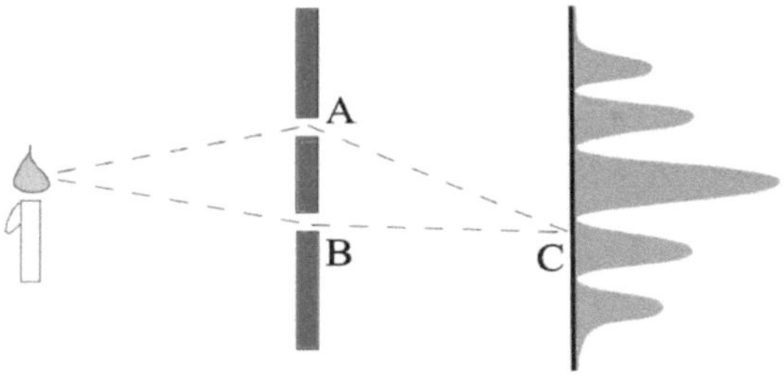

Figure 6.5: Quantum interference in the experiment with two holes.

Suppose that the source emits electrons, which pass through two holes and form an image on the screen, as shown in Figure 6.5. Wave packets can reach the point C on the screen in two mutually exclusive ways: either through the hole A, or through the hole B. Both types of wave packets contribute to the total wave function at the point C.[37] Their complex phase factors $\exp(\frac{i}{\hbar}\phi)$ should add up when calculating the probability amplitude for detecting an electron at the point C, so

$$\psi \propto \frac{1}{\sqrt{2}}\left(e^{\frac{i}{\hbar}\phi_{AC}} + e^{\frac{i}{\hbar}\phi_{BC}}\right).$$

36 Notice that the integrand has the form ("kinetic energy" – "potential energy"), which is known in classical mechanics as the *Lagrangian*.

37 For simplicity, we assume that both packets have equal amplitudes.

The observed probability is then obtained by taking the square of the module of the amplitude,

$$|\psi|^2 \propto \frac{1}{2}\left|e^{\frac{i}{\hbar}\phi_{AC}} + e^{\frac{i}{\hbar}\phi_{BC}}\right|^2 = 1 + \cos\left(\frac{\phi_{AC} - \phi_{BC}}{\hbar}\right).$$

In our case, this calculation is especially simple, since there is no external potential ($V(\boldsymbol{r}) = 0$). The momentum (and velocity) of each wave packet remains constant ($p_0^2(t) = \text{const}$) along its trajectory, so the action integral (6.111) is proportional to the travel time of the wave packet from the hole to the screen. The phase $\delta\phi$ accumulated by the wave packet during the travel time Δt equals

$$\delta\phi = \Delta t \cdot \frac{p_0^2}{2m}.$$

Hence the phase difference between the two paths is

$$\phi_{AC} - \phi_{BC} = \frac{r_{AC} - r_{BC}}{v_0} \cdot \frac{p_0^2}{2m} = (r_{AC} - r_{BC}) \cdot \frac{p_0}{2}.$$

In other words, the nature of the interference (constructive or destructive) at the point C depends on the difference between distances AC and BC. Two paths interfere constructively (the probability is maximal), if

$$\phi_{AC} - \phi_{BC} = 2\pi n\hbar, \quad n = \ldots, -1, 0, 1, 2, \ldots,$$

i. e., for those points on the screen where

$$r_{AC} - r_{BC} = \frac{4\pi\hbar n}{p_0}.$$

At points where the distance difference is $4\pi\hbar(n + 1/2)/p_0$, the interference is destructive (the probability is minimal).

7 Scattering

Physics is becoming so unbelievably complex that it is taking longer and longer to train a physicist. It is taking so long, in fact, to train a physicist to the place where he understands the nature of physical problems that he is already too old to solve them.
Eugene P. Wigner

Finding solutions of the time-dependent Schrödinger equation (6.87) is incredibly difficult, even for the simplest models. However, nature had mercy on us and created a very important class of experiments, where a description of dynamics by equation (6.87) is completely unnecessary, because it is too detailed. Here we are talking about scattering experiments, which are the topic of this chapter.

A typical scattering experiment is designed in such a way that free particles (or their bound states, such as atoms or nuclei) are prepared at a great distance from each other and directed into collision. Then experimentalists study the properties of free particles or stable bound states leaving the collision region. In these experiments, as a rule, it is impossible to see evolution during the interaction process: particle reactions occur almost instantaneously, and we can observe only reactants and products moving freely before and after the collision. In such situations, the theory is not required to describe the true dynamics of the particles during the short interval of interaction. It is sufficient to understand only the mapping of free states before the collision to free states after the collision. This mapping is described by the so-called S-operator, which we are going to discuss in the next section.

In Section 7.2 we will consider the situation of scattering equivalence when two different Hamiltonians have exactly the same scattering properties.

7.1 Scattering operators

7.1.1 Physical meaning of S-operator

Let us consider a scattering experiment in which the free states of the reactants were prepared in the distant past at $t = -\infty$. The collision itself occurred within a short interval $[\eta', \eta]$ near time zero.[1] Free states of the collision products are registered in the distant future, $t = \infty$, so that the inequalities $\infty \gg \eta > 0 > \eta' \gg -\infty$ are satisfied. For simplicity, we assume that the two colliding particles do not form bound states

1 The short interaction time interval (and the applicability of the scattering theory) is guaranteed if three conditions are satisfied. First, the interaction between particles is short-range or, in a more general setting, cluster-separable. Second, the states of particles are describable by localized wave packets, for example, as in Subsection 6.6.1. Third, the velocities (or momenta) of the particles are high enough.

https://doi.org/10.1515/9783110492132-007

either before or after the collision. Therefore, at asymptotically distant times, the exact time evolution of the system is well approximated by the noninteracting operators $U_0(\eta' \leftarrow -\infty)$ and $U_0(\infty \leftarrow \eta)$.[2] Then the full evolution operator from the remote past to the distant future is (here we use properties (6.84) and (6.85))

$$\begin{aligned}
U(\infty \leftarrow -\infty) &\approx U_0(\infty \leftarrow \eta)U(\eta \leftarrow \eta')U_0(\eta' \leftarrow -\infty) \\
&= U_0(\infty \leftarrow \eta)U_0(\eta \leftarrow 0)[U_0(0 \leftarrow \eta)U(\eta \leftarrow \eta')U_0(\eta' \leftarrow 0)] \\
&\quad \times U_0(0 \leftarrow \eta')U_0(\eta' \leftarrow -\infty) \\
&= U_0(\infty \leftarrow 0)S_{\eta,\eta'}U_0(0 \leftarrow -\infty),
\end{aligned} \tag{7.1}$$

where we denote

$$S_{\eta,\eta'} \equiv U_0(0 \leftarrow \eta)U(\eta \leftarrow \eta')U_0(\eta' \leftarrow 0). \tag{7.2}$$

Equation (7.1) means that it is possible to formulate a simplified description for the time evolution in collision processes. In this description, the evolution is always free, except for a sudden change of state at the time $t = 0$. This change is described by the unitary operator $S_{\eta,\eta'}$. Approximation (7.1) becomes more accurate if we increase the time interval $[\eta', \eta]$, during which the exact time evolution is taken into account, i. e., in the limits $\eta' \to -\infty$ and $\eta \to \infty$.[3] Therefore, the exact formula for the time evolution from $-\infty$ to ∞ has the form

$$U(\infty \leftarrow -\infty) = U_0(\infty \leftarrow 0)SU_0(0 \leftarrow -\infty), \tag{7.3}$$

where the S-operator (or *scattering operator*) is defined by

$$\begin{aligned}
S &= \lim_{\eta' \to -\infty, \eta \to \infty} S_{\eta,\eta'} = \lim_{\eta' \to -\infty, \eta \to \infty} U_0(0 \leftarrow \eta)U(\eta \leftarrow \eta')U_0(\eta' \leftarrow 0) \\
&= \lim_{\eta' \to -\infty, \eta \to \infty} e^{\frac{i}{\hbar}H_0\eta} e^{-\frac{i}{\hbar}H(\eta-\eta')} e^{-\frac{i}{\hbar}H_0\eta'} \\
&= \lim_{\eta \to \infty} S(\eta),
\end{aligned} \tag{7.4}$$

where

$$S(\eta) \equiv \lim_{\eta' \to -\infty} e^{\frac{i}{\hbar}H_0\eta} e^{-\frac{i}{\hbar}H(\eta-\eta')} e^{-\frac{i}{\hbar}H_0\eta'}. \tag{7.5}$$

To better understand how scattering theory describes the time evolution, we turn to Figure 7.1. In this figure, we have plotted the state of the physical system (represented

2 Here we denote by $U_0(t \leftarrow t_0) \equiv \exp(-\frac{i}{\hbar}H_0(t-t_0))$ the time evolution operator associated with the noninteracting Hamiltonian H_0. The interaction evolution operator will be denoted by $U(t \leftarrow t_0) \equiv \exp(-\frac{i}{\hbar}H(t-t_0))$. In the Schrödinger representation, this operator acts on state vectors, as in (6.86).

3 Of course, we assume that the right-hand side of (7.2) converges in these limits. The question of convergence will be discussed briefly in Subsection 7.1.4.

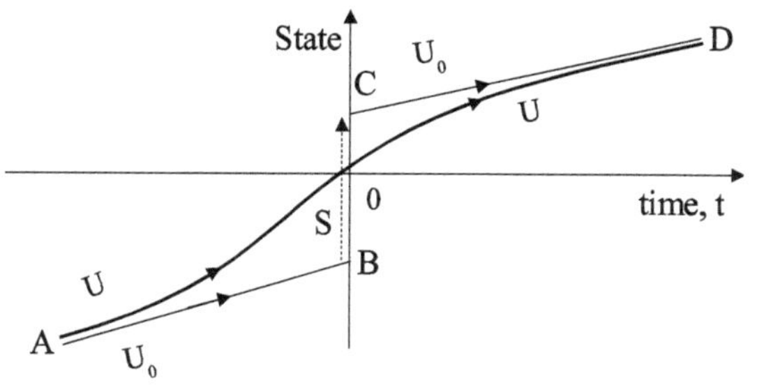

Figure 7.1: Schematic representation of the scattering process.

abstractly as a point on the vertical axis) as a function of time. The exact development of the system, governed by the complete evolution operator U, is shown by the thick line $A \to D$. In asymptotic regions (when the time t is either very negative or very positive), the interaction between the colliding parts of the system is weak. In these regions, the exact time evolution can be rather well approximated by the free operator U_0. These free "trajectories" are shown in the figure by two thin straight lines with arrows: one for very positive times $C \to D$ and the other for very negative times $A \to B$. The thick line (the exact interacting time evolution) asymptotically approaches these thin lines (free evolution) in the remote past (near A) and in the distant future (near D).

If we extrapolate the future and past free evolutions to the time $t = 0$, we will realize that there is a gap B–C between these extrapolated states. The S-operator is defined precisely in such a way as to close this gap, i. e., to connect the free extrapolated states B and C, as shown by the dashed arrow in the figure.

So, in the theory of scattering, the exact time evolution $A \to D$ is approximated in three stages: first the system develops freely up to the time instant $t = 0$, i. e., from A to B. Then there is a sharp jump $B \to C$, represented by the S-operator. Finally, the system again goes into the free evolution mode $C \to D$. As can be seen from the figure, this description of the scattering process is absolutely accurate, as long as we are interested only in the mapping of asymptotically free states from the remote past (A) into asymptotically free states in the distant future (D).

It should also be clear that the scattering operator S contains information about particle interactions in an averaged form integrated over the infinite time interval $t \in (-\infty, \infty)$. This operator is not designed to describe the interacting time evolution during the short interval of collision ($t \approx 0$). For these purposes, we would need the complete interacting time evolution operator U.

In applications, we are mainly interested in matrix elements of the S-operator

$$S_{i \to f} = \langle f|S|i\rangle, \tag{7.6}$$

where $|i\rangle$ is the initial state of the colliding particles and $|f\rangle$ is their final state. Such matrix elements are called the *S-matrix*. Formulas relating the S-matrix to observable quantities, such as scattering cross sections, can be found in any textbook on scattering theory [34, 89].

An important property of the S-operator is its “Poincaré invariance,” i. e., zero commutators with generators of the noninteracting representation of the Poincaré group [95, 44],

$$[S, H_0] = [S, \boldsymbol{P}_0] = [S, \boldsymbol{J}_0] = [S, \boldsymbol{K}_0] = 0. \tag{7.7}$$

In particular, the commutator $[S, H_0] = 0$ implies that in (7.3) we can interchange U_0 and S, so that the full interacting time evolution can be represented as the product of the free evolution operator and the S-operator in any order:

$$U(\infty \leftarrow -\infty) = SU_0(\infty \leftarrow -\infty) = U_0(\infty \leftarrow -\infty)S. \tag{7.8}$$

7.1.2 S-operator in perturbation theory

There are many methods for calculating the S-operator. In this book, we will mainly use perturbation theory. To derive the perturbation theory series for the S-operator, we first note that the operator $S(t)$ in (7.5) satisfies the equation

$$\begin{aligned}
\frac{d}{dt}S(t) &= \frac{d}{dt}\lim_{t'\to-\infty} e^{\frac{i}{\hbar}H_0t}e^{-\frac{i}{\hbar}H(t-t')}e^{-\frac{i}{\hbar}H_0t'} \\
&= \lim_{t'\to-\infty}\left(e^{\frac{i}{\hbar}H_0t}\left(\frac{i}{\hbar}H_0\right)e^{-\frac{i}{\hbar}H(t-t')}e^{-\frac{i}{\hbar}H_0t'} + e^{\frac{i}{\hbar}H_0t}\left(-\frac{i}{\hbar}H\right)e^{-\frac{i}{\hbar}H(t-t')}e^{-\frac{i}{\hbar}H_0t'}\right) \\
&= -\frac{i}{\hbar}\lim_{t'\to-\infty} e^{\frac{i}{\hbar}H_0t}(H-H_0)e^{-\frac{i}{\hbar}H(t-t')}e^{-\frac{i}{\hbar}H_0t'} \\
&= -\frac{i}{\hbar}\lim_{t'\to-\infty} e^{\frac{i}{\hbar}H_0t}Ve^{-\frac{i}{\hbar}H(t-t')}e^{-\frac{i}{\hbar}H_0t'} \\
&= -\frac{i}{\hbar}\lim_{t'\to-\infty} e^{\frac{i}{\hbar}H_0t}Ve^{-\frac{i}{\hbar}H_0t}e^{\frac{i}{\hbar}H_0t}e^{-\frac{i}{\hbar}H(t-t')}e^{-\frac{i}{\hbar}H_0t'} \\
&= -\frac{i}{\hbar}\lim_{t'\to-\infty} V(t)e^{\frac{i}{\hbar}H_0t}e^{-\frac{i}{\hbar}H(t-t')}e^{-\frac{i}{\hbar}H_0t'} \\
&= -\frac{i}{\hbar}V(t)S(t),
\end{aligned} \tag{7.9}$$

where we denote[4]

$$V(t) = e^{\frac{i}{\hbar}H_0t}Ve^{-\frac{i}{\hbar}H_0t}. \tag{7.10}$$

---

4 Note that the t-dependence in $V(t)$ does not mean that we are considering time-dependent interactions. The argument t has no relation with the true *time* dependence of operators in the Heisenberg representation. The latter must be generated by the full interacting Hamiltonian H, not by the free Hamiltonian H_0, as in equation (7.10). To emphasize this difference, in cases like (7.10) we will talk about “t-dependence” instead of “time-dependence”.

By a direct substitution, one can verify that a solution of equation (7.9) with the natural initial condition $S(-\infty) = 1$ is given by the Dyson perturbation series

$$S(t) = 1 - \frac{i}{\hbar} \int_{-\infty}^{t} V(t')dt' - \frac{1}{\hbar^2} \int_{-\infty}^{t} V(t')dt' \int_{-\infty}^{t'} V(t'')dt'' + \cdots .$$

Therefore, the S-operator can be calculated by substituting $t = +\infty$ as the upper limit of t-integrals, so we have

$$S = 1 - \frac{i}{\hbar} \int_{-\infty}^{+\infty} V(t')dt' - \frac{1}{\hbar^2} \int_{-\infty}^{+\infty} V(t')dt' \int_{-\infty}^{t'} V(t'')dt'' + \cdots . \tag{7.11}$$

As a rule, only the first few terms of this series are used in calculations, assuming, therefore, that the interaction V is so weak that it can be regarded as a small perturbation; and the scattering itself is just a small correction to the free propagation of particles. We shall say that a term in the perturbation theory series has *order n* if it contains a product of n factors V. Thus, in (7.11) we have explicit terms in the zero, first and second perturbation orders. We do not want to dwell on mathematical details related to (nontrivial) convergence properties of the expansion (7.11). Throughout this book, we will tacitly assume that all relevant perturbation series do converge.

7.1.3 Convenient notation for *t*-integrals

We shall often use the following symbols for t-integrals:

$$\underline{Y(t)} \equiv -\frac{i}{\hbar} \int_{-\infty}^{t} Y(t')dt'. \tag{7.12}$$

$$\underset{\smile}{Y} \equiv -\frac{i}{\hbar} \int_{-\infty}^{+\infty} Y(t')dt' = \underline{Y(\infty)}. \tag{7.13}$$

In this notation, the perturbation theory series for the S-operator (7.11) has a compact form. We have

$$S = 1 + \underset{\smile}{\Sigma}, \tag{7.14}$$

$$\Sigma(t) = V(t) + V(t)\underline{V(t')} + V(t)\underline{V(t')\underline{V(t'')}} + V(t)\underline{V(t')\underline{V(t'')\underline{V(t''')}}} + \cdots . \tag{7.15}$$

There exist S-operator perturbation expansions other than (7.14)–(7.15). In many cases they are even more convenient. For example, in quantum field theory, the preference is given to the time-ordered perturbation series [95, 65], which uses time ordering[5] of interaction operators in the integrand:

$$
\begin{aligned}
S = 1 &- \frac{i}{\hbar} \int_{-\infty}^{+\infty} dt_1 V(t_1) - \frac{1}{2!\hbar^2} \int_{-\infty}^{+\infty} dt_1 dt_2 T[V(t_1)V(t_2)] \\
&+ \frac{i}{3!\hbar^3} \int_{-\infty}^{+\infty} dt_1 dt_2 dt_3 T[V(t_1)V(t_2)V(t_3)] \\
&+ \frac{1}{4!\hbar^4} \int_{-\infty}^{+\infty} dt_1 dt_2 dt_3 dt_4 T[V(t_1)V(t_2)V(t_3)V(t_4)] + \cdots .
\end{aligned} \tag{7.17}
$$

In the second and third volumes of our book, we will appreciate another perturbative expression,

$$
S = \exp(\underset{\sim}{\Phi}). \tag{7.18}
$$

proposed by Magnus [52, 64, 12]. In this formula, the Hermitian operator $\Phi(t)$ is called the *scattering phase*. It is represented by a series of multiple commutators with t-integrals,

$$
\begin{aligned}
\Phi(t) = V(t) &- \frac{1}{2}\Big[\underline{V(t'), V(t)}\Big] + \frac{1}{6}\Big[\underline{\underline{V(t''), [V(t'), V(t)]}}\Big] \\
&+ \frac{1}{6}\Big[\underline{\underline{[V(t''), V(t')]}, V(t)}\Big] - \frac{1}{12}\Big[\underline{\underline{\underline{V(t'''), [[V(t''), V(t')], V(t)}}}\Big] \\
&- \frac{1}{12}\Big[\underline{\underline{\underline{[V(t'''), [V(t''), V(t')]], V(t)}}}\Big] \\
&- \frac{1}{12}\Big[\underline{\underline{\underline{[V(t'''), V(t'')], [V(t'), V(t)]}}}\Big] + \cdots .
\end{aligned} \tag{7.19}
$$

An important advantage of equation (7.18) is that it explicitly preserves the unitarity of the S-operator in each perturbation order.[6] The three listed perturbation theory series (Dyson, time-ordered and Magnus) are equivalent in the sense that in the limit $n \to \infty$ they converge to the same S-operator. However, in each finite order n their terms can differ.

5 When applied to a product of several t-dependent boson operators, the time ordering symbol T changes the order of the operators so that their t arguments increase from right to left, e. g.,

$$
T\big[A(t_1)B(t_2)\big] = \begin{cases} A(t_1)B(t_2), & \text{if } t_1 > t_2, \\ B(t_2)A(t_1), & \text{if } t_1 < t_2. \end{cases} \tag{7.16}
$$

6 The argument of the exponent is an anti-Hermitian operator $\underset{\sim}{\Phi}$; therefore the exponent itself is unitary.

For brevity, we will often omit t-arguments in operator expressions. Then equations (7.15) and (7.19) simplify, even more, to

$$\Sigma = V + V\underline{V} + V\underline{\underline{V\underline{V}}} + \cdots, \tag{7.20}$$

$$\Phi = V - \frac{1}{2}[\underline{V}, V] + \frac{1}{6}\left[\underline{\underline{V}, [\underline{V}, V]}\right] + \cdots. \tag{7.21}$$

7.1.4 Adiabatic switching of interaction

In formulas for scattering operators (7.15) and (7.19), we encounter t-integrals, *like* $\underline{V(t)}$. Straightforward calculation of such integrals leads to a rather depressing result. To understand, let us introduce the complete basis $|n\rangle$ of eigenvectors of the free Hamiltonian,

$$H_0|n\rangle = E_n|n\rangle, \tag{7.22}$$

$$\sum_n |n\rangle\langle n| = 1, \tag{7.23}$$

and calculate matrix elements of $\underline{V(t)}$ in this basis. Then we have

$$\begin{aligned}\langle n|\underline{V(t)}|m\rangle &\equiv -\frac{i}{\hbar}\int\limits_{-\infty}^{t} \langle n|e^{\frac{i}{\hbar}H_0 t'} V e^{-\frac{i}{\hbar}H_0 t'}|m\rangle dt' = -\frac{i}{\hbar}V_{nm}\int\limits_{-\infty}^{t} e^{\frac{i}{\hbar}(E_n - E_m)t'} dt' \\ &= -V_{nm}\left(\frac{e^{\frac{i}{\hbar}(E_n - E_m)t}}{E_n - E_m} - \frac{e^{\frac{i}{\hbar}(E_n - E_m)(-\infty)}}{E_n - E_m}\right).\end{aligned} \tag{7.24}$$

What can we do with the meaningless term on the right-hand side that contains $(-\infty)$?

This term can be made harmless if we take into account the important fact that the S-operator cannot be applied to all states in the Hilbert space. According to our discussion in Subsection 7.1.1, scattering theory is applicable in its entirety only to *scattering states* $|\Psi\rangle$, in which free particles are far apart in the asymptotic limits $t \to \pm\infty$, so that the time evolution of these states coincides with free evolution in the remote past and in the distant future. Naturally, these assumptions are inapplicable to all states in the Hilbert space. For example, the time evolution of bound states of the full Hamiltonian H does not resemble the free evolution in any time interval. It turns out that if we restrict our theory only to scattering states, then there are no problems with t-integrals.

Indeed, for scattering states $|\Psi\rangle$, the interaction operator effectively vanishes in asymptotic regions, so we can write

$$\lim_{t\to\pm\infty} V e^{-\frac{i}{\hbar}H_0 t}|\Psi\rangle = 0,$$

$$\lim_{t\to\pm\infty} V(t)|\Psi\rangle = \lim_{t\to\pm\infty} e^{\frac{i}{\hbar}H_0 t}(V e^{-\frac{i}{\hbar}H_0 t}|\Psi\rangle) = 0. \tag{7.25}$$

How can we apply this condition to calculations of integrals like (7.24)?

One approach to a rigorous formulation of scattering theory is the explicit consideration of localized wave packets [34]. Then, the cluster separability (= short range) of the interaction V ensures the correct asymptotic behavior of colliding particles and the validity of equation (7.25). However, this approach is rather complicated, and we prefer to keep away from wave packets.

There is another, less rigorous, but shorter way to achieve the same goal – to use a trick known as the *adiabatic switching*. The idea is to add the property (7.25) "by hands." To do this, we multiply $V(t)$ by a real nonnegative function of t that grows slowly from zero (= interaction is "off") at $t = -\infty$ to 1 in the vicinity of $t \approx 0$ (interaction is "on") and then slowly decreases back to zero at $t = +\infty$ (interaction "switches off" again). For example, one convenient choice for such a function is the exponent

$$V(t) = e^{\frac{i}{\hbar}H_0 t} V e^{-\frac{i}{\hbar}H_0 t} e^{-\epsilon|t|}. \tag{7.26}$$

If the parameter ϵ is small and positive, then such a modification of the interaction operator will have no effect on the movement of wave packets and on the S-matrix.[7] For the integral (7.24), we then get

$$\begin{aligned}\langle n|\underline{V(t)}|m\rangle &\approx -V_{nm}\left(\frac{e^{\frac{i}{\hbar}(E_n-E_m)t-\epsilon|t|}}{E_n-E_m} - \frac{e^{\frac{i}{\hbar}(E_n-E_m)(-\infty)-\epsilon(\infty)}}{E_n-E_m}\right)\\ &= -V_{nm}\frac{e^{\frac{i}{\hbar}(E_n-E_m)t-\epsilon|t|}}{E_n-E_m}.\end{aligned}$$

At the end of the calculations, we have to go to the limit $\epsilon \to +0$. Then the t-integral becomes

$$\langle n|\underline{V(t)}|m\rangle \longrightarrow -V_{nm}\frac{e^{\frac{i}{\hbar}(E_n-E_m)t}}{E_n-E_m} \tag{7.27}$$

and the unpleasant expression $e^{i\infty}$ vanishes.

The "adiabatic switching" trick makes possible an alternative derivation of equation (7.11). Let us take equation (6.89) with the initial time in the remote past $t_0 = -\infty$ and the final time in the distant future $t = +\infty$. Then we have

$$\begin{aligned}|\Psi(+\infty)\rangle = \lim_{t\to+\infty} e^{-\frac{i}{\hbar}H_0(t-t_0)}\Bigg(1 - \frac{i}{\hbar}\int_{-\infty}^{\infty} e^{\frac{i}{\hbar}H_0(t'-t_0)} V e^{-\frac{i}{\hbar}H_0(t'-t_0)} dt'\\ - \frac{1}{\hbar^2}\int_{-\infty}^{\infty} e^{\frac{i}{\hbar}H_0(t'-t_0)} V e^{-\frac{i}{\hbar}H_0(t'-t_0)} dt' \int_{-\infty}^{t'} e^{\frac{i}{\hbar}H_0(t''-t_0)} V e^{-\frac{i}{\hbar}H_0(t''-t_0)} dt'' + \cdots\Bigg)\\ \times |\Psi(-\infty)\rangle.\end{aligned}$$

7 Indeed, when the interaction is "off," the wave packets are far from each other, anyway.

Next change the integration variables $t' - t_0 = \tau'$ and $t'' - t_0 = \tau''$ so that[8]

$$
\begin{aligned}
|\Psi(+\infty)\rangle &= U_0(\infty \leftarrow -\infty) \lim_{t\to+\infty}\left(1 - \frac{i}{\hbar}\int_{-\infty}^{\infty} e^{\frac{i}{\hbar}H_0\tau'} V e^{-\frac{i}{\hbar}H_0\tau'} d\tau'\right.\\
&\quad \left. - \frac{1}{\hbar^2}\int_{-\infty}^{\infty} e^{\frac{i}{\hbar}H_0\tau'} V e^{-\frac{i}{\hbar}H_0\tau'} d\tau' \int_{-\infty}^{\tau'} e^{\frac{i}{\hbar}H_0\tau''} V e^{-\frac{i}{\hbar}H_0\tau''} d\tau'' + \cdots \right)|\Psi(-\infty)\rangle \\
&= U_0(\infty \leftarrow -\infty) S |\Psi(-\infty)\rangle.
\end{aligned}
$$

Comparing this formula with equation (7.8), we conclude that the S-factor is exactly as in (7.11).

7.1.5 *T*-matrix

In this subsection we will get acquainted with a useful concept of the T matrix.[9] Let us calculate matrix elements of the S-operator (7.11) in the basis of eigenvectors of the free Hamiltonian (7.22)–(7.23). We have[10]

$$
\begin{aligned}
\langle n|S|m\rangle &= \delta_{nm} - \frac{i}{\hbar}\int_{-\infty}^{\infty} \langle n| e^{\frac{i}{\hbar}H_0 t'} V e^{-\frac{i}{\hbar}H_0 t'} |m\rangle dt' \\
&\quad - \frac{1}{\hbar^2}\int_{-\infty}^{\infty} \langle n| e^{\frac{i}{\hbar}H_0 t'} V e^{-\frac{i}{\hbar}H_0 t'} |k\rangle dt' \int_{-\infty}^{t'} \langle k| e^{\frac{i}{\hbar}H_0 t''} V e^{-\frac{i}{\hbar}H_0 t''} |m\rangle dt'' + \cdots \\
&= \delta_{nm} - \frac{i}{\hbar}\int_{-\infty}^{\infty} e^{\frac{i}{\hbar}(E_n - E_m)t'} V_{nm} dt' \\
&\quad - \frac{1}{\hbar^2}\int_{-\infty}^{\infty} e^{\frac{i}{\hbar}(E_n - E_k)t'} V_{nk} dt' \int_{-\infty}^{t'} e^{\frac{i}{\hbar}(E_k - E_m)t''} V_{km} dt'' + \cdots \\
&= \delta_{nm} - 2\pi i \delta(E_n - E_m) V_{nm} + \frac{i}{\hbar}\int_{-\infty}^{\infty} e^{\frac{i}{\hbar}(E_n - E_k)t'} dt' \frac{e^{\frac{i}{\hbar}(E_k - E_m)t'}}{E_m - E_k} V_{nk} V_{km} + \cdots \\
&= \delta_{nm} - 2\pi i \delta(E_n - E_m) V_{nm} + 2\pi i \delta(E_n - E_m) \frac{1}{E_m - E_k} V_{nk} V_{km} + \cdots \\
&= \delta_{nm} - 2\pi i \delta(E_n - E_m) V_{nk}
\end{aligned}
$$

---

8 For brevity, we do not show the "adiabatic switching" factors explicitly. They force the integrands to vanish at $\pm\infty$ asymptotics. So, they allow us to leave the infinite integration limits ($-\infty$ and ∞) unchanged.

9 I am grateful to Cao Bin for online communications that led to the writing of this subsection.

10 Summation over repeated indices k and l is implied. Equation (7.27) is used for t-integrals.

$$\times \left(\delta_{km} + \frac{1}{E_m - E_k} V_{km} + \frac{1}{E_m - E_k} V_{kl} \frac{1}{E_m - E_l} V_{lm} + \cdots \right)$$
$$= \delta_{nm} - 2\pi i \delta(E_n - E_m) T_{nm}. \tag{7.28}$$

The first term is the unit matrix expressing the free propagation of particles. The matrix in the second term is called the *transition matrix* (or *T-matrix*).[11] For further simplifications we can write

$$\begin{aligned}
T_{nm} &\equiv V_{nk}\left(\delta_{km} + \frac{1}{E_m - E_k} V_{km} + \frac{1}{E_m - E_k} V_{kl} \frac{1}{E_m - E_l} V_{lm} + \cdots \right) \\
&= \langle n|V|k\rangle\langle k|m\rangle + \langle n|V(E_m - E_k)^{-1}|k\rangle\langle k|V|m\rangle \\
&\quad + \langle n|V(E_m - E_k)^{-1}|k\rangle\langle k|V(E_m - E_l)^{-1}|l\rangle\langle l|V|m\rangle + \cdots \\
&= \langle n|V|k\rangle\langle k|m\rangle + \langle n|V(E - H_0)^{-1}|k\rangle\langle k|V|m\rangle \\
&\quad + \langle n|V(E - H_0)^{-1}|k\rangle\langle k|V(E - H_0)^{-1}|l\rangle\langle l|V|m\rangle + \cdots \\
&= \langle n|V|m\rangle + \langle n|V(E - H_0)^{-1}V|m\rangle \\
&\quad + \langle n|V(E - H_0)^{-1}V(E - H_0)^{-1}V|m\rangle + \cdots \\
&= \langle n|V\left(1 + \frac{1}{E - H_0} V + \frac{1}{E - H_0} V \frac{1}{E - H_0} V + \cdots \right)|m\rangle.
\end{aligned}$$

The infinite series in the parenthesis can be summed by the standard formula $(1-x)^{-1} = 1 + x + x^2 + \cdots$, so we have

$$\begin{aligned}
T_{nm} &= \langle n|V \frac{1}{1 - (E - H_0)^{-1}V}|m\rangle = \langle n|V(E - H_0)(E - H_0 - V)^{-1}|m\rangle \\
&= \langle n|V(E - H_0)(E - H)^{-1}|m\rangle = \langle n|T(E)|m\rangle.
\end{aligned}$$

Thus, the T-matrix is represented by matrix elements of the energy-dependent $T(E)$-operator given by formula

$$T(E) \equiv V(E - H_0)(E - H)^{-1}. \tag{7.29}$$

The beauty of this result is that it provides a closed expression for the S-operator that goes beyond perturbation theory. This result is widely used in numerical calculations [71, 13, 49, 25].

7.1.6 *S*-matrix and bound states

The formal expression (7.29) can be used to derive an important connection between poles of the S-matrix and energies of bound states. Our derivation will be formal and

---

11 Taking into account the fact that the T-matrix enters the S-matrix multiplied by the delta-function of energy $\delta(E_n - E_m)$, we have denoted $E = E_m = E_n$.

heuristic. More rigorous reasoning can be found in textbooks on scattering theory [34, 89].

We already mentioned that scattering theory can be formulated only for states that behave asymptotically as free ones. The energies E of such states exceed the energy E_0 of separated reactants at rest, for which we have

$$E_0 = \sum_{a=1}^{N} m_a c^2.$$

Therefore, the operator $T(E)$ introduced in (7.29) has a well-defined meaning only in the energy interval

$$E \in [E_0, \infty). \tag{7.30}$$

There are reasons to believe that this operator is an analytic function of energy E. Therefore, it would be interesting to find out where the poles of this function are located. We can expect the appearance of poles at those values of E, where the denominator of the expression $(E - H_0)(E - H)^{-1}$ in (7.29) vanishes. Hence (at least some of) the poles E_β can be found as solutions of the eigenvalue equation

$$(H - E_\beta)|\Psi_\beta\rangle = 0.$$

Obviously, this is the familiar stationary Schrödinger equation (6.79) for bound states. This means that there is a connection between poles of the T-operator and energies of bound states E_β of the full Hamiltonian H.[12] These energies are always lower than E_0, i. e., formally they are outside the domain of the operator $T(E)$. Therefore, the abovementioned connection (poles of the T-operator) $\leftrightarrow$ (energies of bound states) can be established only in the sense of analytic continuation of the operator $T(E)$ from its natural domain (7.30) to values below E_0.

It is important to emphasize that the possibility of finding the energies of the bound states E_β from the T-operator does not mean that the eigenvectors of these states $|\Psi_\beta\rangle$ can be found in the same way. All bound states are eigenstates of the T-operator, corresponding only to one (infinite) eigenvalue. Therefore, even knowing the exact T-operator, the most we can do is to find a subspace in $\mathcal{H}$ that is the linear span of all bound states. This ambiguity is closely related to the scattering equivalence of Hamiltonians, which we shall consider in the next section.

7.2 Scattering equivalence

The results of the previous section indicate that even exact knowledge of the S-operator does not allow us to completely reconstruct the corresponding Hamiltonian H. In

---

12 Similarly, the S-operator can be also regarded as an analytic function $S(E)$ on the complex energy plane with poles at positions $E = E_\beta$.

other words, many different Hamiltonians can have identical scattering properties. Here we will discuss these issues in more detail, because they will play an important role in Volume 3.

7.2.1 Equivalent Hamiltonians

The S-operator and the Hamiltonian are two fundamentally different ways of describing dynamics. From the Hamiltonian H one can obtain the evolution operator $U(t \leftarrow t_0) \equiv e^{-\frac{i}{\hbar}H(t-t_0)}$, which describes in detail the development of states in all time intervals, both large and small. On the other hand, the S-operator represents only the "integrated" time evolution in the infinite interval. In other words, if we know the state of the system in the remote past $|\Psi(-\infty)\rangle$, the free Hamiltonian H_0 and the scattering operator S, then we can find the final state of the system in the far future (7.8), i. e.,

$$|\Psi(\infty)\rangle = U(\infty \leftarrow -\infty)|\Psi(-\infty)\rangle = U_0(\infty \leftarrow -\infty)S|\Psi(-\infty)\rangle.$$

However, we cannot say anything about the system's evolution in the interacting regime.

Despite its incomplete nature, the information contained in the S-operator is fully sufficient for the analysis of most experiments in high-energy physics. In particular, from the S-operator one can obtain accurate scattering cross sections as well as energies and lifetimes of stable and metastable bound states. This situation gave the impression that an exhaustive theory of elementary particles could be constructed on the basis of the S-operator alone without resorting to the Hamiltonian and wave functions. However, this impression is deceptive, because the description of physics by means of scattering theory is incomplete, and such a theory is applicable only to a limited class of experiments.

Knowing the full interacting Hamiltonian H, we can calculate the corresponding S-operator by formulas (7.14), (7.17) or (7.18). However, the converse is not true: even if the S-operator is fully known, it is impossible to recover the unique underlying Hamiltonian. The same S-operator can be obtained from many different Hamiltonians.

Suppose that two Hamiltonians H and H' are related to each other by a unitary transformation $e^{i\Xi}$, i. e.,

$$H' = e^{i\Xi}He^{-i\Xi}. \tag{7.31}$$

Then they have exactly the same scattering properties[13] if the following condition is satisfied:

13 We say that such Hamiltonians are *scattering-equivalent*.

$$\lim_{t\to\pm\infty} e^{\frac{i}{\hbar}H_0 t}\Xi e^{-\frac{i}{\hbar}H_0 t} = 0. \tag{7.32}$$

Indeed, in the limits $\eta \to +\infty$, $\eta' \to -\infty$, we obtain from (7.5) that the two scattering operators are equal [27], i. e.,[14]

$$\begin{aligned}
S' &= \lim_{\eta'\to-\infty,\eta\to\infty} e^{\frac{i}{\hbar}H_0\eta} e^{-\frac{i}{\hbar}H'(\eta-\eta')} e^{-\frac{i}{\hbar}H_0\eta'} \\
&= \lim_{\eta'\to-\infty,\eta\to\infty} e^{\frac{i}{\hbar}H_0\eta} (e^{i\Xi} e^{-\frac{i}{\hbar}H(\eta-\eta')} e^{-i\Xi}) e^{-\frac{i}{\hbar}H_0\eta'} \\
&= \lim_{\eta'\to-\infty,\eta\to\infty} (e^{\frac{i}{\hbar}H_0\eta} e^{i\Xi} e^{-\frac{i}{\hbar}H_0\eta}) e^{\frac{i}{\hbar}H_0\eta} e^{-\frac{i}{\hbar}H(\eta-\eta')} e^{-\frac{i}{\hbar}H_0\eta'} \\
&\quad \times (e^{\frac{i}{\hbar}H_0\eta'} e^{-i\Xi} e^{-\frac{i}{\hbar}H_0\eta'}) \qquad (7.33) \\
&= \lim_{\eta'\to-\infty,\eta\to\infty} e^{\frac{i}{\hbar}H_0\eta} e^{-\frac{i}{\hbar}H(\eta-\eta')} e^{-\frac{i}{\hbar}H_0\eta'} = S. \qquad (7.34)
\end{aligned}$$

Note that, due to Lemma G.9, the energy spectra of the two equivalent Hamiltonians H and H' also coincide. However, their eigenvectors could be different, and the corresponding description of dynamics (e. g., by equation (6.88)) could also differ. Therefore, two theories predicting the same scattering are not necessarily equivalent in the full physical sense.

7.2.2 Bakamjian construction of point-form dynamics

It turns out that the above statement about scattering equivalent Hamiltonians can be generalized in the sense that even two different forms of dynamics (for example, the instant and point forms) can have the same S-operators. This nontrivial fact will be discussed in Subsection 7.2.4. To prepare for this discussion, here we will build a specific version of the point form of dynamics, using a recipe due to Bakamjian [5]. This method is very similar to the method of Bakamjian–Thomas from Subsection 6.3.2.

We consider a system of $n \geq 2$ massive free particles with noninteracting operators of mass M_0, linear momentum $\boldsymbol{P}_0$, angular momentum $\boldsymbol{J}_0$, center-of-energy position $\boldsymbol{R}_0$ and spin $\boldsymbol{S}_0 = \boldsymbol{J}_0 - [\boldsymbol{R}_0 \times \boldsymbol{P}_0]$. Then we introduce two new operators,

$$\boldsymbol{Q}_0 \equiv \frac{\boldsymbol{P}_0}{M_0 c^2},$$

14 In (7.33) we use condition (7.32) to make replacements

$$\begin{aligned}
&\lim_{\eta\to\infty}\left[\exp\left(\frac{i}{\hbar}H_0\eta\right)\exp(i\Xi)\exp\left(-\frac{i}{\hbar}H_0\eta\right)\right] \\
&= \lim_{\eta'\to-\infty}\left[\exp\left(\frac{i}{\hbar}H_0\eta'\right)\exp(-i\Xi)\exp\left(-\frac{i}{\hbar}H_0\eta'\right)\right] = 1.
\end{aligned}$$

$$\boldsymbol{X}_0 \equiv M_0 c^2 \boldsymbol{R}_0,$$

which satisfy the canonical commutation relations

$$[X_{0i}, Q_{0i}] = [X_{0i}, X_{0j}] = [Q_{0i}, Q_{0j}] = 0,$$
$$[X_{0i}, Q_{0j}] = i\hbar\delta_{ij}.$$

Next, we express the generators $\{H_0, \boldsymbol{P}_0, \boldsymbol{J}_0, \boldsymbol{K}_0\}$ of the noninteracting representation of the Poincaré group through the alternative set of operators $\{M_0, \boldsymbol{Q}_0, \boldsymbol{X}_0, \boldsymbol{S}_0\}$ as follows (compare with (6.37)–(6.38)):

$$\boldsymbol{P}_0 = M_0 c^2 \boldsymbol{Q}_0,$$
$$\boldsymbol{K}_0 = -\frac{1}{2}\left(\sqrt{1 + Q_0^2 c^2}\boldsymbol{X}_0 + \boldsymbol{X}_0\sqrt{1 + Q_0^2 c^2}\right) - \frac{[\boldsymbol{Q}_0 \times \boldsymbol{S}_0]}{1 + \sqrt{1 + Q_0^2 c^2}},$$
$$\boldsymbol{J}_0 = [\boldsymbol{X}_0 \times \boldsymbol{Q}_0] + \boldsymbol{S}_0,$$
$$H_0 = M_0 c^2 \sqrt{1 + Q_0^2 c^2}.$$

Now, a point-form interaction can be introduced by modifying the mass operator

$$M_0 \to M \tag{7.35}$$

so as to satisfy the following conditions[15]:

$$[M, \boldsymbol{Q}_0] = [M, \boldsymbol{X}_0] = [M, \boldsymbol{S}_0] = 0.$$

These conditions guarantee, in particular, that the mass operator M is invariant under transformations from the Lorentz subgroup, i. e.,

$$[M, \boldsymbol{K}_0] = [M, \boldsymbol{J}_0] = 0.$$

The mass modification (7.35) introduces interaction in generators of the translation subgroup,

$$\boldsymbol{P} = M c^2 \boldsymbol{Q}_0,$$
$$H = M c^2 \sqrt{1 + Q_0^2 c^2}, \tag{7.36}$$

while Lorentz subgroup generators $\boldsymbol{K}_0$ and $\boldsymbol{J}_0$ remain interaction-free. So, we succeeded in defining a nontrivial interaction $\{H, \boldsymbol{P}, \boldsymbol{J}_0, \boldsymbol{K}_0\}$ in the point form of dynamics.

15 As in Subsection 6.3.2, these conditions can be fulfilled by defining $M = M_0 + N$, where N is a rotationally invariant function of operators of the relative position and momentum that commute with both $\boldsymbol{Q}_0$ and $\boldsymbol{X}_0$.

7.2.3 Unitary link between point and instant forms of dynamics

The S-matrix equivalence of Hamiltonians established in Subsection 7.2.1 remains valid even if the transformation $e^{i\Xi}$ (7.31) changes the relativistic form of dynamics [78, 79]. Here we are going to demonstrate this equivalence using the example of the Dirac instant and point forms [78].

To begin with, suppose that we are given a Bakamjian point form of dynamics with operators

$$\begin{aligned} M &\neq M_0, \\ \boldsymbol{P} &= \boldsymbol{Q}_0 M c^2, \\ \boldsymbol{J} &= \boldsymbol{J}_0, \\ \boldsymbol{R} &= \boldsymbol{X}_0 M^{-1} c^{-2}, \end{aligned}$$

specified in Subsection 7.2.2. Then we define a unitary operator

$$\Theta = \zeta_0 \zeta^{-1},$$

where

$$\begin{aligned} \zeta_0 &\equiv \exp(-i \ln(M_0 c^2) \mathfrak{Z}), \\ \zeta &\equiv \exp(-i \ln(M c^2) \mathfrak{Z}) \end{aligned}$$

and operator

$$\mathfrak{Z} \equiv \frac{1}{2\hbar}(\boldsymbol{R} \cdot \boldsymbol{P} + \boldsymbol{P} \cdot \boldsymbol{R}) = \frac{1}{2\hbar}(\boldsymbol{Q}_0 \cdot \boldsymbol{X}_0 + \boldsymbol{X}_0 \cdot \boldsymbol{Q}_0)$$

was defined in (4.53). Our goal is to show that the set of operators $\Theta M \Theta^{-1}$, $\Theta \boldsymbol{P} \Theta^{-1}$, $\Theta \boldsymbol{J}_0 \Theta^{-1}$ and $\Theta \boldsymbol{R} \Theta^{-1}$ generates a Poincaré group representation in the Bakamjian–Thomas instant form.

Let us define

$$\boldsymbol{Q}_0(b) \equiv e^{ib\mathfrak{Z}} \boldsymbol{Q}_0 e^{-ib\mathfrak{Z}}, \quad b \in \mathbb{R}.$$

From the commutator

$$[\mathfrak{Z}, \boldsymbol{Q}_0] = i\boldsymbol{Q}_0,$$

it follows that

$$\begin{aligned} \frac{d}{db}\boldsymbol{Q}_0(b) &= i[\mathfrak{Z}, \boldsymbol{Q}_0] = -\boldsymbol{Q}_0, \\ \boldsymbol{Q}_0(b) &= e^{-b}\boldsymbol{Q}_0. \end{aligned}$$

This formula remains valid even if b is not a number but any Hermitian operator commuting with both $\boldsymbol{Q}_0$ and $\boldsymbol{X}_0$. For example, if $b = \ln(M_0 c^2)$, then

$$\zeta_0^{-1}\mathbf{Q}_0\zeta_0 = e^{i\ln(M_0c^2)3}\mathbf{Q}_0e^{-i\ln(M_0c^2)3} = e^{-\ln(M_0c^2)}\mathbf{Q}_0 = \mathbf{Q}_0/(M_0c^2).$$

Similarly, one can show

$$\begin{aligned}\zeta^{-1}\mathbf{Q}_0\zeta &= e^{i\ln(Mc^2)3}\mathbf{Q}_0e^{-i\ln(Mc^2)3} = \mathbf{Q}_0/(Mc^2),\\ \zeta_0^{-1}\mathbf{X}_0\zeta_0 &= e^{i\ln(M_0c^2)3}\mathbf{X}_0e^{-i\ln(M_0c^2)3} = \mathbf{X}_0M_0c^2,\\ \zeta^{-1}\mathbf{X}_0\zeta &= e^{i\ln(Mc^2)3}\mathbf{X}_0e^{-i\ln(Mc^2)3} = \mathbf{X}_0Mc^2,\end{aligned}$$

which implies

$$\begin{aligned}\Theta\mathbf{P}\Theta^{-1} &= \zeta_0\zeta^{-1}(\mathbf{Q}_0Mc^2)\zeta\zeta_0^{-1} = \zeta_0\mathbf{Q}_0\zeta_0^{-1} = \mathbf{Q}_0M_0c^2 = \mathbf{P}_0,\\ \Theta\mathbf{J}_0\Theta^{-1} &= \mathbf{J}_0,\\ \Theta\mathbf{R}\Theta^{-1} &= \zeta_0\zeta^{-1}(\mathbf{X}_0/(Mc^2))\zeta\zeta_0^{-1} = \zeta_0\mathbf{X}_0\zeta_0^{-1} = \mathbf{X}_0/(M_0c^2) = \mathbf{R}_0.\end{aligned}$$

From these formulas, it is clear that the transformation Θ really changes dynamics from the Bakamjian point form to the Bakamjian–Thomas instant form.

7.2.4 Scattering equivalence of forms of dynamics

Let us now verify that the scattering operator S, calculated with the point-form Hamiltonian $H = Mc^2\sqrt{1+c^2Q_0^2}$, is the same as the operator S' calculated with the instant-form Hamiltonian $H' = \Theta H\Theta^{-1}$. Notice that we can rewrite equation (7.4) as

$$S = \Omega^+(H,H_0)\Omega^-(H,H_0),$$

where operators

$$\Omega^\pm(H,H_0) \equiv \lim_{t\to\pm\infty} e^{\frac{i}{\hbar}H_0t}e^{-\frac{i}{\hbar}Ht}$$

are called *Møller wave operators*. Next we use the so-called *Birman–Kato invariance principle* [24], which states that $\Omega^\pm(H,H_0) = \Omega^\pm(f(H),f(H_0))$, where f can be any smooth function with positive derivative. Using the relationship between mass operators in the point (M) and instant (M') forms

$$M = \zeta^{-1}M\zeta = \zeta^{-1}\Theta^{-1}M'\Theta\zeta = \zeta^{-1}\zeta\zeta_0^{-1}M'\zeta_0\zeta^{-1}\zeta = \zeta_0^{-1}M'\zeta_0,$$

we obtain

$$\begin{aligned}\Omega^\pm(H,H_0) &\equiv \Omega^\pm\left(Mc^2\sqrt{1+Q_0^2c^2}, M_0c^2\sqrt{1+Q_0^2c^2}\right) = \Omega^\pm(Mc^2, M_0c^2)\\ &= \Omega^\pm(\zeta_0^{-1}M'\zeta_0c^2, M_0c^2) = \zeta_0^{-1}\Omega^\pm(M'c^2, M_0c^2)\zeta_0\end{aligned}$$

$$
\begin{aligned}
&= \zeta_0^{-1}\Omega^{\pm}\Big(\sqrt{(M')^2c^4 + P_0^2c^2}, \sqrt{M_0^2c^4 + P_0^2c^2}\Big)\zeta_0 \\
&= \zeta_0^{-1}\Omega^{\pm}(H', H_0)\zeta_0,
\end{aligned}
$$

$$
\begin{aligned}
S' &= \Omega^{+}(H', H_0)\Omega^{-}(H', H_0) = \zeta_0\Omega^{+}(H, H_0)\zeta_0^{-1}\zeta_0\Omega^{-}(H, H_0)\zeta_0^{-1} \\
&= \zeta_0\Omega^{+}(H, H_0)\Omega^{-}(H, H_0)\zeta_0^{-1} = \zeta_0 S\zeta_0^{-1}.
\end{aligned}
$$

Then we notice that S commutes with free generators (7.7) and therefore with ζ_0 as well. Hence, $S' = S$ and the transformation Θ conserves the S-matrix. This completes the proof.

In addition to the above results, Sokolov and Shatnii [79] established the mutual scattering equivalence of all three basic forms of dynamics – instant, point and front ones. Then it seems reasonable to assume that the S-operator is not sensitive to the form of dynamics at all.

The scattering equivalence of Hamiltonians and forms of dynamics gives us great advantages in calculations. If we are only interested in scattering amplitudes, energies and lifetimes of bound states,[16] then we can choose the Hamiltonian and the form of dynamics from a wide selection, as convenient. However, as we have already said, the scattering equivalence does not mean full physical equivalence of different theories. In the third volume of our book, we will see that an adequate description of the time evolution and other inertial transformations is possible only within the instant-form framework.

16 That is, the properties directly related to the scattering matrix.

A Delta function

Dirac's *delta function* $\delta(x)$ is defined by the following properties:

(1) $\delta(x) = 0$ for all x, except $x = 0$;

(2) $\delta(0)$ is infinite;

(3) the integral $\int_\Delta \delta(x)dx$ is equal to 1 for any interval Δ that includes the point $x = 0$.

Often the delta function is defined by the following integral:

$$\int_\Delta f(x)\delta(x)dx = f(0),$$

where $f(x)$ is any smooth function. Quite useful is the following integral representation:

$$\frac{1}{2\pi\hbar}\int_{-\infty}^{\infty} e^{\frac{i}{\hbar}kx}dk = \delta(x).$$

The other important property is

$$\delta(bx) = \frac{1}{|b|}\delta(x)$$

for any $b \neq 0$.

The delta function of a vector argument $\boldsymbol{r} = (x, y, z)$ is defined as the product of "scalar" delta functions

$$\delta(\boldsymbol{r}) = \delta(x)\delta(y)\delta(z).$$

It has the integral representation

$$\frac{1}{(2\pi\hbar)^3}\int e^{\frac{i}{\hbar}\boldsymbol{k}\cdot\boldsymbol{r}}d\boldsymbol{k} = \delta(\boldsymbol{r}) \tag{A.1}$$

and the useful property

$$\frac{\partial^2}{\partial\boldsymbol{r}^2}\left(\frac{1}{r}\right) = -4\pi\delta(\boldsymbol{r}), \tag{A.2}$$

where we have introduced a formal notation for the *Laplacian*

$$\frac{d^2}{d\boldsymbol{r}^2} \equiv \frac{\partial^2}{\partial x^2} + \frac{\partial^2}{\partial y^2} + \frac{\partial^2}{\partial z^2}.$$

https://doi.org/10.1515/9783110492132-008

B Orthocomplemented lattices

B.1 Derivation of quantum axioms

Here we continue our derivation of the axioms of orthocomplemented lattices[1] from elementary properties of probability measures $(\phi|\mathcal{X})$ (see Subsection 1.4.2).

The implication relation $\leq$ has three obvious properties.

Lemma B.1 (property 1 from Table 1.2). *Each proposition implies itself:* $\mathcal{X} \leq \mathcal{X}$.

Proof. This follows from the simple fact that for each state ϕ we have $(\phi|\mathcal{X}) \leq (\phi|\mathcal{X})$. □

Lemma B.2 (property 2). *If two propositions imply each other, then they are equal. If* $\mathcal{X} \leq \mathcal{Y}$ *and* $\mathcal{Y} \leq \mathcal{X}$, *then* $\mathcal{X} = \mathcal{Y}$.

Proof. If the conditions of the lemma are satisfied, then $(\phi|\mathcal{X}) \leq (\phi|\mathcal{Y})$ and $(\phi|\mathcal{Y}) \leq (\phi|\mathcal{X})$ for any state ϕ. It then follows that $(\phi|\mathcal{X}) = (\phi|\mathcal{Y})$ and, according to (1.5), $\mathcal{X} = \mathcal{Y}$. □

Lemma B.3 (property 3). *If* $\mathcal{X} \leq \mathcal{Y}$ *and* $\mathcal{Y} \leq \mathcal{Z}$, *then* $\mathcal{X} \leq \mathcal{Z}$.

Proof. From the conditions of the lemma it follows that $(\phi|\mathcal{X}) \leq (\phi|\mathcal{Y}) \leq (\phi|\mathcal{Z})$ for any state ϕ. Therefore, $(\phi|\mathcal{X}) \leq (\phi|\mathcal{Z})$ for any ϕ, and $\mathcal{X} \leq \mathcal{Z}$. □

From equalities (1.1) and (1.2) we also conclude the following.

Corollary B.4 (property 4). $\mathcal{X} \leq \mathcal{I}$ *for any* $\mathcal{X} \in \mathcal{L}$.

Corollary B.5 (property 5). $\emptyset \leq \mathcal{X}$ *for any* $\mathcal{X} \in \mathcal{L}$.

The next postulate defines the operation of logical negation.

Postulate B.6 (definition of orthocomplement). For each proposition $\mathcal{X}$ there exists an *orthocomplemented* proposition $\mathcal{X}^\perp$ such that for all states ϕ

$$(\phi|\mathcal{X}^\perp) = 1 - (\phi|\mathcal{X}). \tag{B.1}$$

Lemma B.7 (property 16). $(\mathcal{X}^\perp)^\perp = \mathcal{X}$.

Proof. From Definition (B.1) we have for any state ϕ

$$(\phi|(\mathcal{X}^\perp)^\perp) = 1 - (\phi|\mathcal{X}^\perp) = 1 - (1 - (\phi|\mathcal{X})) = (\phi|\mathcal{X}).$$ □

Lemma B.8 (property 14). $\mathcal{X} \wedge \mathcal{X}^\perp = \emptyset$.

---

1 Properties 1–17 from Table 1.2.

https://doi.org/10.1515/9783110492132-009

Proof. Suppose that $\mathcal{X} \wedge \mathcal{X}^{\perp} = \mathcal{Y} \neq \emptyset$. Then, according to Postulate 1.4, there exists a state ϕ such that $(\phi|\mathcal{Y}) = 1$. From the definition of $\wedge$ (property 6 in Table 1.2) we obtain

$$\begin{aligned} \mathcal{Y} &\leq \mathcal{X}, \\ \mathcal{Y} &\leq \mathcal{X}^{\perp}, \\ 1 = (\phi|\mathcal{Y}) &\leq (\phi|\mathcal{X}), \\ 1 = (\phi|\mathcal{Y}) &\leq (\phi|\mathcal{X}^{\perp}). \end{aligned}$$

This implies $(\phi|\mathcal{X}) = 1$ and $(\phi|\mathcal{X}^{\perp}) = 1$. Since $\mathcal{X}^{\perp} \leq \mathcal{X}^{\perp}$, we conclude that propositions $\mathcal{X}$ and $\mathcal{X}^{\perp}$ are disjoint. Hence, we can apply Kolmogorov's axiom (1.3), which yields the absurd result

$$(\phi|\mathcal{X} \vee \mathcal{X}^{\perp}) = (\phi|\mathcal{X}) + (\phi|\mathcal{X}^{\perp}) = 1 + 1 = 2.$$

Thus, our supposition was wrong, and $\mathcal{X} \wedge \mathcal{X}^{\perp} = \emptyset$. □

Lemma B.9 (property 17). *If $\mathcal{X} \leq \mathcal{Y}$, then $\mathcal{Y}^{\perp} \leq \mathcal{X}^{\perp}$.*

Proof. If $\mathcal{X} \leq \mathcal{Y}$ then $(\phi|\mathcal{X}) \leq (\phi|\mathcal{Y})$ and $(1 - (\phi|\mathcal{X})) \geq (1 - (\phi|\mathcal{Y}))$ for all states ϕ. But by our definition (B.1) the two parts of the last inequality express probability measures for the propositions $\mathcal{X}^{\perp}$ and $\mathcal{Y}^{\perp}$, respectively. □

Lemma B.10 (property 15). $\mathcal{X} \vee \mathcal{X}^{\perp} = \mathcal{I}$.

Proof. Propositions $\mathcal{X}$ and $\mathcal{X}^{\perp}$ are disjoint. Therefore, by equation (1.3), for any state ϕ we have

$$(\phi|\mathcal{X} \vee \mathcal{X}^{\perp}) = (\phi|\mathcal{X}) + (\phi|\mathcal{X}^{\perp}) = (\phi|\mathcal{X}) + (1 - (\phi|\mathcal{X})) = 1,$$

which proves the lemma. □

B.2 Some lemmas and theorems

From the axioms of orthocomplemented lattices[2] one can prove a number of useful results.

Lemma B.11. *We have*

$$\mathcal{Z} \leq \mathcal{X} \wedge \mathcal{Y} \quad \Rightarrow \quad \mathcal{Z} \leq \mathcal{X}. \tag{B.2}$$

Proof. From property 6 we have $\mathcal{X} \wedge \mathcal{Y} \leq \mathcal{X}$, thus $\mathcal{Z} \leq \mathcal{X} \wedge \mathcal{Y} \leq \mathcal{X}$, and by the transitivity property 3, $\mathcal{Z} \leq \mathcal{X}$. □

2 Properties 1–17 from Table 1.2.

Lemma B.12. *We have*

$$\mathcal{X} \leq \mathcal{Y} \quad \Leftrightarrow \quad \mathcal{X} \wedge \mathcal{Y} = \mathcal{X}. \tag{B.3}$$

Proof. From $\mathcal{X} \leq \mathcal{Y}$ and $\mathcal{X} \leq \mathcal{X}$ by property 7 it follows that $\mathcal{X} \leq \mathcal{X} \wedge \mathcal{Y}$. On the other hand, $\mathcal{X} \wedge \mathcal{Y} \leq \mathcal{X}$ (property 6). From property 2 we then obtain $\mathcal{X} \wedge \mathcal{Y} = \mathcal{X}$. The converse statement follows from property 6, written in the form

$$\mathcal{X} \wedge \mathcal{Y} \leq \mathcal{Y}.$$

Since $\mathcal{X} \wedge \mathcal{Y} = \mathcal{X}$, replacing the left-hand side with $\mathcal{X}$, we obtain the left-hand side of (B.3). □

Lemma B.13. *For each proposition* $\mathcal{Z}$

$$\mathcal{X} \leq \mathcal{Y} \quad \Rightarrow \quad \mathcal{X} \wedge \mathcal{Z} \leq \mathcal{Y} \wedge \mathcal{Z}. \tag{B.4}$$

Proof. This follows from $\mathcal{X} \wedge \mathcal{Z} \leq \mathcal{X} \leq \mathcal{Y}$ and $\mathcal{X} \wedge \mathcal{Z} \leq \mathcal{Z}$ by property 7. □

Proofs of equalities

$$\mathcal{X} \wedge \mathcal{X} = \mathcal{X}, \tag{B.5}$$

$$\emptyset \wedge \mathcal{X} = \emptyset, \tag{B.6}$$

$$\mathcal{I} \wedge \mathcal{X} = \mathcal{X}, \tag{B.7}$$

$$\emptyset^{\perp} = \mathcal{I} \tag{B.8}$$

are left as exercises for the reader.

The following observation simplifies substantially proofs of various results in orthocomplemented lattices. If you have a correct expression composed of lattice elements, then you can form a *dual* expression by simultaneous application of the following operations:

(1) swap symbols "$\wedge$" $\rightleftharpoons$ "$\vee$";
(2) change directions of the implication signs "$\leq$" $\rightleftharpoons$ "$\geq$" and "$<$" $\rightleftharpoons$ "$>$";
(3) swap special lattice elements $\emptyset \rightleftharpoons \mathcal{I}$.

Then it is easy to realize that all axioms in Table 1.2 have the property of *duality*. Each axiom is either self-dual or its dual statement is also a valid axiom. Therefore, for each logical (in)equality, its dual is also a true (in)equality. For example, using the duality property, we get from (B.2)–(B.8)

$$\mathcal{X} \vee \mathcal{Y} \leq \mathcal{Z} \quad \Rightarrow \quad \mathcal{X} \leq \mathcal{Z},$$

$$\mathcal{X} \leq \mathcal{Y} \quad \Leftrightarrow \quad \mathcal{X} \vee \mathcal{Y} = \mathcal{Y},$$

$$\mathcal{Y} \leq \mathcal{X} \quad \Rightarrow \quad \mathcal{Y} \vee \mathcal{Z} \leq \mathcal{X} \vee \mathcal{Z},$$

$$\mathcal{X} \vee \mathcal{X} = \mathcal{X},$$
$$\mathcal{I} \vee \mathcal{X} = \mathcal{I},$$
$$\emptyset \vee \mathcal{X} = \mathcal{X},$$
$$\mathcal{I}^{\perp} = \emptyset. \tag{B.9}$$

Theorem B.14. *In an orthocomplemented lattice, all propositions are compatible with each other if and only if the lattice is distributive, i. e., properties 18 and 19 from Table* 1.2 *are fulfilled.*

Proof. If the lattice is distributive, then for each pair of propositions $\mathcal{X}$ and $\mathcal{Y}$ one has[3]

$$(\mathcal{X} \wedge \mathcal{Y}) \vee (\mathcal{X} \wedge \mathcal{Y}^{\perp}) = \mathcal{X} \wedge (\mathcal{Y} \vee \mathcal{Y}^{\perp}) = \mathcal{X} \wedge \mathcal{I} = \mathcal{X}.$$

Changing the places of $\mathcal{X}$ and $\mathcal{Y}$, we get

$$(\mathcal{X} \wedge \mathcal{Y}) \vee (\mathcal{X}^{\perp} \wedge \mathcal{Y}) = \mathcal{Y}.$$

These formulas coincide with our definition of compatibility (1.6)–(1.7), which proves the direct statement of the theorem.

The proof of the converse statement (compatibility → distributivity) is more complicated. Suppose that in our lattice all pairs of propositions are compatible and choose three arbitrary propositions $\mathcal{X}$, $\mathcal{Y}$ and $\mathcal{Z}$. Our task is to prove the validity of the distributive laws

$$(\mathcal{X} \wedge \mathcal{Z}) \vee (\mathcal{Y} \wedge \mathcal{Z}) = (\mathcal{X} \vee \mathcal{Y}) \wedge \mathcal{X}, \tag{B.10}$$
$$(\mathcal{X} \vee \mathcal{Z}) \wedge (\mathcal{Y} \vee \mathcal{Z}) = (\mathcal{X} \wedge \mathcal{Y}) \vee \mathcal{Z}. \tag{B.11}$$

Let us first prove that the seven propositions[4] (see Figure B.1)

$$\mathcal{Q}_1 = \mathcal{X} \wedge \mathcal{Y} \wedge \mathcal{Z},$$
$$\mathcal{Q}_2 = \mathcal{X}^{\perp} \wedge \mathcal{Y} \wedge \mathcal{Z},$$
$$\mathcal{Q}_3 = \mathcal{X} \wedge \mathcal{Y}^{\perp} \wedge \mathcal{Z},$$
$$\mathcal{Q}_4 = \mathcal{X} \wedge \mathcal{Y} \wedge \mathcal{Z}^{\perp},$$
$$\mathcal{Q}_5 = \mathcal{X} \wedge \mathcal{Y}^{\perp} \wedge \mathcal{Z}^{\perp},$$
$$\mathcal{Q}_6 = \mathcal{X}^{\perp} \wedge \mathcal{Y} \wedge \mathcal{Z}^{\perp},$$
$$\mathcal{Q}_7 = \mathcal{X}^{\perp} \wedge \mathcal{Y}^{\perp} \wedge \mathcal{Z}$$

are mutually disjoint, i. e., $\mathcal{Q}_i \le \mathcal{Q}_j^{\perp}$ if $i \neq j$.

3 Here we applied property 15 from Table 1.2 and formula (B.7).

4 Depending on the choice of $\mathcal{X}$, $\mathcal{Y}$ and $\mathcal{Z}$ some of them can be empty.

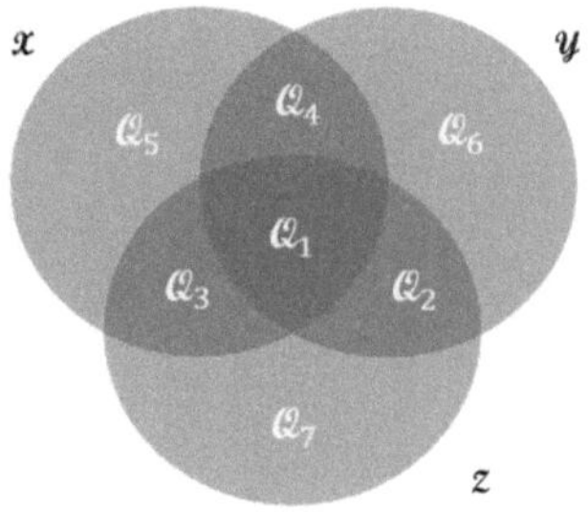

Figure B.1: To the proof of Theorem B.14. Representation of propositions by regions on a plane is for illustration purposes only. In the proof, we do not assume that the lattice of propositions is distributive (= Boolean).

For example, to prove the disjointness of $\mathcal{Q}_3$ and $\mathcal{Q}_5$, we notice that $\mathcal{Q}_3 \le \mathcal{Z}$ and $\mathcal{Q}_5 \le \mathcal{Z}^\perp$.[5] Then by property 17, $\mathcal{Z} \le \mathcal{Q}_5^\perp$ and $\mathcal{Q}_3 \le \mathcal{Z} \le \mathcal{Q}_5^\perp$, and by property 3, $\mathcal{Q}_3 \le \mathcal{Q}_5^\perp$.

Since by our assumption both $\mathcal{X} \wedge \mathcal{Z}$ and $\mathcal{X} \wedge \mathcal{Z}^\perp$ are compatible with $\mathcal{Y}$, we obtain

$$\mathcal{X} \wedge \mathcal{Z} = (\mathcal{X} \wedge \mathcal{Z} \wedge \mathcal{Y}) \vee (\mathcal{X} \wedge \mathcal{Z} \wedge \mathcal{Y}^\perp) = \mathcal{Q}_1 \vee \mathcal{Q}_3,$$
$$\mathcal{X} \wedge \mathcal{Z}^\perp = (\mathcal{X} \wedge \mathcal{Z}^\perp \wedge \mathcal{Y}) \vee (\mathcal{X} \wedge \mathcal{Z}^\perp \wedge \mathcal{Y}^\perp) = \mathcal{Q}_4 \vee \mathcal{Q}_5,$$
$$\mathcal{X} = (\mathcal{X} \wedge \mathcal{Z}) \vee (\mathcal{X} \wedge \mathcal{Z}^\perp) = \mathcal{Q}_1 \vee \mathcal{Q}_3 \vee \mathcal{Q}_4 \vee \mathcal{Q}_5.$$

Similarly

$$\mathcal{Y} \wedge \mathcal{Z} = \mathcal{Q}_1 \vee \mathcal{Q}_2,$$
$$\mathcal{Y} \wedge \mathcal{Z}^\perp = \mathcal{Q}_4 \vee \mathcal{Q}_6,$$
$$\mathcal{Y} = \mathcal{Q}_1 \vee \mathcal{Q}_2 \vee \mathcal{Q}_4 \vee \mathcal{Q}_6,$$
$$\mathcal{Z} = \mathcal{Q}_1 \vee \mathcal{Q}_2 \vee \mathcal{Q}_3 \vee \mathcal{Q}_7.$$

Then, denoting $\mathcal{T} \equiv \mathcal{Q}_1 \vee \mathcal{Q}_2 \vee \mathcal{Q}_3$ and using properties 10, 12 and (B.9), we obtain

$$(\mathcal{X} \wedge \mathcal{Z}) \vee (\mathcal{Y} \wedge \mathcal{Z}) = (\mathcal{Q}_1 \vee \mathcal{Q}_3) \vee (\mathcal{Q}_1 \vee \mathcal{Q}_2) = \mathcal{Q}_1 \vee \mathcal{Q}_2 \vee \mathcal{Q}_3 = \mathcal{T}. \tag{B.12}$$

From properties 7, 8 and the equality $\mathcal{Y} \vee \mathcal{X} = \mathcal{T} \vee \mathcal{Q}_4 \vee \mathcal{Q}_5 \vee \mathcal{Q}_6$ it follows that

$$\mathcal{T} \le (\mathcal{T} \vee \mathcal{Q}_7) \wedge (\mathcal{T} \vee \mathcal{Q}_4 \vee \mathcal{Q}_5 \vee \mathcal{Q}_6) = (\mathcal{X} \vee \mathcal{Y}) \wedge \mathcal{Z}. \tag{B.13}$$

On the other hand, $\mathcal{Q}_4 \le \mathcal{Z}^\perp$, $\mathcal{Q}_5 \le \mathcal{Z}^\perp$ and $\mathcal{Q}_6 \le \mathcal{Z}^\perp$. Therefore, $\mathcal{Q}_4 \vee \mathcal{Q}_5 \vee \mathcal{Q}_6 \le \mathcal{Z}^\perp \le \mathcal{Q}_7^\perp$. Adding here Lemma B.13 and the definition of compatibility, we get

$$(\mathcal{X} \vee \mathcal{Y}) \wedge \mathcal{Z} = (\mathcal{T} \vee \mathcal{Q}_4 \vee \mathcal{Q}_5 \vee \mathcal{Q}_6) \wedge (\mathcal{T} \vee \mathcal{Q}_7) \le (\mathcal{T} \vee \mathcal{Q}_7^\perp) \wedge (\mathcal{T} \vee \mathcal{Q}_7) = \mathcal{T}. \tag{B.14}$$

Then, application of the symmetry property 2 to formulas (B.13) and (B.14) yields

$$(\mathcal{X} \vee \mathcal{Y}) \wedge \mathcal{Z} = \mathcal{T}. \tag{B.15}$$

Comparing equations (B.12) and (B.15), we see that the distributive law (B.10) is indeed satisfied. The other distributive law (B.11) follows from (B.10) by the principle of duality. □

5 Both inequalities follow from property 6 in Table 1.2.

C Groups and vector spaces

C.1 Groups

A *group* is a set where the composition (or *product*) ab of any two elements a and b is defined. This product is also an element of the group, and the following conditions are fulfilled:

(1) associativity,

$$(ab)c = a(bc); \tag{C.1}$$

(2) there exists a unique *identity element* e, such that for each group element a

$$ea = ae = a; \tag{C.2}$$

(3) for each group element a there exists a unique *inverse* element a^{-1}, such that

$$aa^{-1} = a^{-1}a = e. \tag{C.3}$$

In commutative (or *Abelian*) groups $ab = ba$ for any two elements a and b.

For physics, the most interesting are the groups of transformations (rotations, translations, etc.) between inertial frames of reference. These are the Galilei and Poincaré groups considered in Chapter 2.

C.2 Vector spaces

Vector (or linear) space $\mathscr{H}$ is a set of objects (called *vectors*)[1] with two operations: addition of two vectors and multiplication of a vector by a *scalar* (that is, a number). In this book, we will be interested only in vector spaces whose scalars are either complex ($\mathbb{C}$) or real ($\mathbb{R}$) numbers. If $\boldsymbol{x}$ and $\boldsymbol{y}$ are two vectors and a and b are two scalars, then

$$a\boldsymbol{x} + b\boldsymbol{y}$$

is also a vector. By definition, the vector space forms an Abelian (= commutative) group with respect to the addition of vectors. In particular, this means:

(1) commutativity: $\boldsymbol{x} + \boldsymbol{y} = \boldsymbol{y} + \boldsymbol{x}$;
(2) associativity: $(\boldsymbol{x} + \boldsymbol{y}) + \boldsymbol{z} = \boldsymbol{x} + (\boldsymbol{y} + \boldsymbol{z})$;
(3) existence of the group identity (which is denoted by $\mathbf{0}$ and called the *zero vector*): $\boldsymbol{x} + \mathbf{0} = \mathbf{0} + \boldsymbol{x} = \boldsymbol{x}$;

---

1 In our book vectors are indicated in bold ($\boldsymbol{x}$).

https://doi.org/10.1515/9783110492132-010

(4) existence of the opposite (= additive inverse) element, which is denoted by $-\boldsymbol{x}$:

$$\boldsymbol{x} + (-\boldsymbol{x}) = \boldsymbol{0}.$$

Moreover, the following properties of the (scalar)×(vector) product are fulfilled:
(5) associativity of the multiplication by scalars: $a(b\boldsymbol{x}) = (ab)\boldsymbol{x}$;
(6) distributivity of the sum of scalars: $(a + b)\boldsymbol{x} = a\boldsymbol{x} + b\boldsymbol{x}$;
(7) distributivity of the sum of vectors: $a(\boldsymbol{x} + \boldsymbol{y}) = a\boldsymbol{x} + a\boldsymbol{y}$;
(8) multiplication by the scalar unity: $1\boldsymbol{x} = \boldsymbol{x}$.

We encourage the reader to verify that these eight axioms allow us to derive the following useful results for arbitrary scalar a and vector $\boldsymbol{x}$:

$$\begin{aligned} 0\boldsymbol{x} &= a\boldsymbol{0} = \boldsymbol{0}, \\ (-a)\boldsymbol{x} &= a(-\boldsymbol{x}) = -(a\boldsymbol{x}), \\ a\boldsymbol{x} &= \boldsymbol{0} \quad \Rightarrow \quad a = 0 \text{ or } \boldsymbol{x} = \boldsymbol{0}. \end{aligned}$$

An example of a vector space is the set of all columns of n numbers[2]

$$\begin{bmatrix} x_1 \\ x_2 \\ \vdots \\ x_n \end{bmatrix}.$$

The sum of two columns is

$$\begin{bmatrix} x_1 \\ x_2 \\ \vdots \\ x_n \end{bmatrix} + \begin{bmatrix} y_1 \\ y_2 \\ \vdots \\ y_n \end{bmatrix} = \begin{bmatrix} x_1 + y_1 \\ x_2 + y_2 \\ \vdots \\ x_n + y_n \end{bmatrix}.$$

The multiplication of a column by the scalar λ is defined as

$$\lambda \begin{bmatrix} x_1 \\ x_2 \\ \vdots \\ x_n \end{bmatrix} = \begin{bmatrix} \lambda x_1 \\ \lambda x_2 \\ \vdots \\ \lambda x_n \end{bmatrix}.$$

The inverse vector is obtained by changing the signs of all components. The zero vector is the column with all zeros.

---

2 If x_i are real (complex) numbers, then the vector space of the columns is denoted by $\mathbb{R}^n$ ($\mathbb{C}^n$).

A set of nonzero vectors $\{\boldsymbol{x}_i\}$ is called *linearly independent* if the equality

$$\sum_i a_i \boldsymbol{x}_i = \boldsymbol{0}$$

is fulfilled only when $a_i = 0$ for all i. A set of linearly independent vectors $\boldsymbol{x}_i$ is called a *basis*, if, with the addition of any nonzero vector $\boldsymbol{y}$ to this set, it ceases to be linearly independent. In this case the equation

$$a_0 \boldsymbol{y} + \sum_i a_i \boldsymbol{x}_i = \boldsymbol{0}$$

has a solution in which $a_0 \neq 0$.[3] This means that we can express any vector $\boldsymbol{y}$ as a linear combination of basis vectors, i. e.,

$$\boldsymbol{y} = -\sum_i \frac{a_i}{a_0} \boldsymbol{x}_i = \sum_i y_i \boldsymbol{x}_i. \tag{C.4}$$

Notice that each vector $\boldsymbol{y}$ has unique *components* y_i with respect to the basis $\boldsymbol{x}_i$. Indeed, suppose that we have found another set of components y_i' such that

$$\boldsymbol{y} = \sum_i y_i' \boldsymbol{x}_i. \tag{C.5}$$

Then, subtracting (C.5) from (C.4), we get

$$\boldsymbol{0} = \sum_i (y_i' - y_i) \boldsymbol{x}_i$$

and $y_i' = y_i$, because $\boldsymbol{x}_i$ are linearly independent.

One can choose many different bases in a given vector space $\mathscr{H}$. However, the number of vectors in each such basis is the same. This number is called the *dimension* of the vector space (denoted by $\dim \mathscr{H}$). The dimension of the space of n-number columns is n. An example of a basis in this space is given by the set of n vectors

$$\begin{bmatrix} 1 \\ 0 \\ \vdots \\ 0 \end{bmatrix}, \begin{bmatrix} 0 \\ 1 \\ \vdots \\ 0 \end{bmatrix}, \ldots, \begin{bmatrix} 0 \\ 0 \\ \vdots \\ 1 \end{bmatrix}.$$

A *linear subspace* $\mathscr{A}$ is a subset of vectors in $\mathscr{H}$ (denoted $\mathscr{A} \subseteq \mathscr{H}$), which is closed with respect to the operations of vector addition and scalar multiplication. For each set of vectors $\boldsymbol{x}_1, \boldsymbol{x}_2, \ldots$ there exists a subspace called *linear span* $\boldsymbol{x}_1 \uplus \boldsymbol{x}_2 \uplus \cdots$, i. e., the set of linear combinations $\sum_i a_i \boldsymbol{x}_i$ with all possible coefficients a_i. The linear span of a single nonzero vector $\biguplus \boldsymbol{x}$ is a one-dimensional subspace, which is also called the *ray*.

3 Otherwise we would have $a_i = 0$ for all i, which would mean that the set $\{\boldsymbol{x}_i, \boldsymbol{y}\}$ is linearly independent, in contradiction to our assumption.

D Group of rotations

D.1 Basics of 3D space

Let us now turn to our usual three-dimensional position space. It consists of *points*. We choose arbitrarily one such point **0** and call it the *origin*. Now we can connect all other points in space to the origin by arrows (=*vectors*) $\boldsymbol{a}$. These are, indeed, vectors in our definition from Appendix C.2, because we can define their sums and products with real scalars. We will define the sum of two vectors by the parallelogram rule, as shown in Figure D.1. The length of the vector $|\boldsymbol{a}|$ (it is also denoted by a) and the angle α between two vectors $\boldsymbol{a}$ and $\boldsymbol{b}$ are determined in a natural way. Then the vector $\lambda\boldsymbol{a}$ has the same direction as $\boldsymbol{a}$,[1] but its length is equal to $|\lambda|a$. The *scalar product* of two vectors is defined by the formula

$$\boldsymbol{a}\cdot\boldsymbol{b}=\boldsymbol{b}\cdot\boldsymbol{a}=ab\cos\alpha. \tag{D.1}$$

Two nonzero vectors are called *perpendicular* or *orthogonal*, if the angle between them is 90°, so that $\boldsymbol{a}\cdot\boldsymbol{b}=0$.

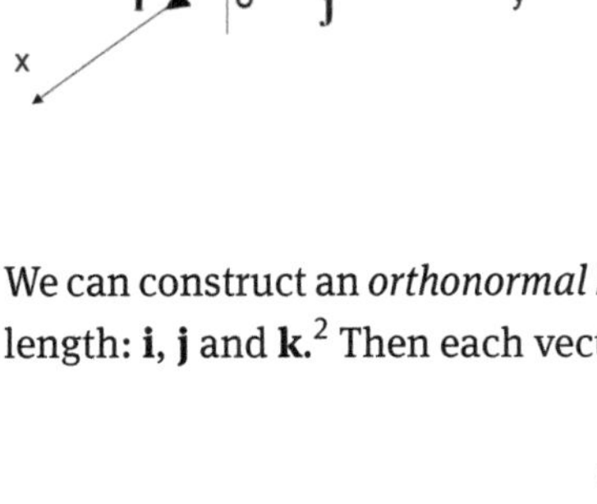

Figure D.1: Some objects in the vector space $\mathbb{R}^3$: the origin **0**; the basis vectors **i**, **j**, **k**; the sum of two vectors $\boldsymbol{a}+\boldsymbol{b}$ is obtained by the parallelogram rule.

We can construct an *orthonormal basis* of three mutually perpendicular vectors of unit length: **i**, **j** and **k**.[2] Then each vector $\boldsymbol{a}$ is representable as a linear combination

$$\boldsymbol{a}=a_x\mathbf{i}+a_y\mathbf{j}+a_z\mathbf{k}$$

1 The direction is reversed if $\lambda<0$.

2 We will choose the triplet of vectors **i**, **j**, **k** which form a *right-oriented basis*, as shown in Figure D.1. Such a basis has the following property: if all three arrows point at you, then you see their ends in the counterclockwise order $\mathbf{i}\to\mathbf{j}\to\mathbf{k}$. The directions **i**, **j**, **k** will be also denoted x, y and z, respectively.

https://doi.org/10.1515/9783110492132-011

or as a column of its components, also called *coordinates* of the vector,[3]

$$\boldsymbol{a} = \begin{bmatrix} a_x \\ a_y \\ a_z \end{bmatrix}.$$

A *transposed* vector is represented as a row

$$\boldsymbol{a}^T = [a_x, a_y, a_z].$$

The scalar product (D.1) can be written in several equivalent forms, i. e.,

$$\boldsymbol{b} \cdot \boldsymbol{a} = \sum_{i=1}^{3} b_i a_i = b_x a_x + b_y a_y + b_z a_z = [b_x, b_y, b_z] \begin{bmatrix} a_x \\ a_y \\ a_z \end{bmatrix} = \boldsymbol{b}^T \boldsymbol{a}, \tag{D.2}$$

where $\boldsymbol{b}^T \boldsymbol{a}$ denotes the usual (row by column) matrix product of the row $\boldsymbol{b}^T$ and the column $\boldsymbol{a}$.

D.2 Scalars and vectors

There are two approaches to inertial transformations, in particular, to rotations: *active* and *passive*. An active rotation moves physical objects, without changing orientations of the basic vectors **i**, **j**, **k**. A *passive rotation* simply changes directions of the basic vectors without touching material things. Therefore, the passive rotation changes vector components, but not the physical vectors themselves. Unless stated otherwise, we will normally use the passive approach to rotations and other inertial transformations.

We will call a quantity a 3-*scalar*, if it does not change under rotations. Distances and angles are examples of scalars.

Let us now find out how rotations change coordinates of vectors in $\mathbb{R}^3$. By definition, rotation keeps the origin and linear combinations of vectors unchanged, so the (linear) action of the rotation on a column-vector should be representable as multiplication by a 3×3 matrix R,

$$a_i' = \sum_{j=1}^{3} R_{ij} a_j, \tag{D.3}$$

or in the matrix form

$$\begin{aligned} \boldsymbol{a}' &= R\boldsymbol{a}, \\ \boldsymbol{b}'^T &= (R\boldsymbol{b})^T = \boldsymbol{b}^T R^T, \end{aligned} \tag{D.4}$$

where R^T denotes the *transposed matrix*.

---

3 Thus, the physical space is identified with the vector space $\mathbb{R}^3$ of all triplets of real numbers. We will denote vector indices either by letters x, y, z, or by numbers 1, 2, 3, as convenient.

D.3 Orthogonal matrices

Rotations preserve lengths, distances and angles, so they should preserve the scalar product

$$\boldsymbol{b} \cdot \boldsymbol{a} = \boldsymbol{b}^T \boldsymbol{a} = (R\boldsymbol{b})^T (R\boldsymbol{a}) = \boldsymbol{b}^T R^T R\boldsymbol{a}.$$

The validity of this equality for all $\boldsymbol{a}$ and $\boldsymbol{b}$ implies that rotation matrices satisfy the following condition:

$$R^T R = I, \tag{D.5}$$

where I is the identity matrix

$$I = \begin{bmatrix} 1 & 0 & 0 \\ 0 & 1 & 0 \\ 0 & 0 & 1 \end{bmatrix}.$$

Multiplying by the inverse matrix R^{-1} on the right, equation (D.5) can be rewritten as

$$R^T = R^{-1}. \tag{D.6}$$

This implies a useful property,

$$R\boldsymbol{b} \cdot \boldsymbol{a} = \boldsymbol{b}^T R^T \boldsymbol{a} = \boldsymbol{b}^T R^{-1} \boldsymbol{a} = \boldsymbol{b} \cdot R^{-1} \boldsymbol{a}. \tag{D.7}$$

In the coordinate notation, condition (D.5) takes the form

$$\sum_{j=1}^{3} R^T_{ij} R_{jk} = \sum_{j=1}^{3} R_{ji} R_{jk} = \delta_{ik} \tag{D.8}$$

where δ_{ij} is the *Kronecker delta*

$$\delta_{ij} = \begin{cases} 1, & \text{if } i = j, \\ 0, & \text{if } i \neq j. \end{cases} \tag{D.9}$$

Matrices satisfying condition (D.6) are called *orthogonal*. Therefore, any rotation has a unique representative in the set of 3×3 orthogonal matrices.

However, not every orthogonal matrix R corresponds to a rotation. To see this, we write

$$1 = \det[I] = \det[R^T R] = \det[R^T] \det[R] = (\det[R])^2,$$

which implies that the determinant of orthogonal matrices is $\det[R] = \pm 1$. Each rotation can be connected in a continuous way to the trivial (0° angle) rotation, which is, of course, represented by the identity matrix with the unit determinant. A continuous transformation cannot abruptly change the determinant from +1 to −1. Therefore, only matrices with unit determinant

$$\det[R] = 1 \tag{D.10}$$

can represent rotations.[4] So, we conclude that rotations are in one-to-one correspondence with orthogonal matrices having the unit determinant.

The concept of a vector is more general than just an arrow sticking out of the origin. We call a 3-*vector* any triplet of numbers $\mathcal{A} = (\mathcal{A}_x, \mathcal{A}_y, \mathcal{A}_z)$, whose transformation under the action of rotations is the same as for vectors–arrows (D.3).

Let us now find out the explicit form of rotation matrices. Each rotation about the z-axis does not change the z-component of 3-vectors. The most general 3×3 matrix satisfying this requirement can be written as

$$R_z = \begin{bmatrix} a & b & 0 \\ c & d & 0 \\ 0 & 0 & 1 \end{bmatrix}$$

and condition (D.10) translates into $ad - bc = 1$. By direct substitution we can check that the inverse matrix is

$$R_z^{-1} = \begin{bmatrix} d & -b & 0 \\ -c & a & 0 \\ 0 & 0 & 1 \end{bmatrix}.$$

According to property (D.6) we should have

$$a = d,$$
$$-b = c.$$

Therefore

$$R_z = \begin{bmatrix} a & -b & 0 \\ b & a & 0 \\ 0 & 0 & 1 \end{bmatrix}.$$

The condition $\det[R_z] = a^2 + b^2 = 1$ implies that we can leave just one parameter φ, such that $a = \cos\varphi$ and $b = \sin\varphi$. Then we have

$$R_z = \begin{bmatrix} \cos\varphi & -\sin\varphi & 0 \\ \sin\varphi & \cos\varphi & 0 \\ 0 & 0 & 1 \end{bmatrix}. \tag{D.11}$$

4 Matrices with $\det[R] = -1$ represent rotations coupled with space inversion (see Subsection 2.2.4).

Obviously, the parameter φ is just the rotation angle. The matrices of rotations about the axes x and y are

$$R_x = \begin{bmatrix} 1 & 0 & 0 \\ 0 & \cos\varphi & -\sin\varphi \\ 0 & \sin\varphi & \cos\varphi \end{bmatrix} \tag{D.12}$$

and

$$R_y = \begin{bmatrix} \cos\varphi & 0 & \sin\varphi \\ 0 & 1 & 0 \\ -\sin\varphi & 0 & \cos\varphi \end{bmatrix}, \tag{D.13}$$

respectively.

D.4 Invariant tensors

A *tensor* of the second *rank*[5] $\mathcal{T}_{ij}$ is defined as a set of nine numbers depending on two indices ($i, j = 1, 2, 3$) and transforming in a linear way (as a vector) with respect to each index,

$$\mathcal{T}'_{ij} = \sum_{kl=1}^{3} R_{ik} R_{jl} \mathcal{T}_{kl}, \tag{D.14}$$

Similarly, one can define tensors of higher ranks, for example $\mathcal{T}_{ijk}$.

There are two *invariant tensors*, playing special roles, because their components do not change under rotations. The first invariant tensor is the Kronecker delta (D.9).[6] Its invariance follows from the orthogonality of R-matrices (D.8),

$$\delta'_{ij} = \sum_{kl=1}^{3} R_{ik} R_{jl} \delta_{kl} = \sum_{k=1}^{3} R_{ik} R_{jk} = \delta_{ij}.$$

The second invariant tensor is the third-rank *Levi-Civita symbol* ϵ_{ijk}, whose only nonzero components are $\epsilon_{xyz} = \epsilon_{zxy} = \epsilon_{yzx} = -\epsilon_{xzy} = -\epsilon_{yxz} = -\epsilon_{zyx} = 1$. We show its invariance by applying an arbitrary rotation R to ϵ_{ijk}. Then

$$\begin{aligned} \epsilon'_{ijk} &= \sum_{lmn=1}^{3} R_{il} R_{jm} R_{kn} \epsilon_{lmn} \\ &= R_{i1}R_{j2}R_{k3} + R_{i3}R_{j1}R_{k2} + R_{i2}R_{j3}R_{k1} - R_{i2}R_{j1}R_{k3} - R_{i3}R_{j2}R_{k1} - R_{i1}R_{j3}R_{k2}. \end{aligned} \tag{D.15}$$

5 Scalars and vectors are sometimes called tensors of the zero and 1st rank, respectively.

6 Which can be regarded as a unit 3×3 matrix.

The right-hand side has the following properties:

(1) It vanishes if any two indices coincide: $i = j$ or $i = k$ or $j = k$; for example,

$$\begin{aligned}\epsilon'_{112} &= R_{11}R_{12}R_{23} + R_{13}R_{11}R_{22} + R_{12}R_{13}R_{21}\\ &\quad - R_{12}R_{11}R_{23} - R_{13}R_{12}R_{21} - R_{11}R_{13}R_{22} = 0.\end{aligned}$$

(2) It does not change under cyclic permutation of the indices ijk.

(3) We have $\epsilon'_{123} = \det[R] = 1$.

These are the same properties that were used in the definition of the Levi-Civita symbol. Hence, the right-hand side (D.15) must have the same components as ϵ_{ijk}, so

$$\epsilon'_{ijk} = \epsilon_{ijk}.$$

Using the invariant tensors δ_{ij} and ϵ_{ijk}, we can perform transformations between scalar, vector and tensor quantities, as shown in Table D.1. For example, any antisymmetric tensor $\mathcal{A}_{ij}$ has three independent components, and it can be represented as

$$\mathcal{A}_{ij} = \sum_{k=1}^{3} \epsilon_{ijk} V_k,$$

where V_k are components of some 3-vector.

Table D.1: Transitions between quantities of different ranks, using invariant tensors.

Scalar S	$\to$	$S\delta_{ij}$ (symmetric tensor of 2nd rank)
Scalar S	$\to$	$S\epsilon_{ijk}$ (antisymmetric tensor of 3rd rank)
Vector V_i	$\to$	$\sum_{k=1}^{3} \epsilon_{ijk} V_k$ (antisymmetric tensor of 2nd rank)
Tensor $\mathcal{T}_{ij}$	$\to$	$\sum_{ij=1}^{3} \delta_{ij} \mathcal{T}_{ji}$ (scalar)
Tensor $\mathcal{T}_{ij}$	$\to$	$\sum_{jk=1}^{3} \epsilon_{ijk} \mathcal{T}_{kj}$ (vector)
2 vectors A_i, B_i	$\to$	$\sum_{ij=1}^{3} \delta_{ij} A_i B_j$ (scalar)
2 vectors A_i, B_i	$\to$	$\sum_{jk=1}^{3} \epsilon_{ijk} A_j B_k$ (vector)
3 vectors A_i, B_i, C_i	$\to$	$\sum_{ijk=1}^{3} \epsilon_{ijk} A_i B_j C_k$ (scalar)

Using invariant tensors, one can build scalars and vectors from any two given vectors $\boldsymbol{A}$ and $\boldsymbol{B}$. The scalar is built with the help of the Kronecker delta

$$\boldsymbol{A} \cdot \boldsymbol{B} = \sum_{ij=1}^{3} \delta_{ij} A_i B_j \equiv A_1 B_1 + A_2 B_2 + A_3 B_3.$$

This is just the familiar scalar product (D.2). The vector is constructed using the Levi-Civita tensor

$$[\boldsymbol{A} \times \boldsymbol{B}]_i \equiv \sum_{jk=1}^{3} \epsilon_{ijk} A_j B_k.$$

It is called the *vector product* of $\boldsymbol{A}$ and $\boldsymbol{B}$, and it has the following components:

$$[\boldsymbol{A}\times\boldsymbol{B}]_x = A_yB_z - A_zB_y,$$
$$[\boldsymbol{A}\times\boldsymbol{B}]_y = A_zB_x - A_xB_z,$$
$$[\boldsymbol{A}\times\boldsymbol{B}]_z = A_xB_y - A_yB_x$$

and properties:

$$[\boldsymbol{A}\times\boldsymbol{B}] = -[\boldsymbol{B}\times\boldsymbol{A}],$$
$$\big[\boldsymbol{A}\times[\boldsymbol{B}\times\boldsymbol{C}]\big] = \boldsymbol{B}(\boldsymbol{A}\cdot\boldsymbol{C}) - \boldsymbol{C}(\boldsymbol{A}\cdot\boldsymbol{B}). \tag{D.16}$$

The *mixed product* is a scalar obtained from three vectors by means of the Levi-Civita invariant tensor

$$[\boldsymbol{A}\times\boldsymbol{B}]\cdot\boldsymbol{C} = \sum_{ijk=1}^{3}\epsilon_{ijk}A_iB_jC_k.$$

It has the properties

$$[\boldsymbol{A}\times\boldsymbol{B}]\cdot\boldsymbol{C} = [\boldsymbol{B}\times\boldsymbol{C}]\cdot\boldsymbol{A} = [\boldsymbol{C}\times\boldsymbol{A}]\cdot\boldsymbol{B}, \tag{D.17}$$
$$[\boldsymbol{A}\times\boldsymbol{B}]\cdot\boldsymbol{B} = 0.$$

D.5 Vector parametrization of rotations

The matrix notation for rotations (D.3) is convenient for describing transformations of vector and tensor components. However, it is often necessary to characterize a rotation in a more physical way, namely, in terms of its axis and angle. In other words, any rotation can be uniquely represented by one vector $\boldsymbol{\varphi} = \varphi_x\mathbf{i} + \varphi_y\mathbf{j} + \varphi_z\mathbf{k}$,[7] such that its direction coincides with the direction of the rotation axis and the vector's length $\varphi \equiv |\boldsymbol{\varphi}|$ is equal to the rotation angle.

The effect of the rotation $\boldsymbol{\varphi}$ on a 3-vector $\boldsymbol{x}$ will be denoted simply as $\boldsymbol{\varphi x}$. Let us now establish the connection between matrix and vector representations of rotations. Our goal is to find the matrix $R_{\boldsymbol{\varphi}}$ corresponding to the rotation $\boldsymbol{\varphi}$.

As shown in Figure D.2, each vector $\boldsymbol{P}$ in $\mathbb{R}^3$ can be uniquely decomposed into two parts: $\boldsymbol{P} = \boldsymbol{P}_\parallel + \boldsymbol{P}_\perp$. The first part

$$\boldsymbol{P}_\parallel \equiv \left(\boldsymbol{P}\cdot\frac{\boldsymbol{\varphi}}{\varphi}\right)\frac{\boldsymbol{\varphi}}{\varphi} \tag{D.18}$$

7 Or simply by a triplet of real numbers $\{\varphi_x, \varphi_y, \varphi_z\}$.

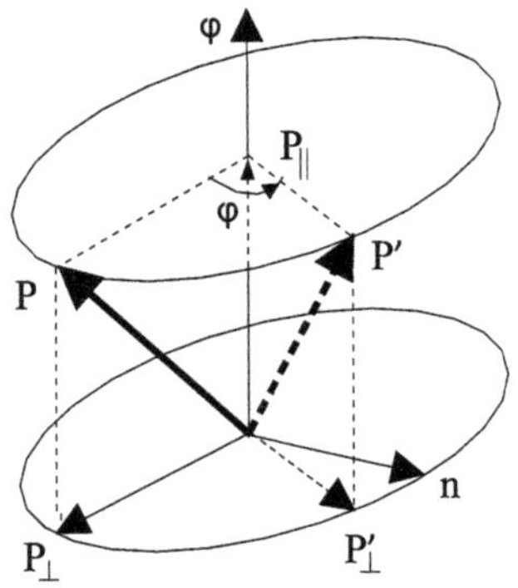

Figure D.2: Transformation of vector components $\boldsymbol{P} = \boldsymbol{P}_{\|} + \boldsymbol{P}_{\perp}$ under active rotation by the angle $\boldsymbol{\varphi}$.

is parallel to the rotation axis, and the second part

$$\boldsymbol{P}_{\perp} = \boldsymbol{P} - \boldsymbol{P}_{\|} \tag{D.19}$$

is perpendicular to it. The rotation does not affect the parallel part of the vector, so after the *active* rotation our vector becomes

$$\boldsymbol{P}' \equiv \boldsymbol{P}_{\|} + \boldsymbol{P}'_{\perp}. \tag{D.20}$$

If $\boldsymbol{P}_{\perp} = 0$, then the rotation leaves $\boldsymbol{P}$ untouched. Let us now find the active rotation of a nonzero perpendicular component $\boldsymbol{P}_{\perp} \neq 0$.

Denote by

$$\boldsymbol{n} = -\frac{[\boldsymbol{P}_{\perp} \times \boldsymbol{\varphi}]}{\varphi} = -\frac{[\boldsymbol{P} \times \boldsymbol{\varphi}]}{\varphi}$$

the vector, which is orthogonal to both $\boldsymbol{\varphi}$ and $\boldsymbol{P}_{\perp}$ and equal to the latter in length. Notice that the triplet of vectors $(\boldsymbol{P}_{\perp}, \boldsymbol{n}, \boldsymbol{\varphi})$ forms a right-oriented system, just like the basis triplet $(\mathbf{i}, \mathbf{j}, \mathbf{k})$. Therefore, the result of rotation through the angle φ in the plane formed by vectors $\boldsymbol{P}_{\perp}$ and $\boldsymbol{n}$ should be the same as the rotation about the axis $\mathbf{k}$ in the plane of vectors $\mathbf{i}$ and $\mathbf{j}$, so we have

$$\boldsymbol{P}'_{\perp} = \boldsymbol{P}_{\perp} \cos\varphi + \boldsymbol{n} \sin\varphi = \boldsymbol{P}_{\perp} \cos\varphi - \frac{[\boldsymbol{P} \times \boldsymbol{\varphi}]}{\varphi} \sin\varphi. \tag{D.21}$$

Joining equations (D.18)–(D.21), we obtain

$$\boldsymbol{P}' = \left(\boldsymbol{P} \cdot \frac{\boldsymbol{\varphi}}{\varphi}\right)\frac{\boldsymbol{\varphi}}{\varphi} + \boldsymbol{P}_{\perp} \cos\varphi - \frac{[\boldsymbol{P} \times \boldsymbol{\varphi}]}{\varphi} \sin\varphi.$$

The formula for the *passive* rotation is obtained by changing the sign of $\boldsymbol{\varphi}$. We have

$$\boldsymbol{\varphi}\boldsymbol{P} = \boldsymbol{P} \cos\varphi + \left(\boldsymbol{P} \cdot \frac{\boldsymbol{\varphi}}{\varphi}\right)\frac{\boldsymbol{\varphi}}{\varphi}(1 - \cos\varphi) + \left[\boldsymbol{P} \times \frac{\boldsymbol{\varphi}}{\varphi}\right] \sin\varphi \tag{D.22}$$

or in the component notation

$$(\boldsymbol{\varphi P})_x = (P_x\varphi_x + P_y\varphi_y + P_z\varphi_z)\frac{\varphi_x}{\varphi^2}(1-\cos\varphi) + P_x\cos\varphi + (P_y\varphi_z - P_z\varphi_y)\frac{\sin\varphi}{\varphi},$$
$$(\boldsymbol{\varphi P})_y = (P_x\varphi_x + P_y\varphi_y + P_z\varphi_z)\frac{\varphi_y}{\varphi^2}(1-\cos\varphi) + P_y\cos\varphi + (P_z\varphi_x - P_x\varphi_z)\frac{\sin\varphi}{\varphi},$$
$$(\boldsymbol{\varphi P})_z = (P_x\varphi_x + P_y\varphi_y + P_z\varphi_z)\frac{\varphi_z}{\varphi^2}(1-\cos\varphi) + P_z\cos\varphi + (P_x\varphi_y - P_y\varphi_x)\frac{\sin\varphi}{\varphi}.$$

This means that a general orthogonal rotation matrix has the following entries:

$$(R_{\boldsymbol{\varphi}})_{ij} = \cos\varphi\delta_{ij} - \sum_{k=1}^{3}\varphi_k\epsilon_{ijk}\frac{\sin\varphi}{\varphi} + \varphi_i\varphi_j\frac{1-\cos\varphi}{\varphi^2}$$

or explicitly[8]

$$R_{\boldsymbol{\varphi}} = \begin{bmatrix} \mathfrak{c} + l_x^2(1-\mathfrak{c}) & l_xl_y(1-\mathfrak{c}) - l_z\mathfrak{s} & l_xl_z(1-\mathfrak{c}) + l_y\mathfrak{s} \\ l_xl_y(1-\mathfrak{c}) + l_z\mathfrak{s} & \mathfrak{c} + l_y^2(1-\mathfrak{c}) & l_yl_z(1-\mathfrak{c}) - l_x\mathfrak{s} \\ l_xl_z(1-\mathfrak{c}) - l_y\mathfrak{s} & l_yl_z(1-\mathfrak{c}) + l_x\mathfrak{s} & \mathfrak{c} + l_z^2(1-\mathfrak{c}) \end{bmatrix}, \quad \text{(D.23)}$$

where we denoted $\mathfrak{c} \equiv \cos\varphi$, $\mathfrak{s} \equiv \sin\varphi$ and $\boldsymbol{l} \equiv \boldsymbol{\varphi}/\varphi$.

To find the reverse link from matrices to rotation vectors, we start with an arbitrary orthogonal matrix R having unit determinant $\det[R] = 1$ and try to find the corresponding vector $\boldsymbol{\varphi}$. Obviously, this vector does not change under the transformation R, so it is an eigenvector of the matrix R with the eigenvalue of 1, thus

$$R\boldsymbol{\varphi} = \boldsymbol{\varphi}.$$

Each orthogonal 3×3 matrix with unit determinant has eigenvalues $(1, e^{i\varphi}, e^{-i\varphi})$.[9] Hence, the eigenvalue 1 is nondegenerate, and the direction of the vector $\boldsymbol{\varphi}$ is defined up to a sign. Next we have to find the length of this vector, i. e., the rotation angle φ. The trace of the matrix R is given by the sum of its eigenvalues,

$$\mathrm{Tr}(R) = 1 + e^{i\varphi} + e^{-i\varphi} = 1 + 2\cos\varphi.$$

Therefore, we can define a mapping from the set of rotation matrices R to the corresponding rotation vectors $\boldsymbol{\varphi}$ by the following rules:

- The direction of the rotation vector $\boldsymbol{\varphi}$ coincides with the direction of the unique eigenvector of R with eigenvalue 1.

8 Matrices (D.11)–(D.13) are obtained as particular cases of (D.23) for $\boldsymbol{l} = (0,0,1)$, $\boldsymbol{l} = (1,0,0)$ and $\boldsymbol{l} = (0,1,0)$, respectively.

9 This result can be verified using explicit representations (D.11)–(D.13) and (D.23).

– The length of the vector (= rotation angle) φ is equal to[10]

$$\varphi = \arccos \frac{\mathrm{Tr}(R) - 1}{2}. \tag{D.24}$$

Finding the additional steps for determining the sign of the vector $\boldsymbol{\varphi}$ is left as an exercise for the reader.

D.6 Group properties of rotations

Rotations form a mathematical Lie group. The identity element of this group is the rotation by the zero angle $\mathbf{0}$. This rotation leaves all vectors unchanged and is represented by the identity matrix $R_{\mathbf{0}} = I$. The symbol $\boldsymbol{\phi} \circ \boldsymbol{\varphi}$ will denote the product (composition) of two rotations: first $\boldsymbol{\varphi}$ and then $\boldsymbol{\phi}$. This composition corresponds to the product of the corresponding orthogonal matrices $R_{\boldsymbol{\phi}}$ and $R_{\boldsymbol{\varphi}}$, which is also an orthogonal matrix, because

$$(R_{\boldsymbol{\phi}} R_{\boldsymbol{\varphi}})^T = R_{\boldsymbol{\varphi}}^T R_{\boldsymbol{\phi}}^T = R_{\boldsymbol{\varphi}}^{-1} R_{\boldsymbol{\phi}}^{-1} = (R_{\boldsymbol{\phi}} R_{\boldsymbol{\varphi}})^{-1}.$$

For each rotation $\boldsymbol{\varphi}$ there exists an inverse rotation $-\boldsymbol{\varphi}$ such that

$$(-\boldsymbol{\varphi}) \circ \boldsymbol{\varphi} = \boldsymbol{\varphi} \circ (-\boldsymbol{\varphi}) = \mathbf{0}.$$

Obviously, the inverse rotation is represented by the inverse matrix $R_{-\boldsymbol{\varphi}} = R_{\boldsymbol{\varphi}}^{-1} = R_{\boldsymbol{\varphi}}^T$. This matrix is also orthogonal:

$$(R_{\boldsymbol{\varphi}}^{-1})^T = (R_{\boldsymbol{\varphi}}^T)^T = R_{\boldsymbol{\varphi}} = (R_{\boldsymbol{\varphi}}^{-1})^{-1}.$$

The associativity law

$$\boldsymbol{\varphi}_1 \circ (\boldsymbol{\varphi}_2 \circ \boldsymbol{\varphi}_3) = (\boldsymbol{\varphi}_1 \circ \boldsymbol{\varphi}_2) \circ \boldsymbol{\varphi}_3$$

follows from the associativity of matrix multiplication.

Rotations about different axes, as a rule, do not commute with each other. However, two rotations $\varphi\boldsymbol{n}$ and $\psi\boldsymbol{n}$ about the same axis[11] do commute. Moreover, our choice of the vector parametrization of rotations leads to the following important relation:

$$(\varphi\boldsymbol{n}) \circ (\psi\boldsymbol{n}) = (\varphi + \psi)\boldsymbol{n}. \tag{D.25}$$

---

10 As expected, this formula does not depend on the basis because the trace of a matrix is basis-independent (see Lemma G.5).

11 Here $\boldsymbol{n}$ is an arbitrary unit vector.

For example, for two rotations about the z-axis we can write

$$
\begin{aligned}
R_{(0,0,\varphi)}R_{(0,0,\psi)} &= \begin{bmatrix} \cos\varphi & -\sin\varphi & 0 \\ \sin\varphi & \cos\varphi & 0 \\ 0 & 0 & 1 \end{bmatrix} \begin{bmatrix} \cos\psi & -\sin\psi & 0 \\ \sin\psi & \cos\psi & 0 \\ 0 & 0 & 1 \end{bmatrix} \\
&= \begin{bmatrix} \cos(\varphi+\psi) & -\sin(\varphi+\psi) & 0 \\ \sin(\varphi+\psi) & \cos(\varphi+\psi) & 0 \\ 0 & 0 & 1 \end{bmatrix} \\
&= R_{(0,0,\varphi+\psi)}.
\end{aligned}
$$

We will say that rotations about one axis $\boldsymbol{n}$ form a *one-parameter subgroup* of the rotation group.

D.7 Generators of rotations

Rotations by small angles (near the identity element $1 \equiv \mathbf{0}$ of the group) can be formally represented by Taylor expansions as follows:

$$\boldsymbol{\varphi} = 1 + \sum_{i=1}^{3} \varphi^i t_i + \frac{1}{2} \sum_{ij=1}^{3} \varphi^i \varphi^j t_{ij} + \cdots .$$

At small values of φ we have

$$\boldsymbol{\varphi} \approx 1 + \sum_{i=1}^{3} \varphi^i t_i. \tag{D.26}$$

The objects t_i are called *generators* or *infinitesimal rotations*. Generators can be formally represented as derivatives of elements of one-parameter subgroups with respect to their parameters φ^i, i. e.,

$$t_i = \lim_{\varphi^i \to 0} \frac{d}{d\varphi^i} \boldsymbol{\varphi}. \tag{D.27}$$

For example, in the matrix notation, the generator of rotations about the z-axis is given by the matrix

$$\mathcal{J}_z = \lim_{\varphi \to 0} \frac{d}{d\varphi} R_z(\varphi) = \lim_{\varphi \to 0} \frac{d}{d\varphi} \begin{bmatrix} \cos\varphi & -\sin\varphi & 0 \\ \sin\varphi & \cos\varphi & 0 \\ 0 & 0 & 1 \end{bmatrix} = \begin{bmatrix} 0 & -1 & 0 \\ 1 & 0 & 0 \\ 0 & 0 & 0 \end{bmatrix}. \tag{D.28}$$

Similarly, for generators of rotations about the axes x and y, we obtain from (D.12) and (D.13)

$$\mathcal{J}_x = \begin{bmatrix} 0 & 0 & 0 \\ 0 & 0 & -1 \\ 0 & 1 & 0 \end{bmatrix}, \quad \mathcal{J}_y = \begin{bmatrix} 0 & 0 & 1 \\ 0 & 0 & 0 \\ -1 & 0 & 0 \end{bmatrix}. \tag{D.29}$$

Using the additivity property (D.25) and the expansion (D.26), we can express an arbitrary rotation $\boldsymbol{\varphi}$ as an exponential function of the generator

$$\boldsymbol{\varphi} = \lim_{N\to\infty}\left\{N\frac{\boldsymbol{\varphi}}{N}\right\} = \lim_{N\to\infty}\left\{\frac{\boldsymbol{\varphi}}{N}\right\}^N = \lim_{N\to\infty}\left(1+\sum_{i=1}^{3}\frac{\varphi^i}{N}t_i\right)^N = \exp\left(\sum_{i=1}^{3}\varphi^i t_i\right). \tag{D.30}$$

For example, the exponent of the matrix $\mathcal{J}_z$

$$\begin{aligned} e^{\mathcal{J}_z\varphi} &= 1+\varphi\mathcal{J}_z+\frac{1}{2!}\varphi^2\mathcal{J}_z^2+O(\varphi^3)\\ &= \begin{bmatrix}1&0&0\\0&1&0\\0&0&1\end{bmatrix}+\begin{bmatrix}0&-\varphi&0\\\varphi&0&0\\0&0&0\end{bmatrix}+\begin{bmatrix}-\frac{\varphi^2}{2}&0&0\\0&-\frac{\varphi^2}{2}&0\\0&0&0\end{bmatrix}+O(\varphi^3)\\ &= \begin{bmatrix}1-\frac{\varphi^2}{2}+\cdots&-\varphi+\cdots&0\\\varphi+\cdots&1-\frac{\varphi^2}{2}+\cdots&0\\0&0&1\end{bmatrix}=\begin{bmatrix}\cos\varphi&-\sin\varphi&0\\\sin\varphi&\cos\varphi&0\\0&0&1\end{bmatrix}\\ &= R_z=(0,0,\varphi) \end{aligned} \tag{D.31}$$

is the matrix of rotation about the z-axis, as expected.

One can notice that three generators (D.28)–(D.29) form a basis in the linear space of antisymmetric 3×3 matrices. A generator of a rotation about an arbitrary axis is also an antisymmetric matrix, which is a linear combination of the generators. It is not difficult to verify that the exponent of any such linear combination of the matrices $\mathcal{J}_x$, $\mathcal{J}_y$ and $\mathcal{J}_z$ is an orthogonal 3×3 matrix with unit determinant, i. e., it represents a certain rotation.

Therefore, the three abstract objects t_i[12] form a basis in the vector space of generators of the rotation group. This vector space is called the *Lie algebra* of the rotation group. General properties of Lie algebras will be discussed in Appendix E.2.

12 In the matrix representation they correspond to the matrices $\mathcal{J}_x$, $\mathcal{J}_y$ and $\mathcal{J}_z$.

E Lie groups and Lie algebras

E.1 Lie groups

Generally speaking, the sets of elements of a group can be either discrete or continuous. Here we are going to discuss a special class of groups, called *Lie groups*, whose *group manifold* is continuous. Moreover, the group composition and the inverse are given by smooth functions on this manifold.

In Appendix D.5 (see also Appendix I.4), we saw that all these properties are satisfied in the group of rotations, which is the prime example of a three-dimensional Lie group.

Analogously to the rotation angles $\boldsymbol{\varphi}$, elements of a general n-dimensional Lie group can be labeled by n continuous parameters (coordinates) η_i. A general element of the group will be denoted by $\{\boldsymbol{\eta}\} = \{\eta_1, \eta_2, \dots \eta_n\}$, so that the operations of product and inverse are smooth functions of these coordinates. We assume that the group coordinates can always be chosen such that the following properties are satisfied:

(1) the identity element has coordinates $\{0, 0, \dots, 0\}$;

(2) $\{\boldsymbol{\eta}\}^{-1} = \{-\boldsymbol{\eta}\}$;

(3) if elements $\{\boldsymbol{\psi}\}$ and $\{\boldsymbol{\varphi}\}$ belong to the same one-parameter subgroup, then coordinates of their products are simply sums of coordinates of the factors

$$\{\boldsymbol{\psi}\}\{\boldsymbol{\varphi}\} = \{\boldsymbol{\psi} + \boldsymbol{\varphi}\}. \tag{E.1}$$

Then one can introduce *infinitesimal transformations* or *generators* t_a $(a = 1, 2, \dots, n)$ and express group elements near the identity as exponential functions of the generators (compare with Appendix D.7)

$$\{\boldsymbol{\eta}\} = \exp\left(\sum_{a=1}^{n} \eta^a t_a\right) = 1 + \sum_{a=1}^{n} \eta^a t_a + \frac{1}{2!} \sum_{bc=1}^{n} \eta^b \eta^c t_{bc} + \cdots. \tag{E.2}$$

Let us introduce the function $\boldsymbol{g}(\boldsymbol{\xi}, \boldsymbol{\zeta})$, which maps a pair of points $\{\boldsymbol{\xi}\}$ and $\{\boldsymbol{\zeta}\}$ in the group manifold to a third point $\boldsymbol{g}(\boldsymbol{\xi}, \boldsymbol{\zeta})$ according to the group multiplication law, i. e.,

$$\{\boldsymbol{\xi}\}\{\boldsymbol{\zeta}\} = \{\boldsymbol{g}(\boldsymbol{\xi}, \boldsymbol{\zeta})\}. \tag{E.3}$$

This function should satisfy the following conditions:

$$\begin{aligned} \boldsymbol{g}(\mathbf{0}, \boldsymbol{\eta}) &= \boldsymbol{g}(\boldsymbol{\eta}, \mathbf{0}) = \boldsymbol{\eta}, \\ \boldsymbol{g}(\boldsymbol{\eta}, -\boldsymbol{\eta}) &= \mathbf{0}, \end{aligned} \tag{E.4}$$

which follow from group properties (C.2) and (C.3), respectively. In order to ensure condition (E.4), the Taylor expansion of $\boldsymbol{g}$ up to second-order terms should look like

$$g^a(\boldsymbol{\xi}, \boldsymbol{\zeta}) = \xi^a + \zeta^a + \sum_{bc=1}^{n} f^a_{bc} \xi^b \zeta^c + \cdots, \tag{E.5}$$

https://doi.org/10.1515/9783110492132-012

where f_{bc}^a are real coefficients. Now, substitute expansions (E.2) and (E.5) into (E.3) to obtain

$$\left(1+\sum_{a=1}^{n}\xi^a t_a+\frac{1}{2}\sum_{bc=1}^{n}\xi^b\xi^c t_{bc}+\cdots\right)\left(1+\sum_{a=1}^{n}\zeta^a t_a+\frac{1}{2}\sum_{bc=1}^{n}\zeta^b\zeta^c t_{bc}+\cdots\right)$$
$$=1+\sum_{a=1}^{n}\left(\xi^a+\zeta^a+\sum_{bc=1}^{n}f_{bc}^a\xi^b\zeta^c+\cdots\right)t_a$$
$$+\frac{1}{2}\sum_{ab=1}^{n}(\xi^a+\zeta^a+\cdots)(\xi^b+\zeta^b+\cdots)t_{ab}+\cdots.$$

Numerical factors in front of the terms $1, \xi, \zeta, \xi^2, \zeta^2$ match exactly on both sides of this equation. However, matching the factors in front of $\xi\zeta$ is possible only under the nontrivial condition

$$\frac{1}{2}(t_{bc}+t_{cb})=t_b t_c-\sum_{a=1}^{n}f_{bc}^a t_a.$$

The left-hand side is symmetric with respect to swapping indices $b \rightleftharpoons c$. Therefore, the right-hand side must be symmetric as well, so we have

$$t_b t_c-\sum_{a=1}^{n}f_{bc}^a t_a-t_c t_b+\sum_{a=1}^{n}f_{cb}^a t_a=0. \tag{E.6}$$

According to (E.6), the *commutator* of two generators defined by the formula

$$[t_b,t_c]\equiv t_b t_c-t_c t_b$$

is a linear combination of other generators, i. e.,

$$[t_b,t_c]=\sum_{a=1}^{n}C_{bc}^a t_a, \tag{E.7}$$

where the real parameters $C_{bc}^a=f_{bc}^a-f_{cb}^a$ are called *structure constants* of the Lie group.

Theorem E.1. *Lie group generators satisfy the Jacobi identity*

$$[t_a,[t_b,t_c]]+[t_b,[t_c,t_a]]+[t_c,[t_a,t_b]]=0. \tag{E.8}$$

Proof. First we write the associative law (C.1) in the form[1]

$$0=g^a(\boldsymbol{\zeta},\boldsymbol{g}(\boldsymbol{\xi},\boldsymbol{\eta}))-g^a(\boldsymbol{g}(\boldsymbol{\zeta},\boldsymbol{\xi}),\boldsymbol{\eta})$$

1 The burden of writing summation signs is becoming unbearable, so we accept Einstein's rule, according to which these signs can be omitted and sums are performed over all pairs of repeated indices. Moreover, we leave only terms of the second and lower orders in expansions (E.5).

$$
\begin{aligned}
&\approx \zeta^a + g^a(\boldsymbol{\xi},\boldsymbol{\eta}) + f^a_{bc}\zeta^b g^c(\boldsymbol{\xi},\boldsymbol{\eta}) - g^a(\boldsymbol{\zeta},\boldsymbol{\xi}) - \eta^a - f^a_{bc} g^b(\boldsymbol{\zeta},\boldsymbol{\xi})\eta^c \\
&\approx \zeta^a + \xi^a + \eta^a + f^a_{bc}\xi^b\eta^c + f^a_{bc}\zeta^b(\xi^c + \eta^c + f^c_{xy}\xi^x\eta^y) \\
&\quad - \zeta^a - \xi^a - f^a_{xy}\zeta^x\xi^y - \eta^a - f^a_{bc}(\zeta^b + \xi^b + f^b_{xy}\zeta^x\xi^y)\eta^c \\
&= f^a_{bc}\xi^b\eta^c + f^a_{bc}\zeta^b\xi^c + f^a_{bc}\zeta^b\eta^c + f^a_{bc}f^c_{xy}\zeta^b\xi^x\eta^y \\
&\quad - f^a_{xy}\zeta^x\xi^y - f^a_{bc}\eta^c\zeta^b - f^a_{bc}\eta^c\xi^b - f^a_{bc}f^b_{xy}\eta^c\zeta^x\xi^y \\
&= f^a_{bc}f^c_{xy}\zeta^b\xi^x\eta^y - f^a_{cy}f^c_{bx}\zeta^b\xi^x\eta^y \\
&= (f^c_{xy}f^a_{bc} - f^c_{bx}f^a_{cy})\zeta^b\xi^x\eta^y .
\end{aligned}
$$

The group elements $\{\boldsymbol{\zeta}\}$, $\{\boldsymbol{\xi}\}$ and $\{\boldsymbol{\eta}\}$ were chosen arbitrarily, which means that

$$
f^x_{ab}f^y_{cx} - f^x_{ca}f^y_{xb} = 0 \tag{E.9}
$$

for all combinations of indices a, b, c, y. Next we turn to the left-hand side of the Jacobi identity (E.8),

$$
\begin{aligned}
[t_a,[t_b,t_c]] + [t_b,[t_c,t_a]] + [t_c,[t_a,t_b]] &= [t_a, C^x_{bc}t_x] + [t_b, C^x_{ca}t_x] + [t_c, C^x_{ab}t_x] \\
&= (C^x_{bc}C^y_{ax} + C^x_{ca}C^y_{bx} + C^x_{ab}C^y_{cx})t_y .
\end{aligned}
$$

The expression inside the parentheses is

$$
\begin{aligned}
&(f^x_{bc} - f^x_{cb})(f^y_{ax} - f^y_{xa}) + (f^x_{ca} - f^x_{ac})(f^y_{bx} - f^y_{xb}) + (f^x_{ab} - f^x_{ba})(f^y_{cx} - f^y_{xc}) \\
&\quad = f^x_{bc}f^y_{ax} - f^x_{bc}f^y_{xa} - f^x_{cb}f^y_{ax} + f^x_{cb}f^y_{xa} + f^x_{ca}f^y_{bx} - f^x_{ca}f^y_{xb} \\
&\qquad - f^x_{ac}f^y_{bx} + f^x_{ac}f^y_{xb} + f^x_{ab}f^y_{cx} - f^x_{ab}f^y_{xc} - f^x_{ba}f^y_{cx} + f^x_{ba}f^y_{xc} \\
&\quad = (f^x_{bc}f^y_{ax} - f^x_{ab}f^y_{xc}) + (f^x_{ca}f^y_{bx} - f^x_{bc}f^y_{xa}) - (f^x_{cb}f^y_{ax} - f^x_{ac}f^y_{xb}) \\
&\qquad - (f^x_{ba}f^y_{cx} - f^x_{cb}f^y_{xa}) + (f^x_{ab}f^y_{cx} - f^x_{ca}f^y_{xb}) - (f^x_{ac}f^y_{bx} - f^x_{ba}f^y_{xc}) .
\end{aligned} \tag{E.10}
$$

According to (E.9), all parentheses on the right-hand side of (E.10) vanish, which proves the theorem. □

E.2 Lie algebras

A *Lie algebra* is a vector space with real scalars $\lambda \in \mathbb{R}$ and with an additional operation called the *Lie bracket* and denoted by $[\boldsymbol{A},\boldsymbol{B}]_L$. This operation maps a pair of vectors $\boldsymbol{A}$ and $\boldsymbol{B}$ to a third vector. By definition, the Lie bracket satisfies the following requirements:

$$
\begin{aligned}
[\boldsymbol{A},\boldsymbol{B}]_L &= -[\boldsymbol{B},\boldsymbol{A}]_L, \\
[\boldsymbol{A},\boldsymbol{B}+\boldsymbol{C}]_L &= [\boldsymbol{A},\boldsymbol{B}]_L + [\boldsymbol{A},\boldsymbol{C}]_L,
\end{aligned}
$$

$$[\boldsymbol{A}, \lambda\boldsymbol{B}]_L = [\lambda\boldsymbol{A}, \boldsymbol{B}]_L = \lambda[\boldsymbol{A}, \boldsymbol{B}]_L, \quad \text{for any } \lambda \in \mathbb{R},$$
$$0 = [\boldsymbol{A}, [\boldsymbol{B}, \boldsymbol{C}]_L]_L + [\boldsymbol{B}, [\boldsymbol{C}, \boldsymbol{A}]_L]_L + [\boldsymbol{C}, [\boldsymbol{A}, \boldsymbol{B}]_L]_L. \tag{E.11}$$

The last equality is referred to as the *Jacobi identity*.

From our discussions in Appendix E.1 it is clear that generators of a Lie group form a Lie algebra if the commutator is used for the Lie bracket.[2] Consider, for example, the group of rotations. In the matrix representation, the generators are linear combinations of matrices (D.28)–(D.29), i. e., arbitrary antisymmetric matrices satisfying $A^T = -A$. The commutator of two such matrices,

$$[A, B] = AB - BA,$$

is also an antisymmetric matrix, because

$$(AB - BA)^T = B^T A^T - A^T B^T = BA - AB = -(AB - BA).$$

Therefore, the linear space of 3×3 antisymmetric matrices is a Lie algebra.

The structure constants of this Lie algebra can be obtained by a direct calculation from explicit formulas (D.28)–(D.29),

$$[\mathcal{J}_x, \mathcal{J}_y]_L = \mathcal{J}_z,$$
$$[\mathcal{J}_x, \mathcal{J}_z]_L = -\mathcal{J}_y,$$
$$[\mathcal{J}_y, \mathcal{J}_z]_L = \mathcal{J}_x.$$

In a more compact notation

$$[\mathcal{J}_i, \mathcal{J}_j]_L = \sum_{k=1}^{3} \epsilon_{ijk} \mathcal{J}_k,$$

which coincides with formulas (2.18) and (2.41).

We will find useful the following property of commutators in the matrix representation:

$$\begin{aligned}[A, BC] &= ABC - BCA \\ &= ABC - BAC + BAC - BCA \\ &= (AB - BA)C + B(AC - CA) \\ &= [A, B]C + B[A, C].\end{aligned} \tag{E.12}$$

2 Note that the Lie bracket in an abstract Lie algebra cannot be written as the commutator, because the product $\boldsymbol{AB}$ of two Lie algebra elements is not defined. However, we *can* represent Lie brackets by commutators in a matrix representation of the Lie algebra.

E.3 One-parameter subgroups of Lie groups

Consider an arbitrary Lie group G and a vector $\boldsymbol{t}$ from its Lie algebra A_G. Consider also the set of group elements having the form

$$\{z\} = e^{z\boldsymbol{t}}, \tag{E.13}$$

where parameter z runs through all real numbers $z \in \mathbb{R}$. Using properties 1–3 from Appendix E.1, it is easy to see that the set (E.13) forms a one-parameter subgroup in G. Indeed, this set contains the identity element (when $z = 0$), the group product is defined as

$$\{z_1\}\{z_2\} = e^{z_1\boldsymbol{t}}e^{z_2\boldsymbol{t}} = e^{(z_1+z_2)\boldsymbol{t}} = \{z_1 + z_2\}$$

and the inverse element is

$$\{z\}^{-1} = e^{-z\boldsymbol{t}} = \{-z\}.$$

In the vicinity of the identity element, every point of the group manifold belongs to some one-parameter subgroup. Therefore, each element can be represented in the form (E.13) with some z and $\boldsymbol{t}$.

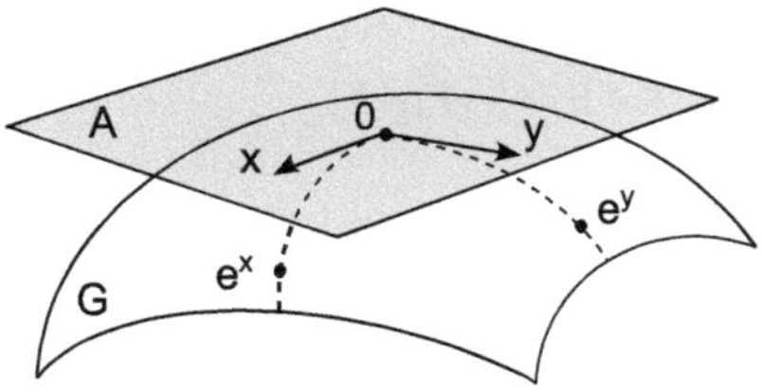

Figure E.1: The connection between a Lie group G and its Lie algebra A.

The connection between a Lie group and its Lie algebra is illustrated in Figure E.1. As we have said already, a Lie group is an n-dimensional surface G with two additional structures (the product and the inverse functions) defined on it. The n-dimensional linear space A corresponding to the Lie algebra of G can be represented as the tangent space at the identity element $e \equiv \{\mathbf{0}\}$. Elements of the Lie algebra (vectors $\boldsymbol{x}, \boldsymbol{y} \in A$) are formal derivatives (D.27) of one-parameter subgroups (shown by broken lines in the figure) in G.

Near the identity, each element of the Lie group can be represented as an exponential $\exp(\boldsymbol{x})$ (E.13) of a (tangent) vector $\boldsymbol{x}$ from the Lie algebra. Our notation (the exponent of a vector) may seem strange, but it accurately reflects the relationship between Lie groups and algebras. In the case of the group of rotations, this relationship is established in Appendix D.7.

There is a unique Lie algebra A_G for each Lie group G. However, there are many Lie groups having the same Lie algebra. These groups have identical properties near the identity element, but their global topologies may be different.

A Lie *subalgebra* is a subspace $B \subseteq A_G$ which is closed with respect to the Lie bracket, i. e., if $\boldsymbol{x}, \boldsymbol{y} \in B$, then $[\boldsymbol{x}, \boldsymbol{y}]_L \in B$.

E.4 Baker–Campbell–Hausdorff formula

We already know that the product of two elements $\exp(\boldsymbol{x})$ and $\exp(\boldsymbol{y})$ of the Lie group is another element of the group. This element should also have an exponential representation. Hence, for any two vectors $\boldsymbol{x}$ and $\boldsymbol{y}$ from the Lie algebra, we can write

$$\exp(\boldsymbol{x})\exp(\boldsymbol{y}) = \exp(\boldsymbol{z}),$$

where $\boldsymbol{z}$ is also a vector in the Lie algebra. The Baker–Campbell–Hausdorff theorem [97] gives the connection between these three vectors in the form of an infinite series,

$$\begin{aligned}
\boldsymbol{z} = \boldsymbol{x} + \boldsymbol{y} &+ \frac{1}{2}[\boldsymbol{x},\boldsymbol{y}]_L + \frac{1}{12}[[\boldsymbol{x},\boldsymbol{y}]_L,\boldsymbol{y}]_L + \frac{1}{12}[[\boldsymbol{y},\boldsymbol{x}]_L,\boldsymbol{x}]_L \\
&+ \frac{1}{24}[[[\boldsymbol{y},\boldsymbol{x}]_L,\boldsymbol{x}]_L,\boldsymbol{y}]_L - \frac{1}{720}[[[[\boldsymbol{x},\boldsymbol{y}]_L,\boldsymbol{y}]_L,\boldsymbol{y}]_L,\boldsymbol{y}]_L \\
&+ \frac{1}{360}[[[[\boldsymbol{x},\boldsymbol{y}]_L,\boldsymbol{y}]_L,\boldsymbol{y}]_L,\boldsymbol{x}]_L + \frac{1}{360}[[[[\boldsymbol{y},\boldsymbol{x}]_L,\boldsymbol{x}]_L,\boldsymbol{x}]_L,\boldsymbol{y}]_L \\
&- \frac{1}{120}[[[[\boldsymbol{x},\boldsymbol{y}]_L,\boldsymbol{y}]_L,\boldsymbol{x}]_L,\boldsymbol{y}]_L - \frac{1}{120}[[[[\boldsymbol{y},\boldsymbol{x}]_L,\boldsymbol{x}]_L,\boldsymbol{y}]_L,\boldsymbol{x}]_L \cdots .
\end{aligned}$$

This means that Lie brackets contain all information about the multiplication law in the vicinity of the group identity. In many cases it is much more convenient to deal with generators and their Lie brackets than directly with group elements and their products.

In applications we will often meet the following useful identity:

$$\exp(a\boldsymbol{x})\boldsymbol{y}\exp(-a\boldsymbol{x}) = \boldsymbol{y} + a[\boldsymbol{x},\boldsymbol{y}]_L + \frac{a^2}{2!}[\boldsymbol{x},[\boldsymbol{x},\boldsymbol{y}]_L]_L + \frac{a^3}{3!}[\boldsymbol{x},[\boldsymbol{x},[\boldsymbol{x},\boldsymbol{y}]_L]_L]_L + O(a^4), \tag{E.14}$$

where $a \in \mathbb{R}$. This formula can be proved by noticing that both sides are solutions of the same differential equation,

$$\frac{d\boldsymbol{y}(a)}{da} = [\boldsymbol{x},\boldsymbol{y}(a)]_L,$$

with the same initial condition $\boldsymbol{y}(0) = \boldsymbol{y}$.

F Hilbert space

F.1 Internal product

The *inner product space* $\mathscr{H}$ is defined as a vector space over complex scalars,[1] where a mapping from ordered pairs of vectors to complex numbers is defined. This mapping $(|y\rangle, |x\rangle) \in \mathbb{C}$ is called the *inner product*, and it has the following properties:

$$(|x\rangle, |y\rangle) = (|y\rangle, |x\rangle)^*, \tag{F.1}$$

$$(|z\rangle, \alpha|x\rangle + \beta|y\rangle) = \alpha(|z\rangle, |x\rangle) + \beta(|z\rangle, |y\rangle), \tag{F.2}$$

$$(|x\rangle, |x\rangle) \in \mathbb{R}, \tag{F.3}$$

$$(|x\rangle, |x\rangle) \geq 0, \tag{F.4}$$

$$(|x\rangle, |x\rangle) = 0 \quad \Leftrightarrow \quad |x\rangle = \mathbf{0}, \tag{F.5}$$

where α and β are complex numbers.

The *Hilbert space* $\mathscr{H}$ is an inner product space, where several additional properties are defined: the convergence of sequences of vectors, the closedness of subspaces and the completeness. We will not discuss these technical details in our book.

F.2 Orthonormal bases

Two vectors $|x\rangle$ and $|y\rangle$ in the Hilbert space $\mathscr{H}$ are called *orthogonal*, if their inner product vanishes, i. e., $(|x\rangle, |y\rangle) = 0$. The vector $|x\rangle$ is called *unimodular*, if $(|x\rangle, |x\rangle) = 1$. In a Hilbert space, we can construct an *orthonormal basis* that consists of mutually orthogonal unimodular vectors $|e_i\rangle$ satisfying the equality[2]

$$(|e_i\rangle, |e_j\rangle) = \delta_{ij}. \tag{F.6}$$

Suppose that in this basis two vectors $|x\rangle$ and $|y\rangle$ have components x_i and y_i, respectively:

$$|x\rangle = x_1|e_1\rangle + x_2|e_2\rangle + \cdots + x_n|e_n\rangle,$$
$$|y\rangle = y_1|e_1\rangle + y_2|e_2\rangle + \cdots + y_n|e_n\rangle.$$

1 Vectors in $\mathscr{H}$ will be denoted $|x\rangle$.

2 All formulas in this appendix are written for finite-dimensional Hilbert spaces. As a rule, Hilbert spaces of physical systems are infinite-dimensional. Moreover, they are inseparable (i. e., with an uncountable basis). However, in our book we ignore all these difficulties, assuming that the properties proved in the finite-dimensional case are extrapolatable to more realistic Hilbert spaces.

https://doi.org/10.1515/9783110492132-013

Then, using (F.1), (F.2) and (F.6), we can express the inner product through the vector components

$$\begin{aligned}(|x\rangle, |y\rangle) &= (x_1|e_1\rangle + x_2|e_2\rangle + \cdots + x_n|e_n\rangle, y_1|e_1\rangle + y_2|e_2\rangle + \cdots + y_n|e_n\rangle) \\ &= x_1^* y_1 + x_2^* y_2 + \cdots + x_n^* y_n = \sum_i x_i^* y_i. \end{aligned} \tag{F.7}$$

F.3 Bra and ket vectors

The notation $(|x\rangle, |y\rangle)$ for the internal product is rather cumbersome. Instead, it is convenient to use the so-called *bra-ket* formalism, which was invented by Dirac and greatly simplifies manipulations with objects in the Hilbert space. Let us call vectors in the Hilbert space *ket* vectors. Then we define a *linear functional* $\langle f| : \mathscr{H} \to \mathbb{C}$ as a linear mapping from ket vectors $|x\rangle \in \mathscr{H}$ to complex numbers,

$$\langle f|(\alpha|x\rangle + \beta|y\rangle) = \alpha\langle f|x\rangle + \beta\langle f|y\rangle.$$

Since any linear combination $\alpha\langle f| + \beta\langle g|$ of two functionals $\langle f|$ and $\langle g|$ is also a linear functional, all such functionals form a vector space (denoted $\mathscr{H}^*$). Vectors in $\mathscr{H}^*$ will be called *bra* vectors. We can define an inner product in $\mathscr{H}^*$ so that it becomes a Hilbert space. For this, let us choose an orthonormal basis $|e_i\rangle$ in $\mathscr{H}$. Then, each functional $\langle f|$ defines a set of complex numbers f_i – the values of this functional on the basis vectors

$$f_i = \langle f|e_i\rangle.$$

These numbers define the functional uniquely, i. e., if two functionals $\langle f|$ and $\langle g|$ are different, then their values are different at least on one basic vector $|e_k\rangle$: $f_k \neq g_k$.[3] Now we can define the inner product of two bra vectors $\langle f|$ and $\langle g|$ by the formula

$$(\langle f|, \langle g|) = \sum_i f_i g_i^*$$

and verify that it satisfies all properties (F.1)–(F.5) of the inner product.

The bra Hilbert space $\mathscr{H}^*$ is called *dual* to the ket space $\mathscr{H}$. Note that each ket vector $|y\rangle \in \mathscr{H}$ defines a unique linear functional $\langle y|$, which acts on all $|x\rangle \in \mathscr{H}$ by the formula

$$\langle y|x\rangle \equiv (|y\rangle, |x\rangle).$$

3 Otherwise, using the property of linearity, one could prove that these two functionals have the same values on all vectors in $\mathscr{H}$, that is, $\langle f| = \langle g|$.

This bra vector $\langle y|$ is called dual to the ket vector $|y\rangle$. In other words, in order to calculate the inner product of $|y\rangle$ and $|x\rangle$, we should find a bra vector (functional) dual to $|y\rangle$ and then find its value on $|x\rangle$. This means that the inner product is obtained by combining the bra and ket vectors $\langle y|x\rangle$, thus forming a closed *bra-c-ket*.

While bra vectors in $\mathscr{H}^*$ define linear functionals on $\mathscr{H}$, ket vectors $|x\rangle \in \mathscr{H}$ define *antilinear* functionals on bras by the formula $\langle y|x\rangle$, i. e.,

$$(\alpha\langle y| + \beta\langle z|)|x\rangle = \alpha^* \langle y|x\rangle + \beta^* \langle z|x\rangle.$$

Applying the same arguments as before, we see that if $\langle y|$ is a bra vector, then there exists a unique ket $|y\rangle$ such that for any $\langle x| \in \mathscr{H}^*$ we have

$$\langle x|y\rangle = (\langle x|, \langle y|). \tag{F.8}$$

Thus, we have established an isomorphism of the two Hilbert spaces $\mathscr{H}$ and $\mathscr{H}^*$. This statement is known as the *Riesz theorem.*

Lemma F.1. *If the kets $|e_i\rangle$ form an orthonormal basis in $\mathscr{H}$, then the dual bras $\langle e_i|$ also form an orthonormal basis in $\mathscr{H}^*$.*

Proof. Suppose that $\langle e_i|$ does not form a basis. Then there is a nonzero bra vector $\langle z| \in \mathscr{H}^*$ orthogonal to all $\langle e_i|$. However, the values of the functional $\langle z|$ on all basis ket vectors $|e_i\rangle$ are equal to zero. Therefore $\langle z| = \mathbf{0}$, in contradiction to our assumption. The orthonormality of $\langle e_i|$ follows from equations (F.8) and (F.6). We have

$$(\langle e_i|, \langle e_j|) = \langle e_i|e_j\rangle = (|e_i\rangle, |e_j\rangle) = \delta_{ij}.$$ □

It is convenient to use the following notation for the components x_i of the vector $|x\rangle$ in the basis $|e_i\rangle$:

$$\langle e_i|x\rangle = \langle e_i|(x_1|e_1\rangle + x_2|e_2\rangle + \cdots + x_n|e_n\rangle) = x_i.$$

Now we can write

$$|x\rangle = \sum_i |e_i\rangle x_i = \sum_i |e_i\rangle\langle e_i|x\rangle. \tag{F.9}$$

The bra vector $\langle y|$ dual to the ket $|y\rangle$ has complex conjugate components in the dual basis

$$\langle y| = \sum_i y_i^* \langle e_i|. \tag{F.10}$$

You can verify this by checking that the functional on the right-hand side, when applied to the vector $|x\rangle \in \mathscr{H}$, gives

$$\sum_i y_i^* \langle e_i|x\rangle = \sum_i y_i^* x_i = (|y\rangle, |x\rangle) = \langle y|x\rangle.$$

F.4 Tensor product of Hilbert spaces

If two Hilbert spaces $\mathscr{H}_1$ and $\mathscr{H}_2$ are given, then we can construct the third Hilbert space $\mathscr{H}$, which is called the *tensor product* of $\mathscr{H}_1$ and $\mathscr{H}_2$ and is denoted by $\mathscr{H} = \mathscr{H}_1 \otimes \mathscr{H}_2$. For each pair of basis ket vectors $|i\rangle \in \mathscr{H}_1$ and $|j\rangle \in \mathscr{H}_2$ there exists one basic ket in $\mathscr{H}$ denoted $|i\rangle \otimes |j\rangle$. Other vectors in $\mathscr{H}$ are linear combinations of the basis kets $|i\rangle \otimes |j\rangle$ with complex coefficients.

The inner product of two basis vectors $|i_1\rangle \otimes |j_1\rangle \in \mathscr{H}$ and $|i_2\rangle \otimes |j_2\rangle \in \mathscr{H}$ is defined as $\langle i_1|i_2\rangle\langle j_1|j_2\rangle$. This inner product extends to all linear combinations of basis vectors by linearity.

G Operators

G.1 Linear operators

Linear transformations (also known as linear *operators*) of vectors in the Hilbert space

$$T|x\rangle = |x'\rangle$$

play an important role in quantum formalism. The linearity of such transformations is expressed by the equality

$$T(\alpha|x\rangle + \beta|y\rangle) = \alpha T|x\rangle + \beta T|y\rangle$$

for any two complex numbers α, β and any two vectors $|x\rangle$, $|y\rangle$. If we have an operator T and a basis $|e_i\rangle$, then we can find images of the basis vectors

$$T|e_i\rangle = |e_i'\rangle$$

and the expansion of these images in the basis $|e_i\rangle$

$$|e_i'\rangle = \sum_j t_{ij}|e_j\rangle.$$

The coefficients t_{ij} of this expansion are called *matrix elements* of the operator T in the basis $|e_i\rangle$. In the bra-ket formalism, we obtain a convenient expression for the matrix elements

$$\langle e_j|(T|e_i\rangle) = \langle e_j|e_i'\rangle = \langle e_j|\sum_k t_{ik}|e_k\rangle = \sum_k t_{ik}\langle e_j|e_k\rangle = \sum_k t_{ik}\delta_{jk} = t_{ij}.$$

Knowing matrix elements of the operator T and components of the vector $|x\rangle$ in the basis $|e_i\rangle$, we can always find components of the transformed vector $|x'\rangle = T|x\rangle$,

$$\begin{aligned} x_i' &= \langle e_i|x'\rangle = \langle e_i|(T|x\rangle) = \langle e_i|\sum_j (T|e_j\rangle)x_j = \sum_{jk}\langle e_i|e_k\rangle t_{kj}x_j \\ &= \sum_{jk}\delta_{ik}t_{kj}x_j = \sum_j t_{ij}x_j. \end{aligned} \tag{G.1}$$

In the bra-ket notation, the operator T has the form

$$T = \sum_{ij}|e_i\rangle t_{ij}\langle e_j|. \tag{G.2}$$

Indeed, applying the right-hand side of (G.2) to an arbitrary vector $|x\rangle$, we get

$$\sum_{ij}|e_i\rangle t_{ij}\langle e_j|x\rangle = \sum_{ij}|e_i\rangle t_{ij}x_j = \sum_i x_i'|e_i\rangle = |x'\rangle = T|x\rangle.$$

In particular, it follows from (F.9) that the identity operator in $\mathscr{H}$ has the representation

$$I = \sum_i |e_i\rangle\langle e_i| = \sum_{ij}|e_i\rangle\delta_{ij}\langle e_j|. \tag{G.3}$$

https://doi.org/10.1515/9783110492132-014

G.2 Matrices and operators

It is often convenient to represent Hilbert space vectors and operators $\mathscr{H}$ in the matrix notation. Let us fix the orthonormal basis $|e_i\rangle \in \mathscr{H}$ and describe each ket vector $|y\rangle$ by the column of its components,

$$|y\rangle = \begin{bmatrix} y_1 \\ y_2 \\ \vdots \\ y_n \end{bmatrix}.$$

A bra vector $\langle x|$ is represented by a row

$$\langle x| = [x_1^*, x_2^*, \dots, x_n^*]$$

of complex conjugated components in the dual basis $\langle e_i|$. Then the inner product (F.7) is obtained by the usual "row-by-column" product rule,

$$\langle x|y\rangle = [x_1^*, x_2^*, \dots, x_n^*] \begin{bmatrix} y_1 \\ y_2 \\ \vdots \\ y_n \end{bmatrix} = \sum_i x_i^* y_i.$$

Matrix elements of any operator T in (G.2) are conveniently arranged in a square *matrix*

$$T = \begin{bmatrix} t_{11} & t_{12} & \dots & t_{1n} \\ t_{21} & t_{22} & \dots & t_{2n} \\ \vdots & \vdots & \ddots & \vdots \\ t_{n1} & t_{n2} & \dots & t_{nn} \end{bmatrix}.$$

Then the action of the operator T on a vector $|x'\rangle = T|x\rangle$ can be represented as a product of the matrix and the vector-column,

$$\begin{bmatrix} x_1' \\ x_2' \\ \vdots \\ x_n' \end{bmatrix} = \begin{bmatrix} t_{11} & t_{12} & \dots & t_{1n} \\ t_{21} & t_{22} & \dots & t_{2n} \\ \vdots & \vdots & \ddots & \vdots \\ t_{n1} & t_{n2} & \dots & t_{nn} \end{bmatrix} \begin{bmatrix} x_1 \\ x_2 \\ \vdots \\ x_n \end{bmatrix} = \begin{bmatrix} \sum_j t_{1j} x_j \\ \sum_j t_{2j} x_j \\ \vdots \\ \sum_j t_{nj} x_j \end{bmatrix}.$$

So, each operator has a unique matrix, and each $n \times n$ matrix defines a single linear operator. This establishes an isomorphism between *matrices* and *operators*. We will use these two terms interchangeably.

The *identity operator* (G.3) has the unit matrix δ_{ij}, so

$$I = \begin{bmatrix} 1 & 0 & \dots & 0 \\ 0 & 1 & \dots & 0 \\ \vdots & \vdots & \ddots & \vdots \\ 0 & 0 & \dots & 1 \end{bmatrix}.$$

A *diagonal operator* has a diagonal matrix $d_i\delta_{ij}$, so

$$D = \begin{bmatrix} d_1 & 0 & \dots & 0 \\ 0 & d_2 & \dots & 0 \\ \vdots & \vdots & \ddots & \vdots \\ 0 & 0 & \dots & d_n \end{bmatrix}.$$

The action of an operator in the dual space $\mathscr{H}^*$ will be indicated by multiplying bra vector-rows by the operator matrix from the right,

$$[y'_1, y'_2, \dots, y'_n] = [y_1, y_2, \dots, y_n] \begin{bmatrix} s_{11} & s_{12} & \dots & s_{1n} \\ s_{21} & s_{22} & \dots & s_{2n} \\ \vdots & \vdots & \ddots & \vdots \\ s_{n1} & s_{n2} & \dots & s_{nn} \end{bmatrix},$$

or in a symbolic notation,

$$y'_i = \sum_j y_j s_{ji},$$

$$\langle y'| = \langle y|S.$$

Suppose that the operator T with the matrix t_{ij} transforms the ket vector $|x\rangle$ into the ket vector $|y\rangle$, i. e.,

$$y_i = \sum_j t_{ij} x_j. \tag{G.4}$$

What is the matrix of the operator S that connects vectors $\langle x|$ and $\langle y|$ in the dual space $\mathscr{H}^*$? Since $\langle x|$ and $\langle y|$ have components that are complex conjugate to the components of $|x\rangle$ and $|y\rangle$ and S acts on the bra vectors from the right, we can write

$$y_i^* = \sum_j x_j^* s_{ji}.$$

On the other hand, taking the complex conjugation of (G.4), we obtain

$$y_i^* = \sum_j t_{ij}^* x_j^*.$$

Comparing these two expressions, we obtain

$$s_{ij} = t_{ji}^*.$$

This means that the matrix representing the action of T in the bra space $\mathscr{H}^*$ differs from t_{ij} in that rows and columns are interchanged,[1] and matrix elements are replaced by their complex conjugates. This joint "transposition + complex conjugation" operation is called the *Hermitian conjugation*: $T \to T^\dagger$. In particular, we can write

$$\langle x|(T|y\rangle) = (\langle x|T^\dagger)|y\rangle, \tag{G.5}$$

$$\det[T^\dagger] = (\det[T])^*. \tag{G.6}$$

G.3 Functions of operators

The sum of two operators A and B and the product of the operator A by a complex number λ are easily expressed in terms of matrix elements as follows:

$$(A+B)_{ij} = a_{ij} + b_{ij},$$

$$(\lambda A)_{ij} = \lambda a_{ij}.$$

We define the product AB as a transformation in $\mathscr{H}$ obtained as a result of sequential application of the two operators: first B and then A. Obviously, this product is also a linear transformation of vectors, i. e., an operator. The matrix of the operator AB is obtained from the matrices a_{ij} and b_{ij} using the standard "row-by-column" rule

$$(AB)_{ij} = \sum_k a_{ik} b_{kj}.$$

Lemma G.1. *The Hermitian conjugate of the product of two operators is equal to the product of conjugate operators in the reverse order:*

$$(AB)^\dagger = B^\dagger A^\dagger.$$

Proof. We have

$$(AB)^\dagger_{ij} = (AB)^*_{ji} = \sum_k a^*_{jk} b^*_{ki} = \sum_k b^*_{ki} a^*_{jk} = \sum_k (B^\dagger)_{ik} (A^\dagger)_{kj} = (B^\dagger A^\dagger)_{ij}. \qquad \square$$

The *inverse operator* A^{-1} is defined by its properties

$$A^{-1}A = AA^{-1} = I.$$

The corresponding matrix is inverse with respect to the matrix of A.

1 This matrix operation is called the *transposition*. It reflects the matrix with respect to the main diagonal.

Using the basic operations of addition, multiplication and inversion, we can define various functions $f(A)$ of operator A. For example, the exponent is determined by its Taylor series

$$e^A = 1 + A + \frac{1}{2!}A^2 + O(A^3). \tag{G.7}$$

For any two operators A and B the expression

$$[A, B] \equiv AB - BA \tag{G.8}$$

is called the *commutator*.[2] We shall say that two operators A and B *commute* with each other if $[A, B] = 0$. It is clear that all powers of A commute; $[A^n, A^m] = 0$ and also $[A, A^{-1}] = 0$. Consequently, any two functions of the operator A also commute: $[f(A), g(A)] = 0$.

The *trace* of a matrix is the sum of its diagonal elements

$$\mathrm{Tr}(A) = \sum_i A_{ii}.$$

Lemma G.2. *The trace of the product of operators is invariant with respect to the cyclic permutation of factors.*

Proof. Take for example the trace of the product of three operators,

$$\mathrm{Tr}(ABC) = \sum_{ijk} A_{ij} B_{jk} C_{ki}.$$

Table G.1: Properties of operators in the Hilbert space.

	Symbol	**Conditions on matrix elements or eigenvalues**
Transformations/functions of operators		
Complex conjugation	$A \to A^*$	$(A^*)_{ij} = A^*_{ij}$
Transposition	$A \to A^T$	$(A^T)_{ij} = A_{ji}$
Hermitian conjugation	$A \to A^\dagger = (A^*)^T$	$(A^\dagger)_{ij} = A^*_{ji}$
Inversion	$A \to A^{-1}$	inverse eigenvalues
Determinant	$\det[A]$	product of eigenvalues
Trace	$\mathrm{Tr}(A)$	$\sum_i A_{ii}$
Types of operators		
Identity	I	$I_{ij} = \delta_{ij}$
Diagonal	D	$D_{ij} = d_i \delta_{ij}$
Hermitian	$A = A^\dagger$	$A_{ij} = A^*_{ji}$
Anti-Hermitian	$A = -A^\dagger$	$A_{ij} = -A^*_{ji}$
Unitary	$A^{-1} = A^\dagger$	unimodular eigenvalues
Projection	$A = A^\dagger, A^2 = A$	eigenvalues only 0 or 1

2 $\{A, B\} \equiv AB + BA$ is called the *anticommutator*.

By changing the dummy summation indices $i \to k, j \to i$ and $k \to j$ we obtain

$$\mathrm{Tr}(ABC) = \sum_{ijk} A_{ki}B_{ij}C_{jk} = \sum_{ijk} B_{ij}C_{jk}A_{ki} = \mathrm{Tr}(BCA). \qquad \square$$

Table G.1 summarizes some operator properties discussed above.

G.4 Hermitian and unitary operators

In quantum mechanics an important role is played by Hermitian and unitary operators (and their matrices). We call the operator T *Hermitian*, if

$$T = T^{\dagger}. \tag{G.9}$$

For the Hermitian T we can write

$$\begin{aligned} t_{ii} &= t_{ii}^{*}, \\ t_{ij} &= t_{ji}^{*}, \end{aligned} \tag{G.10}$$

i. e., diagonal matrix elements are real, and off-diagonal matrix elements symmetric with respect to the main diagonal are complex conjugates of each other. Moreover, from equations (G.5) and (G.9), we can calculate the inner product of the vectors $\langle x|$ and $T|y\rangle$ as follows:

$$\langle x|(T|y\rangle) = (\langle x|T^{\dagger})|y\rangle = (\langle x|T)|y\rangle \equiv \langle x|T|y\rangle.$$

From this symmetric notation it is clear that the Hermitian T can act both to the right (on the ket $|y\rangle$) and to the left (on the bra $\langle x|$)

The operator U is called *unitary*, if

$$U^{-1} = U^{\dagger}$$

or, equivalently,

$$U^{\dagger}U = UU^{\dagger} = I.$$

Unitary operators preserve the inner product of vectors, i. e.,

$$\langle Ua|Ub\rangle \equiv (\langle a|U^{\dagger})(U|b\rangle) = \langle a|U^{-1}U|b\rangle = \langle a|I|b\rangle = \langle a|b\rangle. \tag{G.11}$$

Lemma G.3. *If F is an Hermitian operator, then $U = e^{iF}$ is a unitary one.*

Proof. We have

$$U^{\dagger}U = (e^{iF})^{\dagger}(e^{iF}) = e^{-iF^{\dagger}}e^{iF} = e^{-iF}e^{iF} = e^{-iF+iF} = e^{0} = I. \qquad \square$$

The operator A is called *antilinear*, if $A(\alpha|x\rangle + \beta|y\rangle) = \alpha^{*}A|x\rangle + \beta^{*}A|y\rangle$ for any complex $\alpha\,\beta$ and any $|x\rangle, |y\rangle \in \mathscr{H}$. An antilinear operator with the additional property $\langle Ay|Ax\rangle = \langle y|x\rangle^{*}$ is called *antiunitary*.

G.5 Linear operators in different orthonormal bases

So far we worked with matrix elements of operators in a fixed orthonormal basis $|e_i\rangle$. However, in another basis, the same operator will be represented by a different matrix. Nevertheless, we are going to show that many of the above properties remain valid in all orthonormal bases (that is, they are basis-independent).

Theorem G.4. *If $|e_i\rangle$ and $|e_i'\rangle$ are two orthonormal bases, then there exists a unitary operator U such that*

$$U|e_i\rangle = |e_i'\rangle \tag{G.12}$$

and conversely, if $|e_i\rangle$ is an orthonormal basis and U is a unitary operator, then (G.12) *is also an orthonormal basis.*

Proof. The basis $|e_i'\rangle$, obtained by applying the unitary operator U to the orthonormal basis $|e_i\rangle$, is orthonormal, because U preserves inner products of vectors (G.11). To prove the direct statement of the theorem, let us form the matrix

$$\begin{bmatrix} \langle e_1|e_1'\rangle & \langle e_1|e_2'\rangle & \dots & \langle e_1|e_n'\rangle \\ \langle e_2|e_1'\rangle & \langle e_2|e_2'\rangle & \dots & \langle e_2|e_n'\rangle \\ \vdots & \vdots & \ddots & \vdots \\ \langle e_n|e_1'\rangle & \langle e_n|e_2'\rangle & \dots & \langle e_n|e_n'\rangle \end{bmatrix}$$

with matrix elements

$$u_{ji} = \langle e_j|e_i'\rangle.$$

The operator U, corresponding to this matrix, can be written as

$$U = \sum_{jk} |e_j\rangle u_{jk}\langle e_k| = \sum_{jk} |e_j\rangle\langle e_j|e_k'\rangle\langle e_k|.$$

Therefore, its action on the vector $|e_i\rangle$,

$$\begin{aligned} U|e_i\rangle &= \sum_{jk} |e_j\rangle\langle e_j|e_k'\rangle\langle e_k|e_i\rangle = \sum_{jk} |e_j\rangle\langle e_j|e_k'\rangle\delta_{ki} \\ &= \sum_{j} |e_j\rangle\langle e_j|e_i'\rangle = I|e_i'\rangle = |e_i'\rangle, \end{aligned}$$

transforms it into the vector $|e_i'\rangle$, as required. This operator is unitary, because

$$\begin{aligned} (UU^\dagger)_{ij} &= \sum_k u_{ik}u_{jk}^* = \sum_k \langle e_i|e_k'\rangle\langle e_j|e_k'\rangle^* = \sum_k \langle e_i|e_k'\rangle\langle e_k'|e_j\rangle \\ &= \langle e_i|I|e_j\rangle = \langle e_i|e_j\rangle = \delta_{ij} = I_{ij}. \end{aligned}$$

□

If F is an operator with matrix elements f_{ij} in the basis $|e_k\rangle$, then its matrix elements f'_{ij} in the basis $|e'_k\rangle = U|e_k\rangle$ can be obtained by the formula

$$f'_{ij} = \langle e'_i|F|e'_j\rangle = (\langle e_i|U^\dagger)F(U|e_j\rangle) = \langle e_i|U^\dagger FU|e_j\rangle = \langle e_i|U^{-1}FU|e_j\rangle. \quad \text{(G.13)}$$

We can look at this equality from two different, but equivalent points of view. On the one hand (the *passive* point of view), we can consider (G.13) as matrix elements of F in the new basis $U|e_i\rangle$. On the other hand (the *active* point of view), they can be perceived as matrix elements of the transformed operator $U^{-1}FU$ in the initial basis $|e_i\rangle$.

Unitary transformation changes the matrix of the operator, but the operator's type remains the same. If the operator F is Hermitian, then after the transformation it remains Hermitian:

$$(F')^\dagger = (U^{-1}FU)^\dagger = U^\dagger F^\dagger (U^{-1})^\dagger = U^{-1}FU = F'.$$

If the operator V is unitary, then for the transformed operator V' we have

$$\begin{aligned}(V')^\dagger V' &= (U^{-1}VU)^\dagger U^{-1}VU = U^\dagger V^\dagger (U^{-1})^\dagger U^{-1}VU \\ &= U^{-1}V^\dagger UU^{-1}VU = U^{-1}V^\dagger VU = U^{-1}U = I,\end{aligned}$$

thus V' is unitary as well.

Lemma G.5. *The trace of an operator is basis-independent.*

Proof. From Lemma G.2 it follows that

$$\mathrm{Tr}(U^{-1}AU) = \mathrm{Tr}(AUU^{-1}) = \mathrm{Tr}(A).$$ □

G.6 Diagonalization of Hermitian and unitary matrices

We know that the choice of a basis in the Hilbert space is a matter of convenience. Therefore, when performing calculations it is desirable to choose a basis in which important operators have the simplest form, e. g., the diagonal one. It turns out that Hermitian and unitary operators can always be brought to the diagonal form by an appropriate choice of basis.

Theorem G.6 (spectral theorem). *For any Hermitian or unitary operator F one can find an orthonormal basis $|e_i\rangle$ where*

$$F|e_i\rangle = f_i|e_i\rangle, \quad \text{(G.14)}$$

where f_i are, generally speaking, complex numbers.

The proof of this theorem can be found in [75].

If the vector $|x\rangle$ satisfies the equation

$$F|x\rangle = f|x\rangle,$$

where f is a complex number, then $|x\rangle$ is called the *eigenvector* of the operator F, and f is its *eigenvalue*. Thus the spectral theorem tells us that one can always build an orthonormal basis from eigenvectors of any Hermitian or unitary operator.

From equation (G.14) it follows that the matrix of the operator F is diagonal in the basis $|e_i\rangle$, so

$$F = \begin{bmatrix} f_1 & 0 & \dots & 0 \\ 0 & f_2 & \dots & 0 \\ \vdots & \vdots & \ddots & \vdots \\ 0 & 0 & \dots & f_n \end{bmatrix} \tag{G.15}$$

and, according to (G.2), any Hermitian or unitary operator can be expressed through its eigenvectors and eigenvalues by the following formula:

$$F = \sum_i |e_i\rangle f_i \langle e_i|. \tag{G.16}$$

Lemma G.7. *Hermitian operators have real eigenvalues.*

Proof. In its diagonal form, the eigenvalues of an Hermitian operator (G.15) are on the diagonal. From (G.10) it follows that these matrix elements are real. □

Lemma G.8. *Unitary operators have unimodular eigenvalues.*

Proof. Using the representation (G.16), we obtain

$$\begin{aligned}
\sum_i |e_i\rangle\langle e_i| = I &= UU^\dagger \\
&= \left(\sum_i |e_i\rangle f_i \langle e_i|\right)\left(\sum_j |e_j\rangle f_j^* \langle e_j|\right) \\
&= \sum_{ij} f_i f_j^* |e_i\rangle\langle e_i|e_j\rangle\langle e_j| \\
&= \sum_{ij} f_i f_j^* |e_i\rangle \delta_{ij} \langle e_j| \\
&= \sum_i |f_i|^2 |e_i\rangle\langle e_i|.
\end{aligned}$$

Therefore, $|f_i|^2 = 1$. □

One benefit of diagonalization is that functions of operators are easily computed in the diagonal form. If the operator A is diagonal i. e.,

$$A = \begin{bmatrix} a_1 & 0 & \dots & 0 \\ 0 & a_2 & \dots & 0 \\ \vdots & \vdots & \ddots & \vdots \\ 0 & 0 & \dots & a_n \end{bmatrix},$$

then the arbitrary function $f(A)$ has the matrix form

$$f(A) = \begin{bmatrix} f(a_1) & 0 & \dots & 0 \\ 0 & f(a_2) & \dots & 0 \\ \vdots & \vdots & \ddots & \vdots \\ 0 & 0 & \dots & f(a_n) \end{bmatrix}.$$

For example, the matrix of the inverse operator is[3]

$$A^{-1} = \begin{bmatrix} a_1^{-1} & 0 & \dots & 0 \\ 0 & a_2^{-1} & \dots & 0 \\ \vdots & \vdots & \ddots & \vdots \\ 0 & 0 & \dots & a_n^{-1} \end{bmatrix}.$$

From Lemma G.8 it follows that there is a basis where the unitary operator U has the form

$$U = \begin{bmatrix} e^{if_1} & 0 & \dots & 0 \\ 0 & e^{if_2} & \dots & 0 \\ \vdots & \vdots & \ddots & \vdots \\ 0 & 0 & \dots & e^{if_n} \end{bmatrix},$$

with real f_i. This implies that each unitary operator is representable as an exponent

$$U = e^{iF},$$

of an Hermitian operator F:

$$F = \begin{bmatrix} f_1 & 0 & \dots & 0 \\ 0 & f_2 & \dots & 0 \\ \vdots & \vdots & \ddots & \vdots \\ 0 & 0 & \dots & f_n \end{bmatrix};$$

see also Lemma G.3.

3 Notice that the inverse operator A^{-1} can be defined only when all eigenvalues of A are nonzero.

Lemma G.9. *Eigenvalues of a unitary or Hermitian operator F are invariant with respect to unitary changes of the basis.*

Proof. If $|\psi_k\rangle$ is an eigenvector of F with the eigenvalue f_k, i. e.,

$$F|\psi_k\rangle = f_k|\psi_k\rangle,$$

then vector $U|\psi_k\rangle$ is an eigenvector of the unitarily transformed operator $F' = UFU^{-1}$ with the same eigenvalue, because

$$F'(U|\psi_k\rangle) = UFU^{-1}(U|\psi_k\rangle) = UF|\psi_k\rangle = Uf_k|\psi_k\rangle = f_k(U|\psi_k\rangle).$$ □

H Subspaces and projections

H.1 Projections

Two subspaces $\mathscr{A}$ and $\mathscr{B}$ in the Hilbert space $\mathscr{H}$ are called *orthogonal* (denoted by $\mathscr{A} \perp \mathscr{B}$), if every vector from $\mathscr{A}$ is orthogonal to every vector from $\mathscr{B}$. The linear span of all vectors orthogonal to $\mathscr{A}$ is called the *orthogonal complement* to the subspace $\mathscr{A}$ and is denoted $\mathscr{A}'$.

If a subspace $\mathscr{A}$ (with dimension $\dim(\mathscr{A}) = m$) is given in the Hilbert space $\mathscr{H}$ (with dimension $\dim(\mathscr{H}) = n > m$), then we can choose an orthonormal basis $|e_i\rangle$ such that m vectors with indices $i = 1, 2, \ldots, m$ lie in $\mathscr{A}$ and $n - m$ vectors with indices $i = m + 1, m + 2, \ldots, n$ lie in the orthogonal complement $\mathscr{A}'$.[1] Then for each vector $|y\rangle \in \mathscr{H}$ we can write

$$|y\rangle = \sum_i^n |e_i\rangle\langle e_i|y\rangle = \sum_{i=1}^m |e_i\rangle\langle e_i|y\rangle + \sum_{i=m+1}^n |e_i\rangle\langle e_i|y\rangle.$$

The first sum lies entirely in $\mathscr{A}$ and is denoted by $|y_\parallel\rangle$. The second sum lies in $\mathscr{A}'$ and is denoted by $|y_\perp\rangle$. This means that we can always decompose $|y\rangle$ into two uniquely defined orthogonal components $|y_\parallel\rangle$ and $|y_\perp\rangle$, i. e.,

$$\begin{aligned} |y\rangle &= |y_\parallel\rangle + |y_\perp\rangle, \\ |y_\parallel\rangle &\in \mathscr{A}, \\ |y_\perp\rangle &\in \mathscr{A}'. \end{aligned}$$

We can also define a linear operator $P_{\mathscr{A}}$, called the *projection* onto the subspace $\mathscr{A}$, which takes each vector $|y\rangle$ into its parallel component, i. e.,

$$P_{\mathscr{A}}|y\rangle = |y_\parallel\rangle.$$

The subspace $\mathscr{A}$ is called the *range* of the projection $P_{\mathscr{A}}$. In the bra-ket notation we can also write

$$P_{\mathscr{A}} = \sum_{i=1}^m |e_i\rangle\langle e_i|,$$

so that in our basis $|e_i\rangle$ $(i = 1, 2, \ldots, n)$ the operator $P_{\mathscr{A}}$ has a diagonal matrix whose first m diagonal matrix elements are 1 and all the other $(i = m+1, \ldots, n)$ diagonal matrix elements are zero. Hence it follows that

$$P_{\mathscr{A}'} = 1 - P_{\mathscr{A}}.$$

---

1 In this case we say that the Hilbert space $\mathscr{H}$ is represented as a *direct sum* ($\mathscr{H} = \mathscr{A} \oplus \mathscr{A}'$) of the orthogonal subspaces $\mathscr{A}$ and $\mathscr{A}'$.

https://doi.org/10.1515/9783110492132-015

A set of projections P_α onto mutually orthogonal subspaces $\mathscr{H}_\alpha$ is called a *resolution of the identity* if

$$1 = \sum_\alpha P_\alpha$$

or, equivalently,

$$\mathscr{H} = \bigoplus_\alpha \mathscr{H}_\alpha.$$

The pair of operators $\{P_\mathscr{A}, P_{\mathscr{A}'}\}$ considered above is an example of such a resolution of the identity: $1 = P_\mathscr{A} + P'_\mathscr{A}$.

Theorem H.1. *An operator P is a projection if and only if it is Hermitian and $P^2 = P$.*

Proof. For the Hermitian P there exists a basis $|e_i\rangle$, in which this operator is diagonal, i. e.,

$$P = \sum_i |e_i\rangle p_i \langle e_i|.$$

Then the condition $P^2 = P$ implies

$$\begin{aligned} 0 = P^2 - P &= \left(\sum_i |e_i\rangle p_i \langle e_i|\right)\left(\sum_j |e_j\rangle p_j \langle e_j|\right) - \sum_i |e_i\rangle p_i \langle e_i| \\ &= \sum_{ij} |e_i\rangle p_i p_j \delta_{ij} \langle e_j| - \sum_i |e_i\rangle p_i \langle e_i| = \sum_i |e_i\rangle (p_i^2 - p_i) \langle e_i|. \end{aligned}$$

Therefore $p_i^2 - p_i = 0$ and either $p_i = 0$ or $p_i = 1$. From this we conclude that P is a projection on the subspace, which is a linear span of all its eigenvectors with the eigenvalue 1.

To prove the converse statement, we note that each projection is an Hermitian operator, because its eigenvalues are real (0 or 1). Moreover, for each vector $|y\rangle$

$$P^2|y\rangle = P(P|y\rangle) = P|y_\parallel\rangle = |y_\parallel\rangle = P|y\rangle,$$

which proves that $P^2 = P$. □

H.2 Commuting operators

Lemma H.2. *Two subspaces $\mathscr{A}$ and $\mathscr{B}$ are orthogonal if and only if $P_\mathscr{A} P_\mathscr{B} = P_\mathscr{B} P_\mathscr{A} = 0$.*

Proof. Assume that

$$P_\mathscr{A} P_\mathscr{B} = P_\mathscr{B} P_\mathscr{A} = 0 \tag{H.1}$$

and that there is a vector $|y\rangle \in \mathscr{B}$, which is not orthogonal to $\mathscr{A}$. Then $P_{\mathscr{A}}|y\rangle = |y_{\|}\rangle \neq \mathbf{0}$. From these properties we obtain

$$P_{\mathscr{A}}P_{\mathscr{B}}|y\rangle = P_{\mathscr{A}}|y\rangle = |y_{\|}\rangle = P_{\mathscr{A}}|y_{\|}\rangle,$$
$$P_{\mathscr{B}}P_{\mathscr{A}}|y\rangle = P_{\mathscr{B}}|y_{\|}\rangle.$$

From the commutativity of $P_{\mathscr{A}}$ and $P_{\mathscr{B}}$ we further have

$$P_{\mathscr{A}}|y_{\|}\rangle = P_{\mathscr{A}}P_{\mathscr{B}}|y\rangle = P_{\mathscr{B}}P_{\mathscr{A}}|y\rangle = P_{\mathscr{B}}|y_{\|}\rangle,$$
$$P_{\mathscr{A}}P_{\mathscr{B}}|y_{\|}\rangle = P_{\mathscr{A}}P_{\mathscr{A}}|y_{\|}\rangle = P_{\mathscr{A}}|y_{\|}\rangle \neq \mathbf{0}.$$

This means that we have found a vector $|y_{\|}\rangle$ for which $P_{\mathscr{A}}P_{\mathscr{B}}|y_{\|}\rangle \neq \mathbf{0}$, in contradiction to our original assumption (H.1). Consequently, the assumption $|y\rangle \neq 0$ was incorrect, which means that $\mathscr{A} \perp \mathscr{B}$.

The converse statement is proved as follows. For each vector $|x\rangle$ the projection $P_{\mathscr{A}}|x\rangle$ is in the subspace $\mathscr{A}$. If $\mathscr{A}$ and $\mathscr{B}$ are orthogonal, then the next projection $P_{\mathscr{B}}P_{\mathscr{A}}|x\rangle$ must result in a zero vector. The same arguments show that $P_{\mathscr{A}}P_{\mathscr{B}}|x\rangle = \mathbf{0}$ and $P_{\mathscr{A}}P_{\mathscr{B}} = P_{\mathscr{B}}P_{\mathscr{A}}$. □

Lemma H.3. *If $\mathscr{A} \perp \mathscr{B}$, then $P_{\mathscr{A}} + P_{\mathscr{B}}$ is a projection on the direct sum of subspaces $\mathscr{A} \oplus \mathscr{B}$.*

Proof. Let us build an orthonormal basis $|e_i\rangle$ in $\mathscr{A} \oplus \mathscr{B}$ such that the first $\dim(\mathscr{A})$ vectors are in $\mathscr{A}$ and the next $\dim(\mathscr{B})$ vectors are in $\mathscr{B}$. Then

$$P_{\mathscr{A}} + P_{\mathscr{B}} = \sum_{i=1}^{\dim(\mathscr{A})} |e_i\rangle\langle e_i| + \sum_{j=1}^{\dim(\mathscr{B})} |e_j\rangle\langle e_j| = P_{\mathscr{A}\oplus\mathscr{B}}.$$ □

Lemma H.4. *If $\mathscr{A}$ is a subspace in $\mathscr{B}$ ($\mathscr{A} \subseteq \mathscr{B}$), then*

$$P_{\mathscr{A}}P_{\mathscr{B}} = P_{\mathscr{B}}P_{\mathscr{A}} = P_{\mathscr{A}}.$$

Proof. If $\mathscr{A} \subseteq \mathscr{B}$, then there exists a subspace $\mathscr{C} \subseteq \mathscr{B}$ such that $\mathscr{C} \perp \mathscr{A}$ and $\mathscr{B} = \mathscr{A} \oplus \mathscr{C}$.[2] In accordance with Lemmas H.2 and H.3,

$$P_{\mathscr{A}}P_{\mathscr{C}} = P_{\mathscr{C}}P_{\mathscr{A}} = 0,$$
$$P_{\mathscr{B}} = P_{\mathscr{A}} + P_{\mathscr{C}},$$
$$P_{\mathscr{A}}P_{\mathscr{B}} = P_{\mathscr{A}}(P_{\mathscr{A}} + P_{\mathscr{C}}) = P_{\mathscr{A}}^2 = P_{\mathscr{A}},$$
$$P_{\mathscr{B}}P_{\mathscr{A}} = (P_{\mathscr{A}} + P_{\mathscr{C}})P_{\mathscr{A}} = P_{\mathscr{A}}.$$ □

2 This subspace is $\mathscr{C} = \mathscr{A}' \cap \mathscr{B}$ – the linear span of all vectors in $\mathscr{B}$ that are orthogonal to $\mathscr{A}$.

If there are three mutually orthogonal subspaces $\mathscr{X}$, $\mathscr{Y}$ and $\mathscr{Z}$ such that $\mathscr{A} = \mathscr{X} \oplus \mathscr{Y}$ and $\mathscr{B} = \mathscr{X} \oplus \mathscr{Z}$, then the subspaces $\mathscr{A}$ and $\mathscr{B}$ (and also the projections $P_{\mathscr{A}}$ and $P_{\mathscr{B}}$) are called *compatible*. By noting that $\mathscr{X} = \mathscr{A} \cap \mathscr{B}$, $\mathscr{Y} = \mathscr{A} \cap \mathscr{B}'$ and $\mathscr{Z} = \mathscr{B} \cap \mathscr{A}'$, it is easy to see that this definition of compatibility is consistent with the quantum-logical definition (1.6)–(1.7), so we have

$$\mathscr{A} = (\mathscr{A} \cap \mathscr{B}) \oplus (\mathscr{A} \cap \mathscr{B}'),$$
$$\mathscr{B} = (\mathscr{B} \cap \mathscr{A}) \oplus (\mathscr{B} \cap \mathscr{A}').$$

Lemma H.5. *Two subspaces $\mathscr{A}$ and $\mathscr{B}$ are compatible if and only if their projections commute, i. e.,*

$$[P_{\mathscr{A}}, P_{\mathscr{B}}] = 0.$$

Proof. First we prove that if $[P_{\mathscr{A}}, P_{\mathscr{B}}] = 0$, then $P_{\mathscr{A}} P_{\mathscr{B}} = P_{\mathscr{B}} P_{\mathscr{A}} = P_{\mathscr{A} \cap \mathscr{B}}$ is a projection on the intersection of the two subspaces.

We have

$$(P_{\mathscr{A}} P_{\mathscr{B}})^2 = P_{\mathscr{A}} P_{\mathscr{B}} P_{\mathscr{A}} P_{\mathscr{B}} = P_{\mathscr{A}}^2 P_{\mathscr{B}}^2 = P_{\mathscr{A}} P_{\mathscr{B}}.$$

The operator $P_{\mathscr{A}} P_{\mathscr{B}}$ is Hermitian, because

$$(P_{\mathscr{A}} P_{\mathscr{B}})^\dagger = P_{\mathscr{B}}^\dagger P_{\mathscr{A}}^\dagger = P_{\mathscr{B}} P_{\mathscr{A}} = P_{\mathscr{A}} P_{\mathscr{B}}.$$

Then, by Theorem H.1, $P_{\mathscr{A}} P_{\mathscr{B}}$ is a projection.

Then two options are possible: either $\mathscr{A}$ and $\mathscr{B}$ are orthogonal ($\mathscr{A} \perp \mathscr{B}$), or they are not orthogonal. In the former case, Lemma H.2 implies $P_{\mathscr{A}} P_{\mathscr{B}} = P_{\mathscr{B}} P_{\mathscr{A}} = 0$, i. e., the direct statement of our lemma.

In the latter case, we denote $\mathscr{C} = \mathscr{A} \cap \mathscr{B}$ (the subspace $\mathscr{C}$ can, of course, be empty). We can always write $\mathscr{A} = \mathscr{C} \oplus \mathscr{X}$ and $\mathscr{B} = \mathscr{C} \oplus \mathscr{Y}$, where $\mathscr{X} \subseteq \mathscr{A}$ and $\mathscr{Y} \subseteq \mathscr{B}$ are some subspaces (possibly zero ones).[3] Then

$$P_{\mathscr{A}} = P_{\mathscr{C}} + P_{\mathscr{X}},$$
$$P_{\mathscr{B}} = P_{\mathscr{C}} + P_{\mathscr{Y}},$$
$$[P_{\mathscr{C}}, P_{\mathscr{X}}] = 0,$$
$$[P_{\mathscr{C}}, P_{\mathscr{Y}}] = 0.$$

To prove the compatibility of $\mathscr{A}$ and $\mathscr{B}$ we are left to show that $\mathscr{X}$ and $\mathscr{Y}$ orthogonal. This follows from the commutator

$$\begin{aligned} 0 = [P_{\mathscr{A}}, P_{\mathscr{B}}] &= [P_{\mathscr{C}} + P_{\mathscr{X}}, P_{\mathscr{C}} + P_{\mathscr{Y}}] \\ &= [P_{\mathscr{C}}, P_{\mathscr{C}}] + [P_{\mathscr{C}}, P_{\mathscr{Y}}] + [P_{\mathscr{X}}, P_{\mathscr{C}}] + [P_{\mathscr{X}}, P_{\mathscr{Y}}] = [P_{\mathscr{X}}, P_{\mathscr{Y}}] \end{aligned}$$

and Lemma H.2.

3 See the proof of Lemma H.4.

Let us now prove the converse statement. From the assumed compatibility of $\mathscr{A}$ and $\mathscr{B}$ it follows that

$$\begin{aligned} P_{\mathscr{A}} &= P_{\mathscr{X}} + P_{\mathscr{Y}}, \\ P_{\mathscr{B}} &= P_{\mathscr{X}} + P_{\mathscr{Z}}, \\ P_{\mathscr{X}} P_{\mathscr{Y}} &= P_{\mathscr{X}} P_{\mathscr{Z}} = P_{\mathscr{Y}} P_{\mathscr{Z}} = 0, \\ [P_{\mathscr{A}}, P_{\mathscr{B}}] &= [P_{\mathscr{X}} + P_{\mathscr{Y}}, P_{\mathscr{X}} + P_{\mathscr{Z}}] = 0. \end{aligned}$$ □

Theorem H.6. *Subspaces in the Hilbert space have the property of orthomodularity* (1.8), *i. e., if $\mathscr{A} \subseteq \mathscr{B}$, then the two subspaces $\mathscr{A}$ and $\mathscr{B}$ are compatible.*

Proof. By Lemma H.4, if $\mathscr{A} \subseteq \mathscr{B}$, then $[P_{\mathscr{A}}, P_{\mathscr{B}}] = 0$. Therefore, by Lemma H.5 the subspaces $\mathscr{A}$ and $\mathscr{B}$ are compatible. □

If two or more (noncollinear) eigenvectors of the Hermitian operator F correspond to the same eigenvalue f, then such eigenvalue is called *degenerate*. Any linear combination of eigenvectors with eigenvalue f is also an eigenvector of F with the same eigenvalue. The linear span of all such eigenvectors is called the *eigensubspace* of the operator F. As usual, with this subspace we can associate a projection P_f. Then the operator F takes the form[4]

$$F = \sum_f f P_f, \tag{H.2}$$

where the index f runs through all distinct eigenvalues of F and P_f are called *spectral projections* of the operator F. This means that each Hermitian operator defines a certain resolution of the identity $I = \sum_f P_f$. Conversely, if a set of projections P_f specifies a resolution of the identity and f are real numbers, then equation (H.2) defines a unique Hermitian operator.

Lemma H.7. *If two Hermitian operators F and G commute, then all spectral projections of F commute with G.*

Proof. Let P_f be one of the spectral projections of the operator F. Take an arbitrary nonzero vector $|x\rangle$ in the range of P_f, that is,

$$\begin{aligned} P_f|x\rangle &= |x\rangle, \\ F|x\rangle &= f|x\rangle, \end{aligned}$$

for some real f. First, we prove that the vector $G|x\rangle$ also lies in the range of P_f. Indeed, using the commutativity of F and G, we obtain

$$F(G|x\rangle) = GF|x\rangle = Gf|x\rangle = f(G|x\rangle).$$

4 This formula is a generalization of (G.16).

This means that the operator G leaves eigenspaces F invariant. Then, for any vector $|x\rangle$, its images $P_f|x\rangle$ and $GP_f|x\rangle$ lie in the range of P_f. Hence

$$P_f G P_f = G P_f.$$

Applying the Hermitian conjugation to both sides of this equation, we have

$$P_f G P_f = P_f G.$$

Subtracting these equalities, we obtain

$$[G, P_f] = GP_f - P_f G = 0. \qquad \square$$

Theorem H.8. *Two Hermitian operators F and G commute if and only if all their spectral projections commute.*

Proof. Let us write spectral decompositions of the two operators

$$F = \sum_f f P_f, \tag{H.3}$$

$$G = \sum_g g Q_g. \tag{H.4}$$

If $[P_f, Q_g] = 0$ for all eigenvalues f, g, then obviously $[F, G] = 0$. To prove the converse statement, notice that by Lemma H.7 each spectral projection P_f commutes with G. Since P_f is Hermitian, from the same lemma it follows that each spectral projection of G commutes with P_f. □

Theorem H.9. *If two Hermitian operators F and G commute, then there is a basis $|e_i\rangle$, where both F and G are diagonal, i. e., $|e_i\rangle$ are common eigenvectors of both F and G.*

Proof. The identity operator can be written in three different ways:

$$I = \sum_f P_f,$$

$$I = \sum_g Q_g,$$

$$I = I \cdot I = \left(\sum_f P_f\right)\left(\sum_g Q_g\right) = \sum_{fg} P_f Q_g,$$

where P_f and Q_g are spectral projections of F and G, respectively. Since F and G commute, according to Theorem H.8, the operators $P_f Q_g$ are projections. Moreover, $P_f Q_g$ and $P_k Q_m$ project on mutually orthogonal subspaces if either $f \neq k$ or $g \neq m$. Therefore, the set of projections $P_f Q_g$ forms a resolution of the identity, and the desired basis $|e_i\rangle$ can be obtained by joining bases in the subspaces that are ranges of the projections $P_f Q_g$. □

I Representations of groups and algebras

A *representation* of a group G is a mapping that associates with each element $g \in G$ a square matrix U_g with nonzero determinant[1] and fixed dimension. Obviously, the matrices U_g can be identified with linear transformations (operators) in some vector space, which is called the *representation space* of the group. The composition of elements in the group is represented by the matrix product, and the inverse matrix corresponds to the inverse group element. In other words, this mapping must satisfy the following conditions:

$$U_{g_1}U_{g_2} = U_{g_1g_2},$$
$$U_{g^{-1}} = U_g^{-1},$$
$$U_e = I.$$

Each group in any vector space has a *trivial* representation in which the same unit matrix (identity operator) corresponds to all elements of the group. Finding nontrivial representations of groups is often a difficult task.

I.1 Unitary representations of groups

In Hilbert spaces, one can define a class of *unitary* representations that is especially useful for quantum mechanics. These representations are realized by unitary operators.

Two representations U_g and U'_g of the group G in the Hilbert space $\mathscr{H}$ are called *unitarily equivalent*, if there exists a unitary operator V such that for each $g \in G$

$$U'_g = VU_gV^{-1}. \tag{I.1}$$

Having two representations U_g and V_g of the group G in $\mathscr{H}_1$ and $\mathscr{H}_2$, respectively, we can always construct a third representation W_g of the same group in the Hilbert space $\mathscr{H} = \mathscr{H}_1 \oplus \mathscr{H}_2$ by combining two submatrices into a block-diagonal matrix

$$W_g = \begin{bmatrix} U_g & 0 \\ 0 & V_g \end{bmatrix}. \tag{I.2}$$

The representation W_g is called the *direct sum* of the representations U_g and V_g and is denoted by $W_g = U_g \oplus V_g$.

A representation U_g is called *reducible*, if there exists a unitary transformation (I.1), bringing representation matrices to the block-diagonal form (I.2) for all g at once.

1 Matrices with zero determinant cannot be reversed; therefore they cannot represent group elements.

https://doi.org/10.1515/9783110492132-016

Otherwise, the representation is called *irreducible*. Each unitary representation of a group is either irreducible or decomposable into a direct sum of irreducible representations. If we find all irreducible representations of a group, then we can construct all (reducible) representations by direct summation.

According to *Schur's first lemma* [38], in irreducible unitary representations, *Casimir operators* are represented by identity operators multiplied by constants.

I.2 Stone theorem

Stone's theorem provides important information about unitary representations of one-dimensional Lie groups with group parameter z.

Theorem I.1 (Stone [86]). *If $U_{g(z)}$ is a unitary representation of a one-dimensional Lie group in the Hilbert space $\mathscr{H}$, then there exists a Hermitian operator T in $\mathscr{H}$ such that*[2]

$$U_{g(z)} = e^{-\frac{i}{\hbar}Tz}. \tag{I.3}$$

This theorem is useful not only for one-dimensional groups, but also for Lie groups of arbitrary dimension. The reason is that in any Lie group one can find many *one-parameter subgroups* (see Appendix E.3), for which the Stone theorem can be applied. In the Poincaré group, examples of such one-parameter subgroups are translations along a fixed axis, rotations about a fixed axis and time translations.

A one-parameter subgroup passes through each element of the group near the group's identity element (see Figure E.1). Therefore, Stone's theorem implies that matrices of a unitary representation of any Lie group in $\mathscr{H}$ can be written in the form (I.3), where Hermitian operators T form a *representation of the Lie algebra* in the Hilbert space $\mathscr{H}$. In this case, Lie brackets are represented by commutators.

I.3 Heisenberg algebra

The *Heisenberg algebra* $\mathfrak{h}_n$ has basis elements traditionally denoted as $\mathcal{P}_i$, $\mathcal{R}_i$ ($i = 1, 2, \ldots, n$) and I with the following commutators[3]:

$$\begin{aligned}
[\mathcal{P}_i, \mathcal{P}_j] &= [\mathcal{R}_i, \mathcal{R}_j] = 0, \\
[\mathcal{R}_i, \mathcal{P}_j] &= \delta_{ij} I, \\
[\mathcal{R}_i, I] &= [\mathcal{P}_i, I] = 0.
\end{aligned}$$

2 In the exponent we use the Planck constant $\hbar$, but any other real constant is also acceptable.

3 Also known as the *canonical commutation relations*.

In quantum physics, an important role is played by Hermitian representations of the Heisenberg algebra in Hilbert spaces, i. e., the sets of Hermitian operators $\{R_i, P_i\}$ $(i = 1, 2, \dots, n)$ satisfying the commutation relations

$$[P_i, P_j] = [R_i, R_j] = 0,$$
$$[R_i, P_j] = i\hbar\delta_{ij}I.$$

For such representations, the following theorem is applicable.

Theorem I.2 (Stone–von Neumann [93]). *The Heisenberg algebra $\mathfrak{h}_n$ has a unique (up to unitary equivalence) Hermitian irreducible representation. This is the so-called Schrödinger representation. In the physically important case $n = 3$, the Schrödinger representation is realized by the operators of momentum $\boldsymbol{P}$ and position $\boldsymbol{R}$, as described in Subsection 5.3.2. The vectors in the representation space $\mathscr{H}$ are complex functions $\psi(\boldsymbol{r})$ on $\mathbb{R}^3$, the operator $\boldsymbol{R}$ multiplies these functions by $\boldsymbol{r}$, i. e.,*

$$\hat{\boldsymbol{R}}\psi(\boldsymbol{r}) = \boldsymbol{r}\psi(\boldsymbol{r}),$$

and the operator $\boldsymbol{P}$ differentiates them as follows:

$$\hat{\boldsymbol{P}}\psi(\boldsymbol{r}) = -i\hbar\frac{d\psi(\boldsymbol{r})}{d\boldsymbol{r}}.$$

Corollary I.3. *Any Hermitian representation of the Heisenberg algebra is a direct sum of identical irreducible Schrödinger representations.*

Corollary I.4. *Two Hermitian representations of the Heisenberg algebra in the same Hilbert space are unitarily equivalent.*

Corollary I.5. *In any Hermitian representation of the Heisenberg algebra, the operators P_i and R_i have continuous spectra from $-\infty$ to ∞.*

I.4 Double-valued representations of rotation group

The group of rotations (see Appendix D) has a nontrivial topology. Results of rotations about one axis at angles $\varphi + 2\pi n$, where $n = \dots, -1, 0, 1, 2, \dots$, are physically indistinguishable. Hence, the region of independent rotation vectors $\boldsymbol{\varphi} \in \mathbb{R}^3$ can be described as the interior of a sphere with radius π, with opposite points on the sphere's surface being considered equivalent. This set of rotation vectors will be called the ball Π (see Figure I.1). The identity element $\boldsymbol{\varphi} = \boldsymbol{0}$ is in the center of the ball. We will be interested in one-parameter families of group elements,[4] that form continuous paths in

4 They are not necessarily one-parameter subgroups.

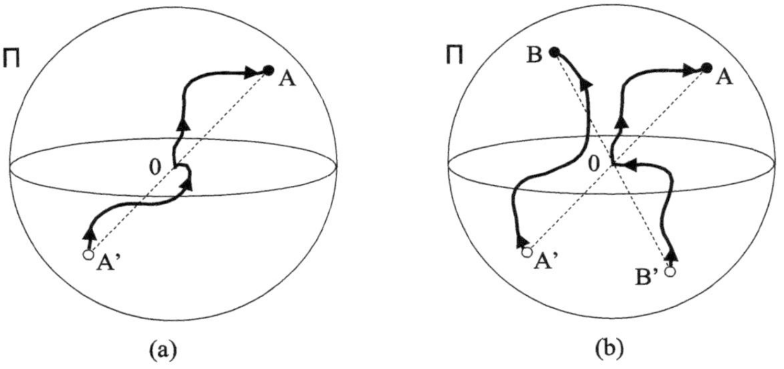

Figure I.1: The manifold of parameters of the rotation group is not simply connected. (a) A closed loop that cannot be shrunk to a single point by a continuous deformation. (b) An example of a loop that can be shrunk to a point.

the group's manifold Π. Since in our topology the opposite points on the sphere are identical, each continuous path crossing this surface re-enters the sphere from the other side. For example, in Figure I.1 (a) we have depicted a closed loop in Π. It starts from the center of the ball $\boldsymbol{\varphi} = \mathbf{0}$, reaches its surface at A and then continues from the opposite point A' back to the center $\mathbf{0}$.

A topological manifold is said to be *simply connected*, if each loop can be continuously deformed to a point. An example of a simply connected manifold is the surface of a sphere. However, the manifold Π of group parameters is not simply connected. The loop shown in Figure I.1 (a) cannot be collapsed to $\mathbf{0}$ by any continuous deformation. It is easy to understand that loops having this property intersect the surface of the ball Π an odd (in our case – one) number of times.

The opposite example is given by the loop $\mathbf{0} \to A \to A' \to B \to B' \to \mathbf{0}$ in Figure I.1 (b). It intersects the surface of the ball Π twice and can be continuously shrunk to the point $\mathbf{0}$. This can be achieved by bringing together the points A' and B (hence also the opposite points A and B' come together), so that the portion $A' \to B$ of the path disappears during the deformation.

Thus, for any rotation $\boldsymbol{\varphi}$ (= for any point inside the ball Π), there are two classes of paths originating from the identity element $\mathbf{0}$ of the group and ending in $\boldsymbol{\varphi}$. These two classes[5] consist of paths that intersect the surface of the ball Π even and odd number of times, respectively. Paths from different classes cannot be continuously deformed into each other.

When we construct a projective representation of the rotation group, then the central charges can be removed by an appropriate choice of numerical constants added to the representatives of generators.[6] Then we can construct a unitary representation

5 They are also called *homotopy classes*.

6 Like in the Poincaré group considered in Subsection 3.2.4.

of the group in which the representatives depend smoothly on the parameters of the group; the identity element **0** is represented by the identity matrix, and by traveling along a small loop in the group manifold from **0** back to **0**, we can be sure that the representative will return to the identity matrix (operator) again. But if we go on a journey along the long path $\mathbf{0} \to A \to A' \to \mathbf{0}$ in Figure I.1 (a), then there is no way to continuously collapse this path to the point **0**, so there is no guarantee that after traveling along this path and returning back to the identity element, we will find its representative equal to I. Instead of I, we can obtain another equivalent operator from the ray of operators containing I. In other words, the representative of the element **0** can differ from I by the phase factor $e^{i\varphi}$ after traveling along a long loop.

On the other hand, after making two passes along this loop $\mathbf{0} \to A \to A' \to \mathbf{0} \to A \to A' \to \mathbf{0}$, we obtain a path that crosses the surface of the ball Π twice and now it *can* be deformed into a point. At the end of such a path we should get the identity matrix representative of **0**. Hence, $e^{2i\varphi} = 1$ or $e^{i\varphi} = \pm 1$. Therefore, two types of unitary representations of the rotation group are possible: single-valued and double-valued. For *single-valued* representations, the representative of the trivial rotation is always I. For *double-valued* representations, the trivial rotation has two representatives, I and $-I$, and the product of two operators in (3.16) can acquire a nontrivial sign, so

$$U_{g_1} U_{g_2} = \pm U_{g_1 g_2}.$$

I.5 Unitary irreducible representations of rotation group

There exist an infinite number of unitary irreducible representations $\mathcal{D}^s$ of the rotation group. These representations are characterized by the values of *spin* $s = 0, 1/2, 1, \ldots$. They are described in detail in textbooks [74], so we will not dwell on them here. In Table I.1 the basic properties of several simple representations are listed: the dimension of the representation space, explicit forms of the three generators S_x, S_y, S_z, their eigenvalues and the value of the Casimir operator S^2.

Representations characterized by an integer spin s are single-valued.

Representations with a half-integer spin are double-valued. For example, in the two-dimensional representation ($s = 1/2$), a rotation by the angle 2π about the z-axis is represented by the minus-identity matrix

$$e^{-\frac{i}{\hbar}S_z 2\pi} = \exp\left(-\frac{2\pi i}{\hbar}\begin{bmatrix} \hbar/2 & 0 \\ 0 & -\hbar/2 \end{bmatrix}\right) = \begin{bmatrix} e^{-i\pi} & 0 \\ 0 & e^{i\pi} \end{bmatrix} = \begin{bmatrix} -1 & 0 \\ 0 & -1 \end{bmatrix} = -I.$$

On the other hand, a rotation by $4\pi = 720°$ is represented, as expected, by the identity matrix

$$e^{-\frac{i}{\hbar}S_z 4\pi} = \begin{bmatrix} 1 & 0 \\ 0 & 1 \end{bmatrix} = I.$$

Table I.1: Unitary irreducible representations $\mathscr{D}^s$ of the rotation group.

Spin	$s = 0$	$s = 1/2$	$s = 1$	$s = 3/2, 2, \dots$
Dimension of $\mathscr{H}$	1	2	3	$2s + 1$
S_x	0	$\begin{bmatrix} 0 & \hbar/2 \\ \hbar/2 & 0 \end{bmatrix}$	$\begin{bmatrix} 0 & 0 & 0 \\ 0 & 0 & -i\hbar \\ 0 & i\hbar & 0 \end{bmatrix}$	see [74]
S_y	0	$\begin{bmatrix} 0 & -i\hbar/2 \\ i\hbar/2 & 0 \end{bmatrix}$	$\begin{bmatrix} 0 & 0 & i\hbar \\ 0 & 0 & 0 \\ -i\hbar & 0 & 0 \end{bmatrix}$	see [74]
S_z	0	$\begin{bmatrix} \hbar/2 & 0 \\ 0 & -\hbar/2 \end{bmatrix}$	$\begin{bmatrix} 0 & -i\hbar & 0 \\ i\hbar & 0 & 0 \\ 0 & 0 & 0 \end{bmatrix}$	see [74]
Spectrum of S_i	0	$\{-\hbar/2, \hbar/2\}$	$\{-\hbar, 0, \hbar\}$	$\{-\hbar s, \dots, \hbar s\}$
$\langle S^2 \rangle$	0	$\frac{3}{4}\hbar^2$	$2\hbar^2$	$\hbar^2 s(s+1)$

J Pseudo-orthogonal representation of Lorentz group

The Lorentz group is a six-dimensional subgroup of the Poincaré group formed by rotations and boosts. Linear representations of the Lorentz group play an important role in many physical problems. In particular, the 4-vector (pseudo-orthogonal) representation is the mathematical basis of the special theory of relativity, which will be summarized in Appendix A of Volume 3.

J.1 Minkowski space–time

Let us first define the vector space on which the 4-vector representation of the Lorentz group acts. This is a four-dimensional real Minkowski vector space $\mathcal{M}$, whose vectors will be marked with the tilde[1]

$$\tilde{\tau} = \begin{bmatrix} ct \\ x \\ y \\ z \end{bmatrix}.$$

We assume that for any two 4-vectors $\tilde{\tau}_1$ and $\tilde{\tau}_2$ in $\mathcal{M}$ the so-called *pseudo-scalar product* is defined, which can be written in several equivalent ways[2]:

$$\begin{aligned} \tilde{\tau}_1 \cdot \tilde{\tau}_2 &\equiv c^2 t_1 t_2 - x_1 x_2 - y_1 y_2 - z_1 z_2 = \sum_{\mu\nu=0}^{3} (\tau_1)_\mu \eta^{\mu\nu} (\tau_2)_\nu \\ &= [ct_1, x_1, y_1, z_1] \begin{bmatrix} 1 & 0 & 0 & 0 \\ 0 & -1 & 0 & 0 \\ 0 & 0 & -1 & 0 \\ 0 & 0 & 0 & -1 \end{bmatrix} \begin{bmatrix} ct_2 \\ x_2 \\ y_2 \\ z_2 \end{bmatrix} \\ &= \tilde{\tau}_1^T \eta \tilde{\tau}_2, \end{aligned} \tag{J.1}$$

where $\eta^{\mu\nu}$ are elements of the so-called *metric tensor*

$$\eta = \begin{bmatrix} 1 & 0 & 0 & 0 \\ 0 & -1 & 0 & 0 \\ 0 & 0 & -1 & 0 \\ 0 & 0 & 0 & -1 \end{bmatrix}. \tag{J.2}$$

---

1 Here c is the speed of light. Unlike most textbooks, we are not going to assign any physical meaning to the Minkowski space–time $\mathcal{M}$. For us $\mathcal{M}$ is just an abstract vector space that has nothing to do with physical space and time. For more discussion on this theme, see Chapter 8 of Volume 3.

2 Here the indices μ and ν range from 0 to 3; $\tau_0 = ct$, $\tau_1 = x$, $\tau_2 = y$, $\tau_3 = z$.

https://doi.org/10.1515/9783110492132-017

For a more compact notation, it is convenient to define 4-vectors with a “raised index” and adopt Einstein’s rule for automatic summation over repeated indices,

$$\tau^{\mu} \equiv \sum_{\nu=0}^{3} \eta^{\mu\nu}\tau_{\nu} \equiv \eta^{\mu\nu}\tau_{\nu} = (ct, -x, -y, -z).$$

Then the pseudo-scalar product can be written

$$\tilde{\tau}_1 \cdot \tilde{\tau}_2 \equiv (\tau_1)_{\mu}(\tau_2)^{\mu} = (\tau_1)^{\mu}(\tau_2)_{\mu}. \tag{J.3}$$

Next we define the *pseudo-scalar square* (or *4-square*) of a 4-vector $\tilde{\tau}$ as

$$\tilde{\tau}^2 \equiv \tilde{\tau} \cdot \tilde{\tau} = \tau_{\mu}\tau^{\mu} = \tau_0^2 - \tau_1^2 - \tau_2^2 - \tau_3^2 = \tau_0^2 - \tau^2,$$

where

$$\tau^2 \equiv (\boldsymbol{\tau} \cdot \boldsymbol{\tau}) = \tau_1^2 + \tau_2^2 + \tau_3^2$$

is the square of the 3-vector part. A 4-vector $(\tau_0, \boldsymbol{\tau})$ is called *spacelike*, if $\tau^2 > \tau_0^2$. *Timelike* 4-vectors have $\tau^2 < \tau_0^2$, and the condition for *lightlike* 4-vectors is $\tau^2 = \tau_0^2$.

J.2 General properties of representation

In the pseudo-orthogonal representation, each element g = (boost) × (rotation) of the Lorentz group is associated with a 4×4 invertible matrix Λ acting on 4-vectors in $\mathcal{M}$ as follows:

$$\tau'_{\mu} = \sum_{\nu=0}^{3} \Lambda_{\mu}^{\ \nu}\tau_{\nu} \equiv (\Lambda\tilde{\tau})_{\mu}.$$

The representation matrices Λ are required to conserve the pseudo-scalar product[3]

$$\tilde{\tau}'_1 \cdot \tilde{\tau}'_2 \equiv \Lambda\tilde{\tau}_1 \cdot \Lambda\tilde{\tau}_2 = \tilde{\tau}_1^T \Lambda^T \eta \Lambda \tilde{\tau}_2 = \tilde{\tau}_1^T \eta \tilde{\tau}_2 = \tilde{\tau}_1 \cdot \tilde{\tau}_2. \tag{J.4}$$

This implies

$$\eta = \Lambda^T \eta \Lambda \tag{J.5}$$

and

$$\Lambda\tilde{\tau}_1 \cdot \tilde{\tau}_2 = \tilde{\tau}_1^T \Lambda^T \eta \tilde{\tau}_2 = \tilde{\tau}_1^T \eta \Lambda^{-1} \tilde{\tau}_2 = \tilde{\tau}_1 \cdot \Lambda^{-1}\tilde{\tau}_2. \tag{J.6}$$

3 Note that this representation is not unitary and the representation space $\mathcal{M}$ is not a Hilbert space.

Another property of the matrices Λ can be obtained by taking the determinant of both sides of equation (J.5),

$$-1 = \det[\eta] = \det[\Lambda^T \eta \Lambda] = \det[\Lambda^T] \det[\eta] \det[\Lambda] = -\det[\Lambda]^2.$$

Therefore $\det[\Lambda] = \pm 1$. Writing equation (J.5) for the component η_{00}, we also obtain

$$1 = \eta_{00} = \sum_{\mu\nu=0}^{3} (\Lambda^T)_0^{\ \mu} \eta_{\mu\nu} \Lambda^\nu_{\ 0} = (\Lambda^0_{\ 0})^2 - (\Lambda^1_{\ 0})^2 - (\Lambda^2_{\ 0})^2 - (\Lambda^3_{\ 0})^2.$$

It then follows that $(\Lambda^0_{\ 0})^2 \geq 1$ and that either $\Lambda^0_{\ 0} \geq 1$ or $\Lambda^0_{\ 0} \leq -1$.

The identity element of the group is represented by the unit matrix I, for which, naturally, $\det[I] = 1$ and $I_{00} = 1$. Since we are only interested in rotations and boosts, which can be connected in a continuous way to the identity element, it is necessary to choose

$$\det[\Lambda] = 1, \tag{J.7}$$

$$\Lambda^0_{\ 0} \geq 1 \tag{J.8}$$

for representatives of all elements g of the Lorentz group. Matrices satisfying equation (J.5) with the additional conditions (J.7)–(J.8) will be called *pseudo-orthogonal*, and our constructed representation of the Lorentz group is also called *pseudo-orthogonal.*

J.3 Matrices of pseudo-orthogonal representation

It is not difficult to establish the explicit form of 4×4 matrices of the pseudo-orthogonal representation in cases of pure boosts and rotations.

The action of boosts on 4-vectors is linear.[4] We have

$$\begin{bmatrix} ct' \\ x' \\ y' \\ z' \end{bmatrix} = \tilde{\boldsymbol{\theta}}^{-1} \begin{bmatrix} ct \\ x \\ y \\ z \end{bmatrix}, \tag{J.9}$$

where the representative of an arbitrary 3-vector of boost has the form

$$\tilde{\boldsymbol{\theta}} = \begin{bmatrix} \cosh\theta & \frac{\theta_x}{\theta}\sinh\theta & \frac{\theta_y}{\theta}\sinh\theta & \frac{\theta_z}{\theta}\sinh\theta \\ \frac{\theta_x}{\theta}\sinh\theta & 1+\chi\theta_x^2 & \chi\theta_x\theta_y & \chi\theta_x\theta_z \\ \frac{\theta_y}{\theta}\sinh\theta & \chi\theta_x\theta_y & 1+\chi\theta_y^2 & \chi\theta_y\theta_z \\ \frac{\theta_z}{\theta}\sinh\theta & \chi\theta_x\theta_z & \chi\theta_y\theta_z & 1+\chi\theta_z^2 \end{bmatrix}. \tag{J.10}$$

4 In our notation 4×4 matrices of the pseudo-orthogonal representation will be marked with the tilde.

Here we have denoted $\chi = (\cosh\theta - 1)\theta^{-2}$. In particular, the boosts along the directions x, y, and z are represented by the following matrices:

$$\tilde{\boldsymbol{\theta}}_x = \begin{bmatrix} \cosh\theta & \sinh\theta & 0 & 0 \\ \sinh\theta & \cosh\theta & 0 & 0 \\ 0 & 0 & 1 & 0 \\ 0 & 0 & 0 & 1 \end{bmatrix}, \tag{J.11}$$

$$\tilde{\boldsymbol{\theta}}_y = \begin{bmatrix} \cosh\theta & 0 & \sinh\theta & 0 \\ 0 & 1 & 0 & 0 \\ \sinh\theta & 0 & \cosh\theta & 0 \\ 0 & 0 & 0 & 1 \end{bmatrix}, \tag{J.12}$$

$$\tilde{\boldsymbol{\theta}}_z = \begin{bmatrix} \cosh\theta & 0 & 0 & \sinh\theta \\ 0 & 1 & 0 & 0 \\ 0 & 0 & 1 & 0 \\ \sinh\theta & 0 & 0 & \cosh\theta \end{bmatrix}. \tag{J.13}$$

Rotations are represented by 4×4 matrices

$$\tilde{\boldsymbol{\varphi}} = \begin{bmatrix} 1 & 0 \\ 0 & R_{\boldsymbol{\varphi}} \end{bmatrix}, \tag{J.14}$$

where $R_{\boldsymbol{\varphi}}$ is the 3×3 rotation matrix (D.23). The general element of the Lorentz group is represented in the form (boost) × (rotation),[5]

$$\Lambda = \tilde{\boldsymbol{\theta}} \circ \tilde{\boldsymbol{\varphi}}. \tag{J.15}$$

Verification of the properties (J.5), (J.7) and (J.8) is left as an exercise for the reader.

J.4 Representation of Lorentz Lie algebra

So far we have discussed the matrix representation of finite transformations from the Lorentz group. Let us now find representatives of infinitesimal transformations of the Lie algebra elements. In accordance with Appendix I.1, matrix representatives of group elements have the exponential form

$$\Lambda = e^{aF},$$

where F is a representative of a Lie algebra element and a is a real constant. Then we can rewrite condition (J.5) as

5 This order of factors is consistent with our convention (2.48).

$$0 = \Lambda^T \eta \Lambda - \eta = e^{aF^T} \eta e^{aF} - \eta = (1 + aF^T + \cdots)\eta(1 + aF + \cdots) - \eta$$
$$= a(F^T \eta + \eta F) + O(a^2).$$

This establishes the following restriction on the matrices F:

$$F^T \eta + \eta F = 0.$$

It is not difficult to find six linearly independent 4×4 matrices that satisfy this condition. The three rotation generators are[6]

$$\mathcal{J}_x = \begin{bmatrix} 0 & 0 & 0 & 0 \\ 0 & 0 & 0 & 0 \\ 0 & 0 & 0 & -1 \\ 0 & 0 & 1 & 0 \end{bmatrix}, \quad \mathcal{J}_y = \begin{bmatrix} 0 & 0 & 0 & 0 \\ 0 & 0 & 0 & 1 \\ 0 & 0 & 0 & 0 \\ 0 & -1 & 0 & 0 \end{bmatrix}, \quad \mathcal{J}_z = \begin{bmatrix} 0 & 0 & 0 & 0 \\ 0 & 0 & -1 & 0 \\ 0 & 1 & 0 & 0 \\ 0 & 0 & 0 & 0 \end{bmatrix}. \tag{J.16}$$

The three generators of boosts are obtained by differentiating the corresponding matrices of finite transformations (J.11)–(J.13). We have

$$\mathcal{K}_x = \frac{1}{c} \lim_{\theta \to 0} \frac{d}{d\theta} \tilde{\boldsymbol{\theta}}_x = \frac{1}{c} \begin{bmatrix} 0 & 1 & 0 & 0 \\ 1 & 0 & 0 & 0 \\ 0 & 0 & 0 & 0 \\ 0 & 0 & 0 & 0 \end{bmatrix}, \tag{J.17}$$

$$\mathcal{K}_y = \frac{1}{c} \lim_{\theta \to 0} \frac{d}{d\theta} \tilde{\boldsymbol{\theta}}_y = \frac{1}{c} \begin{bmatrix} 0 & 0 & 1 & 0 \\ 0 & 0 & 0 & 0 \\ 1 & 0 & 0 & 0 \\ 0 & 0 & 0 & 0 \end{bmatrix}, \tag{J.18}$$

$$\mathcal{K}_z = \frac{1}{c} \lim_{\theta \to 0} \frac{d}{d\theta} \tilde{\boldsymbol{\theta}}_z = \frac{1}{c} \begin{bmatrix} 0 & 0 & 0 & 1 \\ 0 & 0 & 0 & 0 \\ 0 & 0 & 0 & 0 \\ 1 & 0 & 0 & 0 \end{bmatrix}. \tag{J.19}$$

Six matrices (J.16)–(J.19) form a basis for a representation of the Lie algebra of the Lorentz group. The commutation relations (2.41) and (2.45) are easily verified. For example,

$$[\mathcal{J}_x, \mathcal{J}_y] = \begin{bmatrix} 0 & 0 & 0 & 0 \\ 0 & 0 & 0 & 0 \\ 0 & 0 & 0 & -1 \\ 0 & 0 & 1 & 0 \end{bmatrix} \begin{bmatrix} 0 & 0 & 0 & 0 \\ 0 & 0 & 0 & 1 \\ 0 & 0 & 0 & 0 \\ 0 & -1 & 0 & 0 \end{bmatrix} - \begin{bmatrix} 0 & 0 & 0 & 0 \\ 0 & 0 & 0 & 1 \\ 0 & 0 & 0 & 0 \\ 0 & -1 & 0 & 0 \end{bmatrix} \begin{bmatrix} 0 & 0 & 0 & 0 \\ 0 & 0 & 0 & 0 \\ 0 & 0 & 0 & -1 \\ 0 & 0 & 1 & 0 \end{bmatrix}$$

6 Note that (D.28)–(D.29) are 3×3 submatrices in (J.16).

$$= \begin{bmatrix} 0 & 0 & 0 & 0 \\ 0 & 0 & -1 & 0 \\ 0 & 1 & 0 & 0 \\ 0 & 0 & 0 & 0 \end{bmatrix} = \mathcal{J}_z,$$

$$c^2[\mathcal{K}_x, \mathcal{K}_y] = \begin{bmatrix} 0 & 1 & 0 & 0 \\ 1 & 0 & 0 & 0 \\ 0 & 0 & 0 & 0 \\ 0 & 0 & 0 & 0 \end{bmatrix} \begin{bmatrix} 0 & 0 & 1 & 0 \\ 0 & 0 & 0 & 0 \\ 1 & 0 & 0 & 0 \\ 0 & 0 & 0 & 0 \end{bmatrix} - \begin{bmatrix} 0 & 0 & 1 & 0 \\ 0 & 0 & 0 & 0 \\ 1 & 0 & 0 & 0 \\ 0 & 0 & 0 & 0 \end{bmatrix} \begin{bmatrix} 0 & 1 & 0 & 0 \\ 1 & 0 & 0 & 0 \\ 0 & 0 & 0 & 0 \\ 0 & 0 & 0 & 0 \end{bmatrix}$$

$$= \begin{bmatrix} 0 & 0 & 0 & 0 \\ 0 & 0 & 1 & 0 \\ 0 & -1 & 0 & 0 \\ 0 & 0 & 0 & 0 \end{bmatrix} = -\mathcal{J}_z.$$

Representatives of finite rotations (J.14) and boosts (J.10) can be written in the exponential notation

$$\tilde{\boldsymbol{\varphi}} = e^{\mathcal{J}\cdot\boldsymbol{\varphi}},$$
$$\tilde{\boldsymbol{\theta}} = e^{c\mathcal{K}\cdot\boldsymbol{\theta}}.$$

Bibliography

[1] Aerts D, Daubechies I. About the structure-preserving maps of a quantum mechanical propositional system. Helv Phys Acta. 1978;51:637.

[2] Aerts D, Daubechies I. Physical justification for using the tensor product to describe two quantum systems as one joint system. Helv Phys Acta. 1978;51:661.

[3] Ahmad SMW, Wigner EP. Invariant theoretic derivation of the connection between momentum and velocity. Nuovo Cimento A. 1975;28:1.

[4] Arons AB, Peppard MB. Einstein's proposal of the photon concept – a translation of the Annalen der Physik paper of 1905. Am J Phys. 1965;33:367.

[5] Bakamjian B. Relativistic particle dynamics. Phys Rev. 1961;121:1849.

[6] Bakamjian B, Thomas LH. Relativistic particle dynamics. II. Phys Rev. 1953;92:1300.

[7] Ballentine LE. Quantum mechanics: A modern development. Singapore: World Scientific; 1998.

[8] Barsella B, Fabri E. Angular momenta in relativistic many-body theory. Phys Rev. 1962;128:451.

[9] Berg RA. Position and intrinsic spin operators in quantum theory. J Math Phys. 1965;6:34.

[10] Birkhoff G, von Neumann J. The logic of quantum mechanics. Ann Math. 1936;37:823.

[11] Bjorken JD, Drell SD. Relativistic quantum mechanics. New York: McGraw–Hill; 1964.

[12] Blanes S, Casas F, Oteo JA, Ros J. The Magnus expansion and some of its applications. Phys Rep. 2009;470:151. arXiv:0810.5488v1.

[13] Brown GE, Jackson AD, Kuo TTS. Nucleon–nucleon potentials and minimal relativity. Nucl Phys A. 1969;113:481.

[14] Caban P, Rembielinski J. Photon polarization and Wigner's little group. Phys Rev A. 2003;68:042107. arXiv:quant-ph/0304120v2.

[15] Cacciatori S., Gorini V., Kamenshchik A.. Special relativity in the 21st century. 2008. arXiv:gr-qc/0807.3009v1.

[16] Candlin DJ. Physical operators and the representations of the inhomogeneous Lorentz group. Nuovo Cimento. 1965;37:1396.

[17] Chakrabarti AJ. On the canonical relativistic kinematics of *N*-particle systems. J Math Phys. 1964;5:922.

[18] Coester F, Polyzou WN. Relativistic quantum mechanics of particles with direct interactions. Phys Rev D. 1982;26:1348.

[19] Curcuraci L. Why do we need Hilbert spaces? 2017. arXiv:1708.08326v1.

[20] Currie DG, Jordan TF, Sudarshan ECG. Relativistic invariance and Hamiltonian theories of interacting particles. Rev Mod Phys. 1963;35:350.

[21] de Sangro R, Finocchiaro G, Patteri P, Piccolo M, Pizzella G. Measuring propagation speed of Coulomb fields. Eur Phys J C. 2015;75:137. arXiv:1211.2913v2.

[22] DeJonghe III RJ. Rebuilding mathematics on a quantum logical foundation. Chicago: University of Illinois; 2013. PhD Thesis.

[23] Dirac PAM. Forms of relativistic dynamics. Rev Mod Phys. 1949;21:392.

[24] Dollard JD. Interpretation of Kato's invariance principle in scattering theory. J Math Phys. 1976;17:46.

[25] Dubovyk I, Shebeko O. The method of unitary clothing transformations in the theory of nucleon–nucleon scattering. Few-Body Syst. 2010;48:109.

[26] Einstein A. In: Albert Einstein: Philosopher-scientist. Peru: Open Court; 1949.

[27] Ekstein H. Equivalent Hamiltonians in scattering theory. Phys Rev. 1960;117:1590.

[28] Feynman R, Leighton R, Sands M. The Feynman lectures on physics, volume III. California: California Institute of Technology; 2013. http://www.feynmanlectures.caltech.edu/III_toc.html.

https://doi.org/10.1515/9783110492132-018

[29] Fong R, Sucher J. Relativistic particle dynamics and the S-matrix. J Math Phys. 1964;5:456.

[30] Foulis DJ. A half century of quantum logic. What have we learned? http://www.quantonics.com/Foulis_On_Quantum_Logic.html.

[31] Fushchich WI, Nikitin AG. Symmetries of equations of quantum mechanics. New York: Allerton Press. 1994.

[32] Galilei G. Dialogues concerning the two chief world systems. Modern library science series. New York: Modern Library. 2001.

[33] Gleason AM. Measures on the closed subspaces of a Hilbert space. J Math Mech. 1957;6:885.

[34] Goldberger ML, Watson KM. Collision theory. New York: J. Wiley & Sons; 1964.

[35] Halvorson H, Clifton R. No place for particles in relativistic quantum theories? Phil Sci. 2002;69:1. arXiv:quant-ph/0103041v1.

[36] Heisenberg W. Physics and philosophy. New York: Harper and Brothers; 1958.

[37] Hobson A. There are no particles, there are only fields. Am J Phys. 2013;81:211. arXiv:1204.4616v2.

[38] Hsiang WY. Lectures on Lie groups. Singapore: World Scientific; 2000.

[39] Jackson JD. Classical electrodynamics. 3rd ed. New York: J Wiley and Sons; 1999.

[40] Jauch JM. Projective representation of the Poincaré group in a quaternionic Hilbert space. Loebl EM, editor. Group theory and its applications. New York: Academic Press; 1971.

[41] Jordan TF. Identification of the velocity operator for an irreducible unitary representation of the Poincaré group. J Math Phys. 1977;18:608.

[42] Jordan TF. Simple proof of no position operator for quanta with zero mass and nonzero helicity. J Math Phys. 1978;19:1382.

[43] Jordan TF. Simple derivation of the Newton–Wigner position operator. J Math Phys. 1980;21:2028.

[44] Kazes E. Analytic theory of relativistic interactions. Phys Rev D. 1971;4:999.

[45] Keister BD. Forms of relativistic dynamics: What are the possibilities? AIP Conf Proc. 1995;334:164. arXiv:nucl-th/9406032v1.

[46] Keister BD, Polyzou WN. Relativistic Hamiltonian dynamics in nuclear and particle physics. In: Negele JW, Vogt EW, editors. Advances in nuclear physics. Vol. 20. New York: Plenum Press; 1991. http://www.physics.uiowa.edu/~wpolyzou/papers/rev.pdf.

[47] Khrennikov A. The present situation in quantum theory and its merging with general relativity. Found Phys. 2017. https://doi.org/10.1007/s10701-017-0089-0. arXiv:1704.04679v2.

[48] Kofler J, Brukner Č. Classical world because of quantum physics. 2006. arXiv:quant-ph/0609079v3.

[49] Korchin AYu, Mel'nik YuP, Shebeko AV. Angular distributions and polarization of protons in the $d(e, e'p)n$ reaction. Few-Body Syst. 1990;9:211.

[50] Mackey GW. The mathematical foundations of quantum mechanics. New York: WA Benjamin; 1963. See esp. Section 2-2.

[51] Mackey GW. The theory of unitary group representations. Chicago: University of Chicago Press; 2000.

[52] Magnus W. On the exponential solution of differential equations for a linear operator. Commun Pure Appl Math. 1954;7:649.

[53] Malament DB. In defence of dogma: Why there cannot be a relativistic quantum mechanics of (localizable) particles. In: Clifton R, editor. Perspectives on quantum reality. Dordrecht: Springer; 1996.

[54] Matolcsi T. Tensor product of Hilbert lattices and free orthodistributive product of orthomodular lattices. Acta Sci Math (Szeged). 1975;37:263.

[55] Mittelstaedt P. Quantum physics and classical physics – in the light of quantum logic. Int J Theor Phys. 2005;44:771. arXiv:quant-ph/0211021v2.

[56] Monahan AH, McMillan M. Lorentz boost of the Newton–Wigner–Pryce position operator. Phys Rev D. 1997;56:2563.
[57] Moretti V, Oppio M. Quantum theory in quaternionic Hilbert space: How Poincaré symmetry reduces the theory to the standard complex one. 2017. arXiv:1709.09246v2.
[58] Moretti V, Oppio M. Quantum theory in real Hilbert space: How the complex Hilbert space structure emerges from Poincaré symmetry. Rev Math Phys. 2017;29:1750021. arXiv:1611.09029v4.
[59] Mutze U. A no-go theorem concerning the cluster decomposition property of direct interaction scattering theories. J Math Phys. 1978;19:231.
[60] Mutze U. Relativistic quantum mechanics of n-particle systems with cluster-separable interactions. Phys Rev D. 1984;29:2255.
[61] Newton I. Letter to R. Bentley. February 1693.
[62] Newton TD, Wigner EP. Localized states for elementary systems. Rev Mod Phys. 1949;21:400.
[63] Osborn H. Relativistic center-of-mass variables for two-particle system with spin. Phys Rev. 1968;176:1514.
[64] Pechukas P, Light JC. On the exponential form of time-displacement operators in quantum mechanics. J Chem Phys. 1966;44:3897.
[65] Peskin ME, Schroeder DV. An introduction to quantum field theory. Boulder: Westview Press; 1995.
[66] Piron C. Axiomatique quantique. Helv Phys Acta. 1964;37:439.
[67] Piron C. Foundations of quantum physics. Reading: WA Benjamin; 1976.
[68] Polyzou WN. Manifestly covariant, Poincaré invariant quantum theories of directly interacting particles. Phys Rev D. 1985;32:995.
[69] Pryce MHL. The mass-centre in the restricted theory of relativity and its connexion with the quantum theory of elementary particles. Proc R Soc Lond Ser A. 1948;195:62.
[70] Putnam H. Is logic empirical? In: Cohen RS, Wartofsky MW, editors. Boston studies in the philosophy of science. Vol. 5. Dordrecht:D Reidel; 1968. p. 216. Reproduced as "The logic of quantum mechanics" in Mathematics, matter and method; 1975. p. 174–197.
[71] Rescigno TN, McCurdy CW, McKoy V. Discrete basis set approach to nonspherical scattering. Chem Phys Lett. 1974;27:401.
[72] Richman F, Bridges D. A constructive proof of Gleason's theorem. J Funct Anal. 1999;162:287.
[73] Ritus VI. Transformations of the inhomogeneous Lorentz group and the relativistic kinematics of polarized states. Sov Phys JETP. 1961;13:240.
[74] Rose ME. Elementary theory of angular momentum. New York: John Wiley & Sons; 1957.
[75] Rudin W. Functional analysis. New York: McGraw–Hill; 1991.
[76] Sakata S. On a composite model for the new particles. Prog Theor Phys. 1956;16:686.
[77] Schrödinger E. Die gegenwartige Situation in der Quantenmechanik. Naturwissenschaften. 1935;23:807, 823, 844.
[78] Sokolov SN. Physical equivalence of the point and instantaneous forms of relativistic dynamics. Theor Math Phys. 1975;24:799.
[79] Sokolov SN, Shatnii AN. Physical equivalence of the three forms of relativistic dynamics and addition of interactions in the front and instant forms. Theor Math Phys. 1978;37:1029.
[80] Stefanovich EV. Sakata model of hadrons revisited. 2010. arXiv:1010.0458v1.
[81] Stefanovich EV. Sakata model of hadrons revisited. II. Nuclei and scattering. 2015. http://viXra.org/abs/1507.0042v4.
[82] Stefanovich E. Relativistic quantum theory of particles. Vol. 1 Quantum electrodynamics. Saarbrücken: Lambert Academic; 2015.

[83] Stefanovich E. Relativistic quantum theory of particles. Vol. 2 A non-traditional perspective on space, time, particles, fields, and action-at-a-distance. Saarbrücken: Lambert Academic; 2015.
[84] Stefanovich E. Elementary particle theory. Vol. 2: Quantum electrodynamics. De Gruyter Stud Math Phys. Vol. 46. Berlin: De Gruyter; 2018.
[85] Stefanovich E. Elementary particle theory. Vol. 3: Relativistic quantum dynamics. De Gruyter Stud Math Phys. Vol. 47. Berlin: De Gruyter; 2018.
[86] Stone MH. On one-parameter unitary groups in Hilbert space. Ann Math. 1932;33:643.
[87] Straumann N. Unitary representations of the inhomogeneous Lorentz group and their significance in quantum physics. 2008. arXiv:0809.4942v1.
[88] Stueckelberg ECG. Quantum theory in real Hilbert space. Helv Phys Acta. 1960;33:727.
[89] Taylor GI. Interference fringes with feeble light. Proc Camb Philos Soc. 1909;15:114.
[90] Thomas LH. The relativistic dynamics of a system of particles interacting at a distance. Phys Rev. 1952;85:868.
[91] Uhlhorn U. Representation of symmetry transformations in quantum mechanics. Ark Fys. 1963;23:307.
[92] Van Dam H, Wigner EP. Classical relativistic mechanics of interacting point particles. Phys Rev B. 1965;138:1576.
[93] von Neumann J. Die Eindeutigkeit der Schrödingerschen Operationen. Math Ann. 1931;104:570.
[94] Weinberg S. Photons and gravitons in *S*-matrix theory: Derivation of charge conservation and equality of gravitational and inertial mass. Phys Rev. 1964;135:B1049.
[95] Weinberg S. The quantum theory of fields. Vol. 1. Cambridge: University Press; 1995.
[96] Weinberg S. What is quantum field theory, and what did we think it is? 1997. arXiv:hep-th/9702027v1.
[97] Weiss GH, Maradudin AA. The Baker–Hausdorff formula and a problem in crystal physics. J Math Phys. 1962;3:771.
[98] Wigner EP. On unitary representations of the inhomogeneous Lorentz group. Ann Math. 1939;40:149.
[99] Wigner EP. Group theory and its application to the quantum mechanics of atomic spectra. New York: Academic Press; 1959.

Index

De Gruyter Studies in Mathematical Physics

Volume 47
Eugene Stefanovich
Elementary Particle Theory: Volume 3: Relativistic Quantum Dynamics, 2018
ISBN 978-3-11-049090-9, e-ISBN (PDF) 978-3-11-049322-1,
e-ISBN (EPUB) 978-3-11-049139-5

Volume 46
Eugene Stefanovich
Elementary Particle Theory: Volume 2: Quantum Electrodynamics, 2018
ISBN 978-3-11-049089-3, e-ISBN (PDF) 978-3-11-049320-7,
e-ISBN (EPUB) 978-3-11-049143-2

Volume 44
Vladimir V. Kiselev
Collective Effects in Condensed Matter Physics, 2018
ISBN 978-3-11-058509-4, e-ISBN (PDF) 978-3-11-058618-3,
e-ISBN (EPUB) 978-3-11-058513-1

Volume 43
Robert F. Snider
Irreducible Cartesian Tensors, 2017
ISBN 978-3-11-056363-4, e-ISBN (PDF) 978-3-11-056486-0,
e-ISBN (EPUB) 978-3-11-056373-3

Volume 42
Javier Roa
Regularization in Orbital Mechanics: Theory and Practice, 2017
ISBN 978-3-11-055855-5, e-ISBN (PDF) 978-3-11-055912-5,
e-ISBN (EPUB) 978-3-11-055862-3

Volume 41
Esra Russell, Oktay K. Pashaev
Oscillatory Models in General Relativity, 2017
ISBN 978-3-11-051495-7, e-ISBN (PDF) 978-3-11-051536-7,
e-ISBN (EPUB) 978-3-11-051522-0

Volume 40
Joachim Schröter
Minkowski Space: The Spacetime of the Special Relativity Theory Groups, 2017
ISBN 978-3-11-048457-1, e-ISBN (PDF) 978-3-11-048573-8,
e-ISBN (EPUB) 978-3-11-048461-8

Bei Fragen zur Produktsicherheit wenden Sie sich bitte an:
If you have any questions regarding product safety,
please contact:

Walter de Gruyter GmbH
Genthiner Straße 13
10785 Berlin
productsafety@degruyterbrill.com